ELECTRONICS ASSEMBLY
AND
FABRICATION METHODS

S. R. DUARTE
COORDINATOR OF INSTRUCTIONAL MEDIA
Riverside City College, Riverside, California
PROJECT DIRECTOR: SOUTHWEST AMIDS
Division of Vocational Education, University of California at Los Angeles, California

R. L. DUARTE
ELECTRONICS ASSEMBLY SPECIALIST
Manpower Training Consultants, Seal Beach, California
CURRICULUM SPECIALIST
Continuing Education, Santa Ana College
Santa Ana, California

ELECTRONICS ASSEMBLY AND FABRICATION METHODS
SECOND EDITION

McGRAW-HILL BOOK COMPANY
New York San Francisco St. Louis Düsseldorf Johannesburg Kuala Lumpur London
Mexico Montreal New Delhi Panama Rio de Janeiro Singapore Sydney Toronto

This book was set in Caledonia by Continental Graphics. The editors were Alan W. Lowe, Ronald Q. Lewton, and Marge Woodhurst, the designer was Janet Durey Bollow, and the production supervisor was Michael A. Ungersma. The drawings were done by Patrick and Yvonne Patterson.

The printer and binder was Von Hoffmann Press, Inc.

**ELECTRONICS ASSEMBLY
AND
FABRICATION METHODS**
Second Edition

Copyright © 1964, 1973 by McGraw-Hill, Inc. All rights reserved. No part of this publication may be reproduced, stored in a retrieval system, or transmitted, in any form or by any means, electronic, mechanical, photocopying, recording, or otherwise, without the prior written permission of the publisher.

Printed in the United States of America.

Library of Congress Cataloging in Publication Data

Duarte, Salvador R.
   Electronics assembly and fabrication methods.

   1964 ed. published under title: Electronics assembly methods.
   Bibliography: p. 290
   1. Electronic apparatus and appliances.
   2. Assembly-line methods. I. Duarte, R. L., joint author. II. Title.
TK7836.D8   1973   621.3815   72-6495
ISBN 0-07-017880-1

   67890   VHVH   798

TO RICH
    LAURIE
    PAM
    SAL, JR.

CONTENTS

LIST OF EXERCISES   xi

PREFACE   xv

PART I: BASICS FOR ELECTRONICS ASSEMBLY

1   **ELECTRONICS FABRICATION**   3
     1•1   INTRODUCTION   3
     1•2   EMPLOYMENT CONDITIONS   3
     1•3   WORK PRACTICES   5
     1•4   ASSEMBLY-LINE PROCESSES   6

2   **COMPONENT IDENTIFICATION**   13
     2•1   ELECTRICAL CONCEPTS   13
     2•2   RESISTOR RATING   15
     2•3   RESISTOR MARKING   17
     2•4   RESISTOR TYPES   17
     2•5   SCHEMATIC SYMBOLS FOR RESISTORS   18
     2•6   CAPACITOR RATING   19
     2•7   CAPACITOR MARKING   19
     2•8   TYPES OF CAPACITORS   19
     2•9   SCHEMATIC SYMBOLS FOR CAPACITORS   20

	2·10	INDUCTOR RATING 21	
	2·11	INDUCTOR MARKING 21	
	2·12	TYPES OF INDUCTORS 22	
	2·13	COVERINGS FOR INDUCTORS 23	
	2·14	SCHEMATIC SYMBOLS FOR INDUCTORS 23	
	2·15	ACTIVE COMPONENT TYPES 24	
	2·16	SEMICONDUCTOR CLASSIFICATIONS 24	
	2·17	ELECTRON-TUBE CLASSIFICATIONS 28	
	2·18	SYMBOLS FOR SEMICONDUCTORS AND VACUUM TUBES 29	
	2·19	DESIGNATIONS FOR SEMICONDUCTORS 30	
	2·20	DESIGNATIONS FOR ELECTRON TUBES 30	
3	**ELECTRONICS PACKAGING**		50
	3·1	WIRED CIRCUITS 50	
	3·2	PRINTED CIRCUITS 51	
	3·3	INTEGRATED CIRCUITS 52	
4	**DIAGRAMS**		57
	4·1	SCHEMATIC DIAGRAM 58	
	4·2	WIRING DIAGRAM 58	
	4·3	INTERCONNECTION DIAGRAM 58	
	4·4	BLOCK DIAGRAM 61	
	4·5	SYMBOLIC DIAGRAM 62	
	4·6	LAYOUT DIAGRAM 63	
	4·7	PRINTED-CIRCUIT LAYOUT DIAGRAM 63	
	4·8	PICTORIAL DRAWING 63	
	4·9	EXPLODED ILLUSTRATION 63	
	4·10	DRAWING DIAGRAMS 63	
	4·11	BLUEPRINTS 64	
	4·12	LINEAR MEASUREMENT 68	
	4·13	MEASURING DEVICES 68	
5	**TOOLS**		75
	5·1	HAND TOOLS FOR ELECTRICAL ASSEMBLY 75	
	5·2	TOOLS FOR MICROELECTRONICS 79	
	5·3	HAND TOOLS FOR MECHANICAL ASSEMBLY 79	
6	**SOLDERING PRINCIPLES**		89
	6·1	SOLDERING REQUIREMENTS 89	
	6·2	TINNING 90	
	6·3	MECHANICAL AND ELECTRICAL CONNECTION 91	
	6·4	THE SOLDERING PROCESS 92	
	6·5	DAMAGE BY HEAT 94	
	6·6	SOLDERING ON CONNECTOR PINS 95	
	6·7	SOLDERING ON PRINTED-CIRCUIT BOARDS 96	
	6·8	DESOLDERING FROM PRINTED-CIRCUITS BOARDS 97	
	6·9	CLEANING SOLDERED JOINTS 97	
	6·10	MASS-PRODUCTION SOLDERING METHODS 98	
7	**WIRE PREPARATION AND HARNESS ASSEMBLY**		110
	7·1	TYPES OF WIRE 110	
	7·2	TYPES OF WIRE INSULATION 111	
	7·3	WIRE BRAIDS AND SHIELDS 111	
	7·4	MULTIPLE-WIRE CABLES 111	
	7·5	PREPARING WIRES FOR SOLDERING 111	
	7·6	SOLDERLESS CONNECTORS 113	
	7·7	TINNING WITH A SOLDER POT 114	
	7·8	CONNECTING LEADS TO BRAIDED SHIELDS 114	
	7·9	WIRE SIZES 115	
	7·10	WIRE COLOR CODING AND IDENTIFICATION 118	
	7·11	HARNESSING WIRES INTO BUNDLES 118	
	7·12	PRODUCTION TECHNIQUES OF WIRE HARNESSES 118	
	7·13	HARNESS TERMINATION 120	
	7·14	FLAT PRINTED-CIRCUIT CABLE 123	
	7·15	SYMBOLS FOR CABLES AND CONNECTORS 123	
8	**TERMINAL CONNECTIONS**		133
	8·1	TURRET TERMINALS 133	
	8·2	BIFURCATED TERMINALS 135	
	8·3	FLAT TERMINALS 135	
	8·4	HOOK TERMINALS 136	
	8·5	CABLE CONNECTORS 137	
	8·6	SOLDERLESS TERMINALS 138	
	8·7	MOUNTING COMPONENTS ON PRINTED-CIRCUIT BOARDS 138	
	8·8	USE OF SOCKETS AS TERMINALS 141	

9	**HARDWARE AND MECHANICAL ASSEMBLY**	**148**	10	**INSPECTION AND QUALITY CONTROL**	**162**

9·1 PHYSICAL-SUPPORT HARDWARE 148
9·2 ELECTRICAL-SUPPORT HARDWARE 151
9·3 MECHANICAL ASSEMBLY PRACTICES 153

10·1 RELIABILITY 162
10·2 NONDESTRUCTIVE TESTING 163
10·3 VISUAL INSPECTION 163
10·4 CONTINUITY CHECKERS 163
10·5 TEST PANELS 163
10·6 ENVIRONMENTAL TESTING 164

PART II: CONCEPTS OF ELECTRONICS

11 **BASIC ELECTRICAL PRINCIPLES** 171
11·1 ELECTRICITY 172
11·2 RESISTANCE FOR CURRENT CONTROL 173
11·3 BATTERY AS A VOLTAGE SOURCE 174
11·4 UNITS OF QUANTITY 175
11·5 ELECTRICAL TEST EQUIPMENT 175

12 **CIRCUITS FOR VOLTAGE AND CURRENT DISTRIBUTION** 180
12·1 VOLTAGE DISTRIBUTION 180
12·2 CURRENT DISTRIBUTION 181
12·3 SERIES-PARALLEL CIRCUIT COMBINATIONS 182
12·4 REQUIREMENTS FOR CIRCUIT EXPERIMENTATION 183

13 **ALTERNATING-CURRENT FUNDAMENTALS** 186
13·1 FREQUENCY OF ALTERNATION 186
13·2 GRAPHICAL REPRESENTATIONS 187
13·3 PEAK AND EFFECTIVE VALUES 189
13·4 VOLTAGE AND CURRENT RELATIONS 189
13·5 MAGNETISM AND INDUCTORS 190
13·6 ELECTROSTATICS AND CAPACITORS 191
13·7 IMPEDANCE 192

14 **SOLID-STATE ELECTRONICS** 196
14·1 SOLID-STATE DIODES 196
14·2 PN MATERIALS 198
14·3 PN JUNCTIONS 198
14·4 TRANSISTORS 200
14·5 FIELD-EFFECT TRANSISTORS 202
14·6 SPECIAL-APPLICATION DIODES 204
14·7 INTEGRATED CIRCUITS 205
14·8 SEMICONDUCTOR LIMITATIONS 207

15 **ACTIVE CIRCUITS: DESCRIPTIONS** 216
15·1 RECTIFICATION 216
15·2 AMPLIFICATION 217
15·3 OSCILLATION 218
15·4 SWITCHING 219

16 **POWER SUPPLIES** 224
16·1 THE NEED FOR A POWER SUPPLY 224
16·2 FUNCTIONS OF A POWER SUPPLY 226
16·3 THE FULL-WAVE RECTIFIER 229
16·4 THE TRANSISTORIZED POWER SUPPLY 230
16·5 THE BLEEDER RESISTOR 230

17 **LOW-FREQUENCY AMPLIFIERS AND OSCILLATORS** 235
17·1 TYPES OF AMPLIFIERS 235
17·2 APPLICATIONS OF LOW-FREQUENCY AMPLIFIERS 236
17·3 THE FIRST AUDIO-FREQUENCY AMPLIFIER 236
17·4 THE POWER OUTPUT STAGE 238
17·5 THE COMPLETE AUDIO AMPLIFIER SUMMARIZED 239
17·6 LOW-FREQUENCY OSCILLATORS 240

18 **HIGH-FREQUENCY AMPLIFIERS AND OSCILLATORS** 251
18·1 RADIO-FREQUENCY SIGNALS 251
18·2 TUNED CIRCUITS 252
18·3 APPLICATION OF TUNED CIRCUITS 253
18·4 HIGH-FREQUENCY OSCILLATORS 254

19 ELECTRONIC TESTING TECHNIQUES 265
- 19·1 COMPONENT TESTING 265
- 19·2 PRODUCTION TESTING 266
- 19·3 TROUBLESHOOTING 266
- 19·4 SUBSTITUTION TECHNIQUES 266
- 19·5 SIGNAL TRACING 267
- 19·6 SIGNAL INJECTION 267
- 19·7 CIRCUIT LOADING 267

20 ELECTRONIC CALCULATIONS 271
- 20·1 CALCULATING D-C CIRCUITS 271
- 20·2 CALCULATING A-C CIRCUITS 274

APPENDIXES

I ELECTRON TUBES 282
- AI·1 PRINCIPLES OF THE VACUUM TUBE 283
- AI·2 THE THEORY OF THE VACUUM-TUBE DIODE 283
- AI·3 THEORY OF THE TRIODE 285
- AI·4 BATTERY REQUIREMENTS 286
- AI·5 THE TRIODE TUBE IN ACTION 286

II STANDARDS OF INDUSTRY 288
- AII·1 MILITARY AND NASA STANDARDS 288
- AII·2 PROTECTING THE PUBLIC INTEREST 289

III BIBLIOGRAPHY 290

IV AUDIO-VISUAL AIDS AND TECHNIQUES 292

INDEX 296

LIST OF EXERCISES

(SUMMARY JOBS)

1·1	How to evaluate the role of the electronics assembler	9
1·2	How to evaluate safe working conditions	10
1·3	How to evaluate employment conditions	10
1·4	How to arrange for a visit to industry	11
2·1	How to identify resistors	31
2·2	How to read the resistor color code	32
2·3	How to use an ohmmeter	33
2·4	How to determine the wattage rating of a resistor	34
2·5	How to test a variable resistor	35
2·6	How to test a tapped resistor	36
2·7	How to identify capacitors	36
2·8	How to interpret capacitor ratings	37
2·9	How to learn to read the capacitor color code	37
2·10	How to investigate capacitors	38

2·11	How to identify inductors	38		6·9	How to install components on printed-circuit boards	108
2·12	How to investigate transformer and choke construction	39		6·10	How to replace components on printed-circuit boards	109
2·13	How to investigate relay operation	40		7·1	How to identify wire sizes	124
2·14	How to associate schematic symbols of inductors with the actual part	40		7·2	How to strip wire	125
2·15	How to identify semiconductors	41		7·3	How to tin wire with a solder pot	126
2·16	How to identify semiconductor diodes	42		7·4	How to use the wire color code	127
2·17	How to identify color-coded diodes	42		7·5	How to attach solderless terminal lugs	127
2·18	How to interpret power-rectifier markings	43		7·6	How to make a spot tie	128
2·19	How to identify transistors	43		7·7	How to lace cable	128
2·20	How to identify transistor leads and sockets	44		7·8	How to make a wire harness	128
2·21	How to identify transistor cases	45		7·9	How to protect a harness	129
2·22	How to identify integrated-circuit packages	45		7·10	How to check the polarity of a cable connector	129
2·23	How to identify electron tubes	46		7·11	How to make a wire-harness jig board	130
2·24	How to identify vacuum-tube sockets	47		7·12	How to use a wire-harness jig board	131
2·25	How to use a vacuum-tube manual	47		8·1	How to appreciate the use of terminals	143
2·26	How to investigate electron-tube construction	48		8·2	How to identify terminals	143
2·27	How to test vacuum tubes	48		8·3	How to make a terminal practice board	144
3·1	How to investigate a wired chassis	54		8·4	How to wrap wires on terminals	144
3·2	How to investigate a 3-D wired circuit	55		8·5	How to interpret tolerances in the metric system	145
3·3	How to investigate a printed-circuit board	55		8·6	How to bend leads of components	145
3·4	How to investigate an integrated circuit	56		8·7	How to install components on printed-circuit boards	146
4·1	How to make schematic diagrams	69		8·8	How to solder on terminals	146
4·2	How to make a chassis wiring diagram	70		9·1	How to make use of catalogs to identify hardware	157
4·3	How to make a block diagram	71		9·2	How to make a hardware display	157
4·4	How to make a symbolic diagram	71		9·3	How to identify switches	158
4·5	How to make a layout diagram	72		9·4	How to lay out a job	159
4·6	How to read a printed-circuit layout diagram	72		9·5	How to install a potentiometer on a chassis	160
4·7	How to read a foot ruler	73		9·6	How to make a heat shunt for power transistors	160
4·8	How to read a metric scale	73		10·1	How to check the quality of a soldered connection	165
5·1	How to get acquainted with hand tools used for electrical assembly	86		10·2	How to make a pull test on a printed-circuit soldered joint	165
5·2	How to get acquainted with tools and equipment used in microelectronics assembly	87		10·3	How to make a continuity tester	166
5·3	How to get acquainted with tools and equipment used in mechanical assembly	88		10·4	How to make a test panel	167
6·1	How to tin the tip of a soldering iron	101		10·5	How to investigate quality control procedures	168
6·2	How to tin wire with a soldering iron	102		11·1	How to verify the effect of resistance	176
6·3	How to splice wires	103		11·2	How to interpret deflection on an ohmmeter scale	176
6·4	How to make a set of practice soldering boards	104		11·3	How to use a milliammeter	178
6·5	How to practice soldering wire	104		11·4	How to use a voltmeter	178
6·6	How to practice soldering to sheet copper	106				
6·7	How to remove solder from a soldered joint	107				
6·8	How to inspect the quality of a soldered joint	107				

12·1	How to test a series circuit	183
12·2	How to test a parallel circuit	184
12·3	How to test a series-parallel circuit	185
13·1	How to use an oscilloscope to display a-c waveforms	193
13·2	How to test capacitors	194
13·3	How to test transformers	194
14·1	How to check front-to-back ratios in semiconductors	209
14·2	How to investigate basic circuit arrangements of transistors (Including testing considerations)	211
14·3	How to investigate basic biasing circuits for transistors	213
14·4	How to investigate special applications for diodes	213
14·5	How to get acquainted with integrated circuits	214
15·1	How to identify a rectifier circuit	220
15·2	How to identify an amplifier circuit	221
15·3	How to identify an oscillator circuit	222
15·4	How to identify a switching circuit	222
16·1	How to identify components in an a-c power supply	231
16·2	How to identify the leads of a power transformer	231
16·3	How to locate the B-supply outputs	232
16·4	How to investigate various power supply circuits	233
16·5	How to use a bench-type power supply	233
17·1	How to check a loudspeaker with a dry cell	243
17·2	How to make an overall test of an audio amplifier	244
17·3	How to make a voltage analysis	244
17·4	How to use a signal tracer	245
17·5	How to inject a test signal	247
17·6	How to convert an amplifier to an oscillator	248
17·7	How to make a tone generator	249
17·8	How to work with a low-frequency, integrated-circuit amplifier	249
18·1	How to identify a radio-frequency stage	256
18·2	How to make a resistance analysis	257
18·3	How to use an r-f signal generator	258
18·4	How to align an intermediate-frequency amplifier	260
18·5	How to check an r-f oscillator	261
18·6	How to make tracking adjustments	261
18·7	How to produce a beat signal	263
19·1	How to make a troubleshooting procedures chart	268
19·2	How to establish priorities for test equipment	269
20·1	How to calculate for total resistance in series	277
20·2	How to calculate for total resistance in parallel	278
20·3	How to calculate d-c Ohm's law problems	278
20·4	How to calculate inductive reactance	279
20·5	How to determine the impedance of an inductor	279
20·6	How to calculate a-c Ohm's law problems	280

PREFACE

Electronics assumes an ever greater role in American daily life. Applications seem unlimited; new equipment is being designed all the time. More and more people seek employment, security, and opportunities for advancement in the electronics industry. However, they cannot merely go to work and pick up a knowledge of the trade on the job; they must be trained. A knowledge of basic electronics theory, hardware, manufacturing, and/or testing techniques is necessary in all but the most routine type of assembly work. This book provides a working knowledge of the basic theory and of the practical aspects that lay the foundation for those who seek entry into the dynamic and challenging electronics industry.

Fabrication of electronic devices is the product of a technical team composed of engineers, technicians, and electronics assemblers. The engineer originates fundamental ideas of product and proceeds to plan and design construction. The electronics assembler manufactures the product, following specifications set forth by the engineer. The technician checks the finished product to see that it meets specifications and functions properly. He is further responsible for maintenance and repair, so that the equipment stays in operation. In short, the engineer designs the product, the assembler builds the product, and the technician checks the product and maintains it in proper operating condition.

Material in this book will be of most interest to those who wish to become electronics assemblers, inspectors, or test technicians. However, anyone interested in the hardware aspects of electronics can benefit by studying the fundamentals presented here. The book requires no technical background, but is designed to *give* background for various electronics-related job clusters. It will help the technician trainee lay the groundwork for more theoretical aspects of electronics; it will help sales and administrative personnel associated with firms producing electronic-controlled products to become better acquainted with fundamental concepts and with hardware related to electronics; and it will help the consumer, appliance repairman, or hobby enthusiast to understand more about the electronic product he handles.

Over a period of several years, the authors have developed and used this material to train electronics assemblers and job-entry-level technicians at several community colleges in Southern California. Students, many with little or no mathematical and physics preparation in high school, have used this book as a springboard to more advanced courses in electronics technology, and have gone on to become successful technicians in the industry. Countless others, many automated out of mechanical assembly jobs, have taken the electronics assembly course to retrain themselves in order to remain employable. A working knowledge of material in this book provides the necessary background for employment in the field of electronics fabrication.

The first edition was used, with a high degree of success, in public and private schools and colleges throughout the country. The book has now been revised in order to keep current with the state-of-the-art as it concerns electronics manufacturing technology. Revision of the text was also advisable in order to meet the current instructional needs of institutions catering to programs for the disadvantaged, manpower development and/or retraining, and other prevocational training courses designed to help meet the needs of industry. In order to meet *instructional needs*, the book has been separated into two distinct parts and reorganized so that it is applicable to individualized instruction; flexible enough to allow for open-entry/open-exit nonstructured courses; with enough variety and depth to be useful for occupational-cluster or career-ladder programs; and with enough breadth to make it suitable as a basic entry-level core-curriculum for most electronics training programs in high schools, community colleges, trade schools, skill centers, adult education schools, and factory in-service training programs. An Instructor's Guide is available to assist educators with the use of this book.

Many individuals and organizations have graciously contributed assistance and/or materials that help make this text relevant to today's needs. The authors are eternally grateful and hereby acknowledge their assistance:

Thanks to Don Elser, Director of Vocational-Occupational Education, Elko County Schools, Elko, Nevada, and to Douglas M. Stoker, Nevada State Supervisor, Manpower Development and Training, Carson City, Nevada, for allowing coauthor R. L. Duarte to use material from this edition in a special Manpower Development Training Act (MDTA) program designed for American Indians. This experience provided the opportunity to test the applicability of the material in a special class for disadvantaged students. (Results of the evaluation were quite favorable.)

Thanks also to LeRoy Olson, formerly with McGraw-Hill Book Company, for his assistance and guidance in the preparation of the original and revised versions of the manuscript; and to Roy Battershield and Arthur D'Braunstein, both formerly training specialists at Autonetics, Division of North American-Rockwell Corporation, Anaheim, California, for their continued help in ensuring that the authors have kept pace with manufacturing trends in the aerospace industry. Thanks in this category are also extended to members of the College Electronics Faculty Association (California) and to the staff of Hickok Teaching Systems, Woburn, Massachusetts, for contributing many ideas and critiques pertaining to the first edition, which lead to an improved second edition. Finally, we wish to thank the following persons for their detailed reviews and suggestions relating specifically to the manuscript for this edition: Peter Pantelis and

Walter Schmaderbeck, Grumman Aircraft Corporation, Bethpage, New York; Buell Munson, Electronics Department, Orange Coast Community College, Costa Mesa, California; and Edwin E. Pollock, Program Director, Cabrillo College, Aptos, California.

For making available some of the photographs and illustrations used, we also thank the Heath Company, Benton Harbor, Michigan; McCulloch Corporation, Scott Division, Minneapolis, Minnesota; Radio Corporation of America, Electronic Components and Electron Tube Divisions, Harrison, New Jersey; Hunter Tools, Santa Fe Springs, California; and Allied Radio Shack, Electronic Division of Tandy Corporation, Fort Worth, Texas.

S. R. DUARTE
R. L. DUARTE

Part I

BASICS FOR ELECTRONICS ASSEMBLY

1

ELECTRONICS FABRICATION

1·1 INTRODUCTION

Fabrication of electronic devices is normally an assembly-line process. That is, one person does not completely fabricate a unit such as a radio or television receiver by himself. Construction of such devices is done by a *team* of factory workers. One person may be responsible for the drilling of holes in metal parts, another person may be responsible for assembling the cabinet, someone else may be responsible for making the printed-circuit boards, and yet another person may be responsible for soldering the electrical parts together. And when all of these separate jobs have been accomplished, someone else will assemble the various parts into the finished product.

In order that all parts manufactured by the various workers are made so that they will fit together, several persons are employed to inspect the *quality* of work being performed at each workstation. These people are known as *inspectors*.

Another group of workers who contribute to the proper operation of the finished product are the *test technicians*. These persons use electronic test equipment to measure the *quality of operation* of the completed unit. In a sense, they too are inspectors.

1·2 EMPLOYMENT CONDITIONS

With all of the above-mentioned people (and others) working as a team to produce a finished electronic product, there must, of necessity, be cooperation and coordination of effort between them. This cooperation and coordination comes about through proper leadership in the person of a supervisor. Supervisors constantly strive to direct the efforts of all employees in a manner that is pleasing and efficient. An attempt is made to make every work situation pleasing because studies have shown that employees who are satisfied with working conditions are more apt to do good work and remain on the job longer. The longer a person practices his trade, the more efficient he becomes. Further, good working conditions, happy and efficient employees, and proper cooperation and coordination all contribute to keep accidents down to a low level. This in turn adds to the overall efficiency of the *workteam*.

Since one key to successful teamwork lies with the attitude of employees, an attitude closely associated to working conditions, it becomes important to analyze working conditions and related work practices normally found in factories that manufacture electronic products. These conditions can generally be listed under four general classifications: (1) health and safety, (2) working hours, (3) education and advancement, and (4) job security.

HEALTH AND SAFETY

In general, factory supervisory personnel make every effort to ensure safe, healthy working conditions for all employees. This is done in order to minimize absenteeism due to illness or accident. Employees who are at work regularly become more efficient. Not only is a healthy, safe working environment mandatory for maximum efficiency, but factory workers are ensured healthy and safe places to work by state and federal laws which insist on these conditions. Government agencies and labor organizations constantly oversee working conditions in factories in order to ensure compliance with these laws.

Conditions found in the electronics industry and which contribute to a healthy and safe place in which to work are:

1. Plenty of good lighting
2. Adequate ventilation
3. Air conditioning, air purification, and controlled temperature
4. Clean work areas
5. Use of safety guard rails
6. Use of safety glasses and clothing
7. Tools in good condition
8. Machinery in good condition
9. Adequate custodial service
10. Enforcement of safety regulations

The student of electronics is encouraged to form good, safe work habits while in school in order that he can easily adjust to safe practices expected of him in industry. This can be done by observing the following list of safe practices while undergoing training in the classroom or laboratory.

1. Walk, do not run, in the classroom or laboratory.
2. Clear the work area of chairs, benches, and stools not in use.
3. Arrange tools and equipment for easy access.
4. When working with hot soldering irons, place the soldering iron stand at a location that has clear access, that is, where it will not be necessary to reach across any object in order to rest the iron on the stand.
5. When cutting wires, keep the open side of the cutter away from your body and confine the cut wires to your own area.
6. Store all knives and pointed instruments in appropriate containers.
7. Wear safety glasses when cutting wires, soldering, using solvent, using an air hose, working around cathode-ray tubes, or working around rotating machinery. Remember: *You cannot see through a glass eye.*
8. Use the proper tool for the proper job.
9. Do not use worn-out or faulty tools.
10. When soldering a chassis, disconnect all power to the chassis before beginning work.
11. A hot soldering iron should be stored in an appropriate holder and kept as far back on the work bench as practical.
12. A hot soldering iron should not be placed in a tool box.

13. When working with power equipment, make certain that the metal case is connected directly to electrical ground.
14. Women in work area should wear low-heeled, completely enclosed shoes. Slacks or overalls should also be worn to protect the body from hot solder.
15. Loose clothing or hair must be secured before working around moving machinery.
16. All frayed or faulty power cords should be replaced.
17. When making resistance measurements, disconnect all power from the chassis under test.
18. Use an isolation transformer when testing a chassis that utilizes a transformerless power supply.
19. When testing or working on a chassis that has power applied, use one hand at a time, that is, never let both of your hands come in contact with the chassis at the same time.
20. Know the location of all emergency power-disconnect switches.
21. Learn the principles of artificial respiration.
22. A second person should be in the room when someone is working on any device that utilizes electrical power.
23. In case of accident, give whatever assistance is necessary and call for the instructor. If the instructor is not readily available, notify the administration office.
24. If an accident occurs that involves an electrical shock, press the emergency power-disconnect switch, give artificial respiration, and send someone for help. However, *continue to give artificial respiration.*
25. In addition to safety glasses, long sleeves should be worn while working with cathode-ray tubes.
26. There shall be no horseplay of any kind while in the classroom or laboratory.
27. Know the school procedure concerning a fire disaster.
28. Know the school procedure concerning a natural disaster.
29. Know the school procedure concerning atomic blasts and radiation fallout.
30. Learn the telephone number of the administration office.
31. Do not operate any electrical device unless instruction has been given concerning the operation of such equipment.
32. Do not operate machinery without safety guards in place and properly secured.
33. Take your time. *Haste makes waste.*
34. Always work in a well-ventilated classroom or laboratory. Fumes given off during soldering demand an adequate exhaust and ventilation system.

1·3 WORK PRACTICES

WORKING HOURS

Working hours are generally the same nationally since they are governed by federal labor laws. The usual (maximum) is eight hours of work per day, five days of work per week. Overtime pay is awarded on at least time-and-a-half basis for those who work beyond these daily or weekly maximum hours. Although most people are employed during daytime hours, many factories often employ a regular night workshift as well. Working conditions and the length of workshift are generally the same. However, a slightly higher pay scale usually prevails for nighttime employees.

EDUCATION AND ADVANCEMENT

The electronics industry and its related technology constantly changes and improves. Therefore, persons who train for work in any phase of electronics must be prepared to keep retraining in order that they can remain employable. Employers recognize this and therefore usually provide in-service training. In-service training is designed to help an employee to keep abreast of technological changes. A factory worker who does not keep up with his trade cannot work efficiently as part of the work-team mentioned above. Therefore, he soon becomes useless to his employer.

In-service training may be accomplished on the job, at factory schools, or at neighborhood schools or colleges. Although responsibility to keep up with the trade belongs to the employee, the employer often helps by providing time off, paying tuition and books, and at times by providing all three.

Promotion and advancement is closely tied to the person's knowledge of the trade. Therefore, employees who voluntarily take in-service training courses are most apt to be promoted. Longevity on the job alone is seldom sufficient for advancement in this technological age, especially in the electronics industry.

JOB SECURITY

Security of one's job is closely related to knowledge of the job. However, labor organizations and fair labor practice laws oversee that employees are not discriminated upon unjustly. Although many firms that manufacture electronic products cater to national defense and the aerospace industry, and are therefore subject to government contracts, nevertheless the electronics industry as a whole is not to be considered a seasonal industry. Employment in this industry is on a twelve-month basis and does not depend on weather or other similar variables.

1·4 ASSEMBLY-LINE PROCESSES

This text deals primarily with practices encountered in assembly and fabrication of electronic products, especially those practices which are normally used in mass production. Large-scale production of electronic devices allows for employment of individuals who have trained for specific jobs. Some of these jobs are relatively simple and require little training, while other more complicated jobs require extensive preemployment instruction. Therefore, there usually is some job classification available for every person who has had some degree of basic training.

The assembly-line process used for manufacturing electronic devices refers to a station-by-station progression of the work to be done. For example, an employee at one workstation might be responsible for installing a certain bracket on an amplifier chassis and nothing else. Figure 1·1 shows a photograph of an amplifier chassis. Eventually the person assigned to install the bracket at that particular workstation becomes quite good at it, having learned how to handle all required tools efficiently. He will also have learned which hand can best hold

FIG. 1·1 Amplifier chassis (printed circuit).

the bracket while it is being installed and which operations can be combined in order to save time. If the overall operation takes 10 minutes to accomplish, 48 brackets might be installed in an 8-hour workshift. With more experience the worker might be able to install 55 or 60 brackets in a workshift. This not only helps to produce more amplifiers, but it makes the employee more valuable to his employer. This will ensure continued employment and it will pave the way for promotions and increase in salary.

WORKSTATIONS FOR MECHANICAL ASSEMBLY

The above example of where a bracket is to be installed would be a *mechanical assembly* workstation. At this workstation would be found only tools necessary for that particular operation. In addition, materials such as rivets, bolts, nuts, washers, and other similar individual units of *mechanical hardware* would be found. However, only hardware units necessary for the installation of the bracket would be found at this workstation.

Storage of appropriate tools and hardware needed are normally arranged neatly in tool boxes or storage bins in such a way that they will be readily accessible when needed. The work area will be found to be clean and well lighted. This cleanliness and orderliness is necessary in order to attain maximum efficiency. If a person knows exactly where a given tool is, he does not waste time looking for it. Besides, a clean and orderly workstation minimizes the chances of accident. It also contributes to making working conditions pleasant.

ELECTRONICS ASSEMBLY WORKSTATION

Workstations organized specifically for the assembly of electronic devices would have tools such as pliers, wire cutters, wire strippers, soldering irons, lead-bending tools, optical magnifiers, and special vises. The bins would have materials such as resistors, capacitors, inductors, potentiometers, transistors, insulators, wire, solder, and lacing cord. Figure 1·2 shows a typical *electronics assembly workstation*. Workers assigned to work at these stations would have been trained to identify these components and to properly solder them together. They would not be expected to know other job functions such as mechanical assembly or testing.

ELECTRONIC TEST STATION

When all of the brackets and resistors and other component parts have been installed in the chassis, the completed chassis is taken to a *test station*. At this station *test technicians* make a series of electrical and electronic tests to see if the completed unit functions properly. To make these tests he will use electronic instruments such as ohmmeters, voltmeters, milliammeters, signal generators, and signal tracers. If the unit does not function properly, the test technician would then send the faulty device to an *electronics technician* for repairs. It should be stated that job titles in industry vary somewhat from one manufacturing company to another. However, the job titles used in this textbook can be considered fairly accurate and representative. Although the electronics technician is highly trained, having at least 2 years of college education or its equivalent, he is also expected to know many of the assembly and fabrication processes. At the workstations one would find soldering irons, schematic diagrams, color-coded resistors and capacitors, printed circuit boards, and semiconductors; all of the same tools and components that the electronics assembler has to deal with. Therefore, the basic training is the same.

1·5 SUMMARY: CONCEPTS OF ELECTRONICS FABRICATION

In summary, a person seeking employment in a manufacturing plant that produces electronic products will find it to his advantage to know as much of the overall trade as possible. He should have a good idea regarding what the trade expects of him, and he should have an idea of what to expect in terms of working conditions and related work practices. And yet, the fact that a person is considering working in the electronics industry should be an indicator that that person already has a basic knowledge of electronic fabrication concepts. He must have this knowledge since he will become part of a team whose objective is to produce a finished

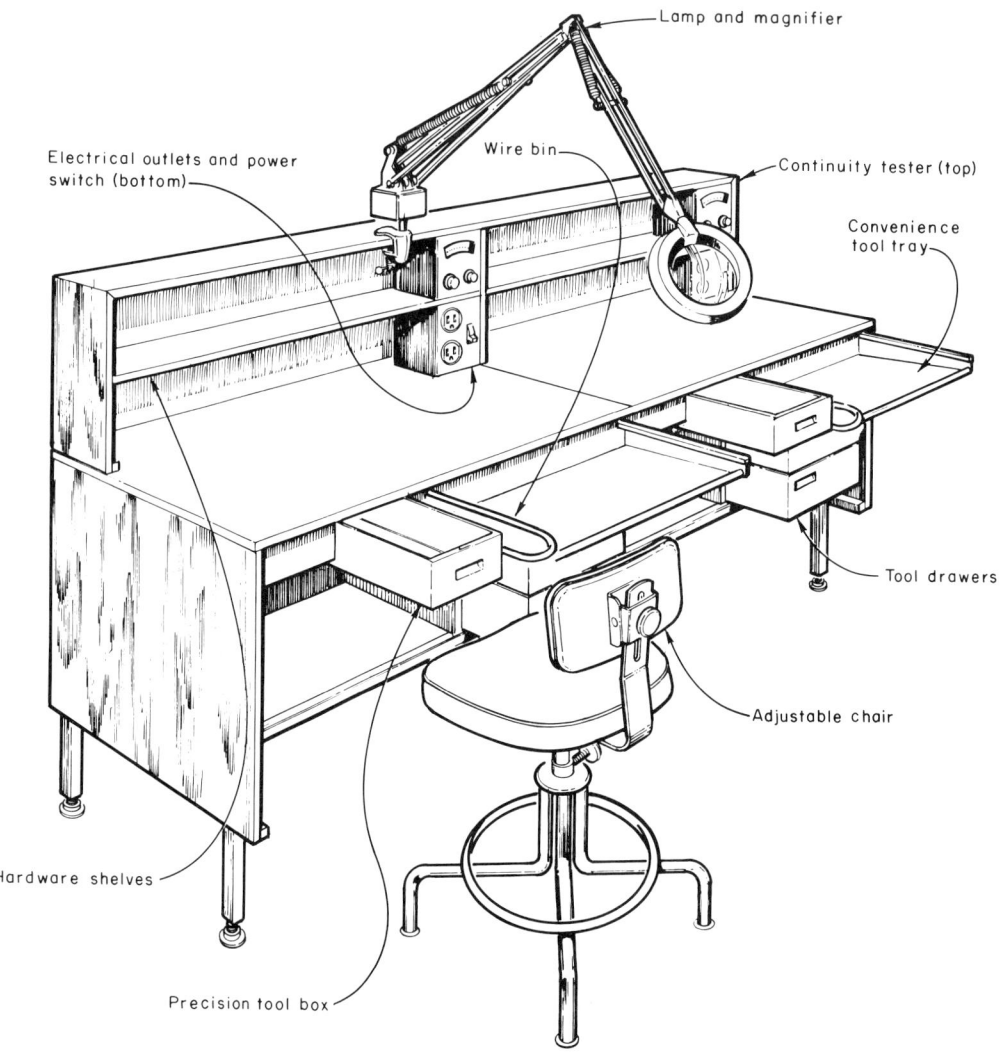

FIG. 1·2 Electronics assembly workstation.

product; and he must therefore be able to hold his end of the effort.

The jobs that follow are designed to give the student an opportunity to expand his knowledge of the subjects covered in this chapter through *practice*. In other words, the student is given an opportunity to put into practice what he has learned. It is through immediate application of the fundamentals that these basics mold into a solid foundation. Besides, if the student encounters difficulty in applying the fundamentals, it is best to know it now before attempting the following chapter. It is only through an awareness of one's weakness that we can attempt to correct the situation.

EXERCISES

JOB 1·1 How to evaluate the role of the electronics assembler

OBJECTIVES

1. To appreciate the role of the electronics assembler
2. To appreciate knowledge required of an electronics assembler
3. To evaluate the student's background

MATERIALS REQUIRED

Supplies
An assortment of electrical and electronic components, preferably in kit form as shown in Fig. 1·3
8½ × 11-inch notebook paper
Pencil

PROCEDURE

1. Carefully examine all parts supplied, both electrical and mechanical.
2. Separate the parts into two groups: electrical and mechanical.
3. On a sheet of notebook paper list only the parts that will serve to handle or control an electrical current. You may sketch the part if you do not know its name.
4. On a separate sheet of paper list only the parts that serve as mechanical-support hardware. Again, draw a sketch if you do not know its name.
5. On a third sheet write a simple explanation regarding who *decides* what parts are to go into a new radio receiver, who will actually *make* the radio

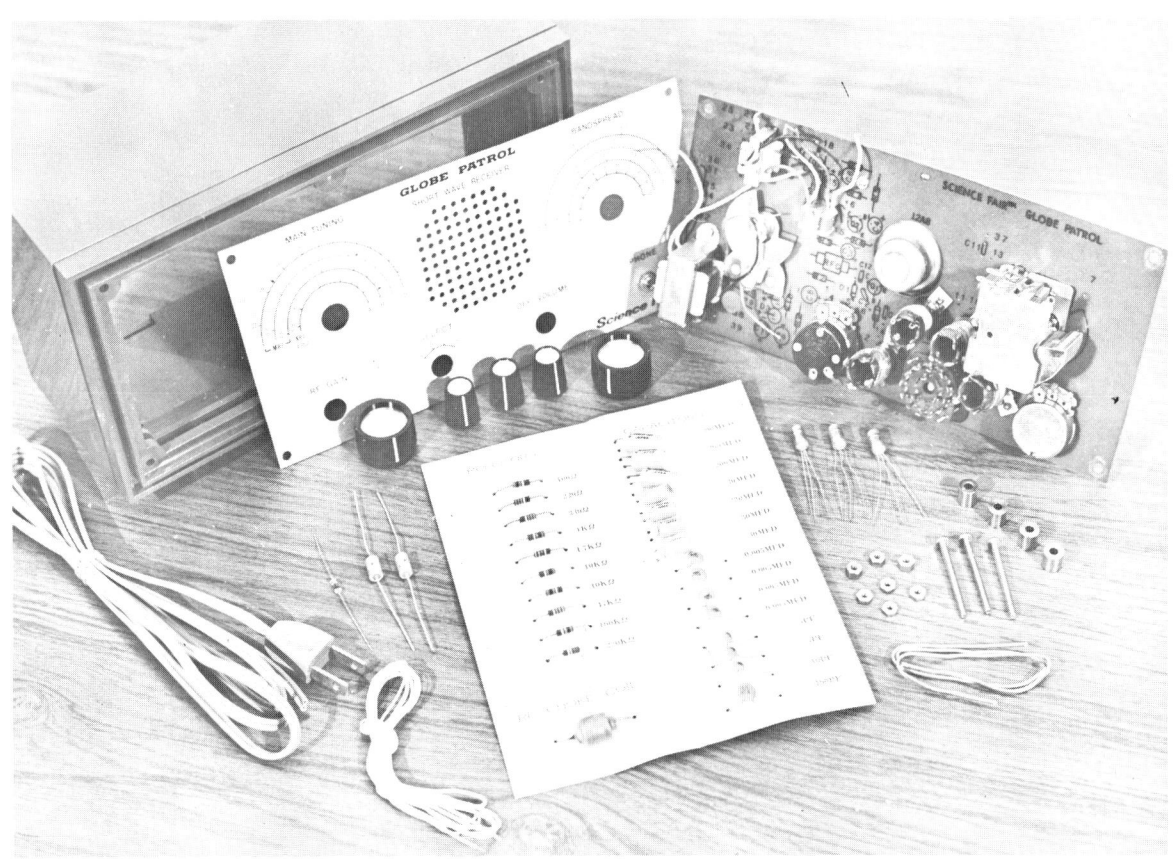

FIG. 1·3 Radio receiver kit.

9

receiver, and who will *test* the receiver when it is finished.
6. On a fourth sheet of notebook paper answer the following questions and submit these answers to your instructor for evaluation. You will follow this same procedure when working the exercises (jobs) at the end of each chapter.

QUESTIONS
1. How many resistors were supplied in the kit of parts?
2. How many capacitors were supplied in the kit of parts?
3. How many inductors were supplied in the kit of parts?
4. Did the kit contain transistors, or vacuum tubes?
5. What purpose does mechanical-support hardware serve in a radio receiver?
6. Explain the difference between roles played by the *electronics engineer,* the *electronics assembler,* and the *electronics technician.*

JOB 1·2 How to evaluate safe working conditions

OBJECTIVES
1. To become aware of unsafe working conditions when they exist
2. To develop the habit of reporting unsafe conditions to supervisors
3. To appreciate safety rules

MATERIALS REQUIRED

Supplies
8½ × 11-inch notebook paper
Pencil

PROCEDURE
1. Section 1·2 of this textbook refers to 10 conditions found in industry which contribute to a healthy and safe place to work. On notebook paper write a brief summary relating to the *importance* of these conditions.
2. Section 1·2 of this textbook refers to 34 safe practices to be observed when working in the classroom or laboratory. On notebook paper write a brief summary statement giving *reasons* for each item on the list. For extra credit, expand the list to cover items overlooked by the authors.
3. Ask your instructor to obtain an invitation for you to visit a shop or laboratory in your school so that you may evaluate their operation in terms of safety. Write a formal report on the *good and bad* safety practices you observed. Submit this report to *your* instructor.

QUESTIONS
1. In your opinion, what are the *three* most important conditions that contribute to a healthy, safe place to work? Why?
2. In your opinion, what are the 10 most important safety practices to be observed? Why?
3. When you visited a neighboring shop or laboratory, what safe practices were most often violated?
4. When working in *your own* shop area, if you see an unsafe condition, or observe another student violating a safety rule, what do *you* intend to do about it?
5. Would you be helping or hurting a friend if you point out to him that he is violating a safety rule? Why?
6. Would you be helping or hurting a friend if you observed him violating a safety rule and, after alerting him of the violation he did nothing about it, you informed your instructor? Why?
7. List violations of possible safety practices by other students which could injure *you* in the event of an accident. Explain each item on your list.

JOB 1·3 How to evaluate employment conditions

OBJECTIVES
1. To become acquainted with the industry you intend to prepare for in terms of employment conditions
2. To consider the future and job security in the industry or trade for which you will be preparing
3. To get an insight as to what will be expected of you when you are ready to apply for employment in this field

MATERIALS REQUIRED

Supplies
Small notebook
Pencil

PROCEDURE

1. Visit a local employment agency and identify yourself as an electronics student assigned the task of evaluating an employment opportunity. Ask for permission to take notes.
2. Ask to see any publications that advertise employment opportunities.
3. Select an advertisement that describes the type of work to be done by the applicant if hired. Ask yourself if you would be happy doing the type of work described. To succeed on a job, one must like what he is doing.
4. Next, try to decide if you have the necessary training and experience to handle the work. You must be practical in selecting a job.
5. Investigate the kind of working conditions associated with this employment opportunity. This will contribute to being happy on the job.
6. Is this place of employment near your home? Will you have to move in order to be near the job? If you have to move, you will have additional expenses that would not otherwise have been necessary.
7. Inquire about public transportation between your home and the place of employment. Private automobiles occasionally break down, and you might have to depend on public transportation to get to work.
8. Find out how many hours per day you are expected to work. Also inquire about the days of the week that you will normally work if you accept employment.
9. Inquire about the workshift on which you will begin employment.
10. Inquire about the maximum wage or salary that is possible to attain. How long a period of time will it be before you can expect to attain maximum pay for doing the type of work you have been hired to do?
11. Ask about the starting wage or salary. Are you reasonably able to meet your financial obligations with the amount of money you will be receiving for the first few months of employment?
12. Inquire about advancement to other more complex areas of employment.
13. Inquire about provisions for sick leave.
14. Inquire about medical group insurance.
15. Inquire about membership in labor organizations.
16. Investigate to find out if the type of product manufactured or serviced is seasonal.
17. What type of retirement system does the company offer to all employees?
18. If you have a physical handicap, it is best to investigate if you can qualify for employment with the company concerned.
19. Is it necessary that the applicant be a citizen of the United States in order to qualify for employment?
20. Inquire about any special requirements that must be met by the applicant in order to be hired.
21. Investigate and analyze any special benefits that are offered by the prospective employer to the employees.
22. Ask yourself, if hired, how long you intend to remain in the employ of the employer.

QUESTIONS

1. What is the most important phase of evaluating an employment opportunity? Why?
2. Explain the purpose of each consideration called for from step 3 through step 22 of this assignment.
3. Why was the matter of wages and salaries delayed until steps 10 and 11 in the procedures?

JOB 1·4 How to arrange for a visit to industry

OBJECTIVES

1. To motivate the student
2. To see firsthand the needs and requirements of industry
3. To appreciate the cooperation between industry and schools

MATERIALS REQUIRED

Equipment
Telephone directory
Typewriter

Supplies
School letterhead stationery
Pen in good condition

PROCEDURE

1. This job is actually a group project. Therefore, select from members of your class a committee to work on arranging for a field trip to industry.
2. Have all members of the class inquire from friends and neighbors for the name and address of a suitable industrial plant in your local area that manufactures electronic products.
3. The class should decide which industrial plant they would like to visit. The instructor should be consulted for advice.
4. Obtain permission from your instructor and from the school administrators to contact the particular firm on behalf of the school, asking for an invitation to visit on a field trip.
5. From the school administrators, get tentative permission to use school transportation in the event that you are successful in arranging a field trip for the class.
6. If approved, get the telephone number of the industrial plant you wish to visit from the telephone directory.
7. Telephone that number and ask for the public relations office.
8. Inquire from the public relations officer the procedure to follow in order to request from his company an invitation for your class to visit on a field trip. State the nature of your class, the number of students enrolled, and your objectives for wanting to visit.
9. If an invitation appears possible, ask for a tentative date and time. Also ask for the name and mailing address of the person to address in writing when requesting the invitation. In addition, find out what additional or specific information they need to know about the class prior to your making the trip. (Security precautions often require that you submit the name of every person who will make the trip.)
10. Next, get a firm commitment from the school to allow the class to visit the plant on the prescribed date and time. The commitment should include the necessary transportation.
11. Having the necessary commitment from the school, write a formal letter to the company to be visited on school letterhead stationery. Type the letter if at all possible. If it is not possible to type it, write the letter neatly and clearly, using a good pen.
12. Assuming you get an invitation in reply to your request, go on the field trip. Allow plenty of time so that you arrive on time.
13. While visiting the plant, do not hesitate to ask questions.
14. Upon return to the school, write a formal thank-you letter on school letterhead stationary. Address it to the same person who sent you the final invitation letter.

QUESTIONS

1. The committee should report to the class the problems and experiences they encountered in arranging the field trip.
2. Every member of the class should write a summary of their observations which pertain to the field trip.
3. Every member of the class should write a statement regarding their desire, or lack of desire, to work in the plant they visited, stating the reasons for their opinion.

2

COMPONENT IDENTIFICATION

One of the first steps that everyone entering the electronics field must take is learning to identify basic parts that he will be working with. He must *recognize* parts, call them by their *proper name*, and be able to associate the part with *symbols* which are used to represent these parts. Further, he must be able to interpret all *markings* on the parts, recognizing *quantitative aspects* which are associated with these markings.

Recognizing parts, learning their proper name, and associating symbols with the part can all be done through memorization. However, in order to interpret markings and recognize the quantitative aspects, the student needs to have a minimum knowledge of basic electrical principles. At least enough to develop a vocabulary in this field.

This chapter deals primarily with identification of *component* parts used in electronics work. However, in order to make identification meaningful, a brief summary of basic electrical concepts must be covered first. The student wishing more information on these principles is referred to Chapter 11 in this book.

2·1 ELECTRICAL CONCEPTS

All components used in electronics work have been designed with the idea that they will exert a certain amount of *control* on an *electric current.* An electric current is a flow of *invisible* particles called *electrons.* Electrons flow readily through some materials, such as copper wire. In other materials, such as rubber, plastic, vinyl or Teflon, electrons find it difficult to move. Materials in which electrons move easy are known as *conductors,* and materials in which electrons find it difficult to move are known as *insulators.* Therefore, a current of electrons will flow through copper wire, but not through the insulation that may surround it. Figure 2·1 illustrates the type of electrical wire often used in electronics work. When there is need for a large amount of electric current, a large diameter wire is used. An electric current is measured in *amperes* and is symbolized by the letter I.

A component known as a *resistor* serves to restrict the flow of an electric current. A resistor lies in the region between a good conductor and a good insulator. Therefore, depending on its value, it can exert some control on the amount of current that will flow. Figure 2·2 shows an assortment of typical resistors used in electronics work. Resistors are rated in *ohm* units. The letter R is used to represent resistance.

When an electric current flows through a resistor, the current will cause it to get hot. In order to protect resistors from this heat, they are made in different *physical* sizes. Therefore, resistors are also rated according to the amount of heat that they can withstand. This is known as the *power rating,* and the *watt* is the basic unit of power.

Electrical *pressure* is needed to force the electric current along a wire or through resistors. This pressure is known as *voltage* and is represented by the letter E. Voltage is made available by batteries, such as shown in Fig. 2·3. Voltage is really the driving force that starts the electrons to move along conductors or resistors. It can be said that the flow of an electric current (I) depends upon how much

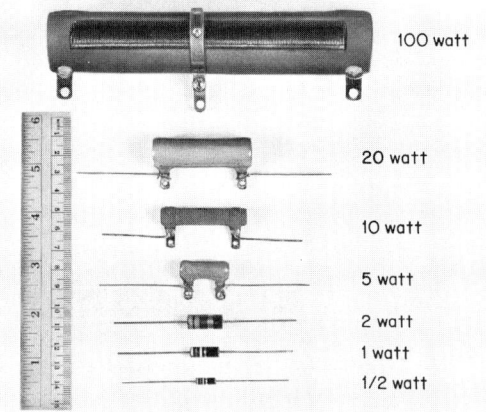

FIG. 2·2 An assortment of resistors with a relative indication of their power rating.

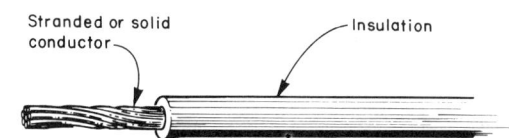

FIG. 2·1 Electrical wire.

FIG. 2·3 An assortment of batteries.

pressure (E) is forced upon it, and on how much resistance (R) it encounters.

It is possible for electricity to be stored for use at a later time. It can be stored *electrostatically* in a *capacitor*, or it can be stored *magnetically* in an *inductor*. Figure 2·4 shows an assortment of capacitors, and Fig. 2·5 shows an assortment of inductors. An inductor can also be used to *alter* the level or amount of electricity. Other components that can also alter the level of electricity are *transistors* and *electron tubes*. Transistors are shown in Fig. 2·6, and electron tubes are shown in Fig. 2·7.

Electricity is an all-encompassing term to which most students should have no trouble relating. The type of electricity available from batteries is known as *direct current*. It is abbreviated *d-c*. The type available in the home or shop from a convenience outlet is referred to as *alternating-current* electricity. It is abbreviated *a-c*. We normally get a-c electricity through a plug as shown in Fig. 2·8.

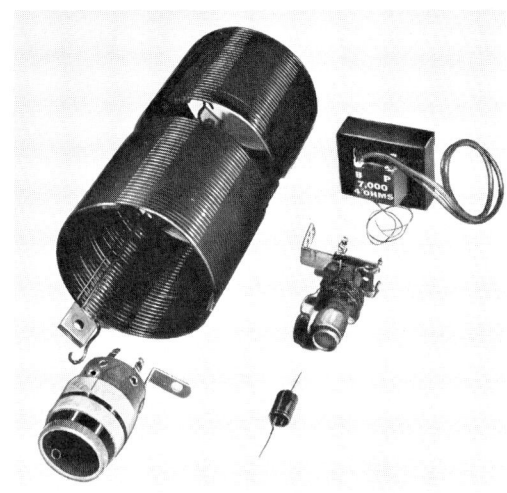

FIG. 2·5 An assortment of inductors.

Electrical devices designed for use with one type of electricity should not be used with the other type.

2·2 RESISTOR RATING

RESISTANCE
Again, a resistor serves to limit an electric current to a usable level. It limits the current by presenting a resistance to the movement of electrons in a closed circuit. The amount of resistance offered by a resistor is expressed in *ohm* units.

WATTAGE
When electrons move as a current within a resistor, they cause a temperature rise in the resistor. This heat can be very low, or it can climb to levels that are quite high. Electrical power is consumed when heating a resistor. Therefore a resistor built to withstand a high temperature level must have a high power rating. (The electrical unit for power is the *watt*.)

In summary, resistors have a resistance rating that is expressed in ohms and a heat-withstanding rating that is expressed in watts.

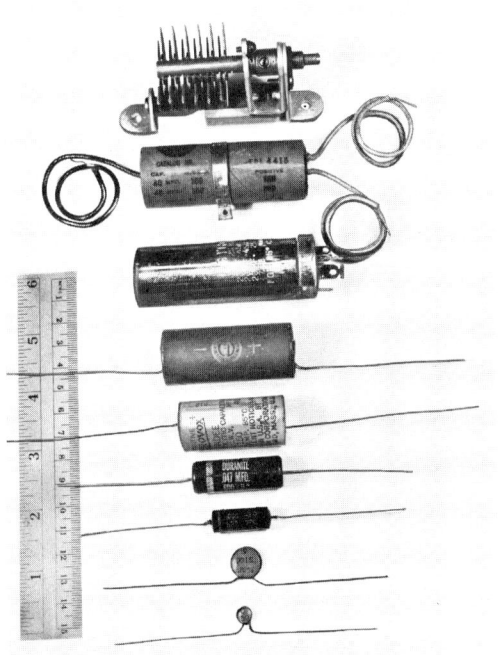

FIG. 2·4 An assortment of capacitors.

FIG. 2·6 An assortment of transistors (compared with an octal-based electron tube).

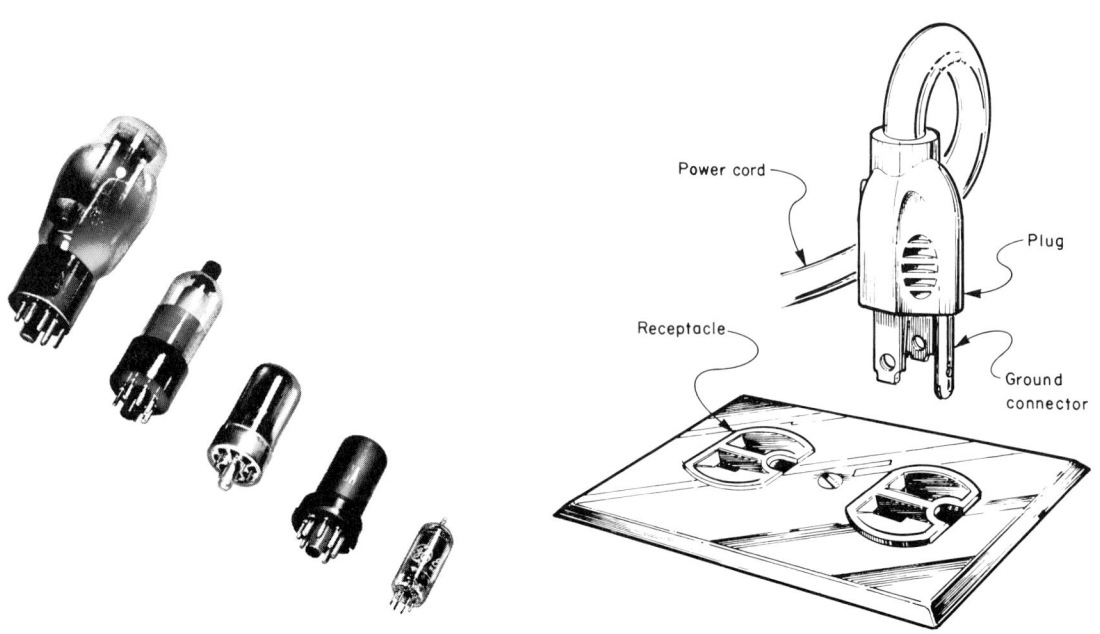

FIG. 2·7 An assortment of electron tubes.

FIG. 2·8 A common source for a-c electricity.

2·3 RESISTOR MARKING

OHMIC VALUES

The resistance rating of a resistor is indicated on the resistor by one of two methods: one utilizes a color code; in the other, the ohmic value is stamped on the resistor. Skill in reading the resistor color code is mandatory when working in electronics. Therefore students should pay detailed attention to Job 2·2. See Fig. 2·9.

WATTAGE RATING

The wattage rating of a resistor is usually a relative condition. Physically small resistors can withstand only low levels of heat generated by their own current. Therefore low-wattage resistors are small and high-wattage resistors are large. Figure 2·2 should help to clarify this point.

2·4 RESISTOR TYPES

POWER RESISTORS

Resistors of the type shown in Fig. 2·9 are used quite frequently. These resistors seldom generate enough heat to be felt by the touch of the hand.

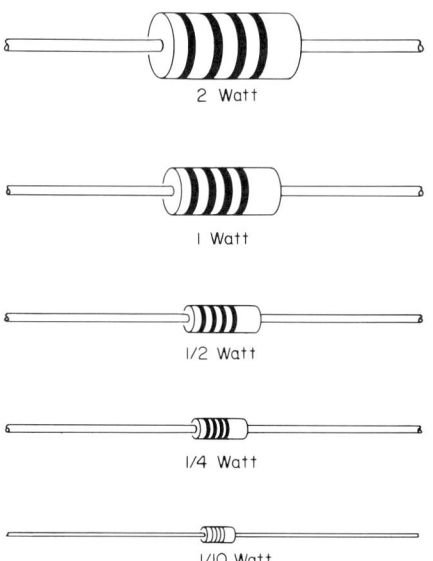

FIG. 2·9 Color-coded resistors are shown in actual size. (Some resistors use five color bands.)

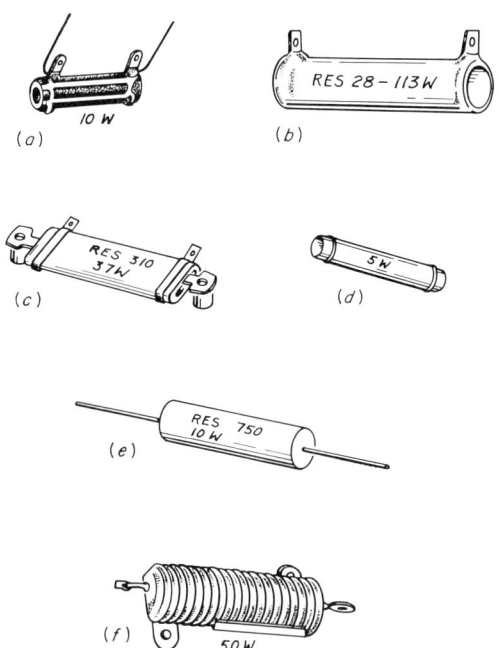

FIG. 2·10 An assortment of power resistors (not drawn to scale).

The resistors illustrated in Fig. 2·10, used less frequently, do generate enough heat to burn the human hand.

Resistors that in normal use generate a great amount of heat are manufactured to withstand this heat. They have high wattage ratings, and are called *power resistors*. Their rating is stamped on them, although in time the heat generated will obliterate any marking.

PRECISION RESISTORS

Many electrical circuits and devices require very exacting resistors. These precision resistors may take an odd shape because of specifications and tolerances that must be observed in their manufacture. A few precision resistors are shown in Fig. 2·11.

Precision resistors are seldom high-power resistors. Any marking or rating is usually stamped on the resistor.

VARIABLE RESISTORS

Since an electric current depends upon the resistance in a closed circuit, it is possible to vary the

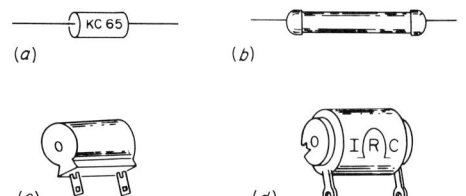

FIG. 2·11 An assortment of precision resistors.

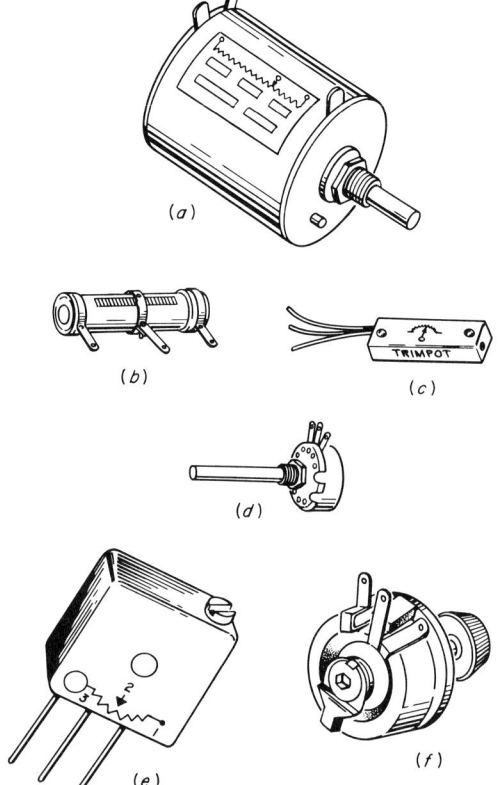

FIG. 2·12 An assortment of variable resistors.

current by varying the resistance. To meet this requirement, variable resistors are manufactured. Some of the most popular are shown in Fig. 2·12.

Variable resistors are rated according to the minimum and maximum resistance limits. They also have power, or wattage, ratings. Resistance variation can be obtained by an operator by turning a knob, or the adjustment may be made with the use of a screwdriver. The volume control in a radio receiver is an example of a variable resistor in use.

A variable resistor used primarily to control current is identified as a *rheostat*. See Fig. 2·12f. A variable resistor used in voltage divider network, where its main function is to control voltage output, is known as a *potentiometer*. See Fig. 2·13.

TAPPED RESISTORS

A resistor similar to that shown in Fig. 2·12b, but whose midconnection is nonadjustable, is identified as a *fixed tapped resistor*. (The resistor in Fig. 2·12b is an adjustable tapped resistor.) Tapped resistors find application in voltage divider circuits or where semipermanent adjustments may be necessary.

2·5 SCHEMATIC SYMBOLS FOR RESISTORS

Symbols used in *schematic* diagrams to represent fixed and variable resistors are shown in Fig. 2·14.

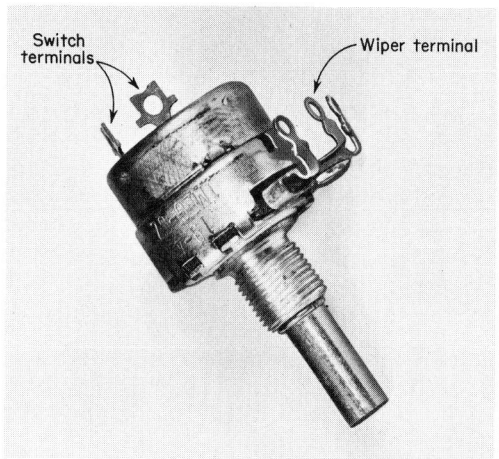

FIG. 2·13 A potentiometer with a switch mounted at the rear.

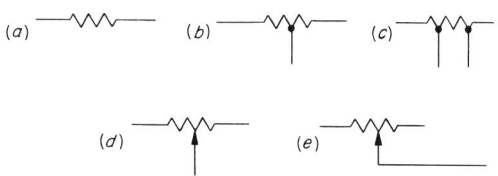

FIG. 2·14 Resistor symbols: (a) Fixed resistor. (b) Fixed, one-tap resistor. (c) Fixed, two-tap resistor. (d) Potentiometer. (e) Rheostat.

Knowledge of these symbols should be mastered by all electrical and electronics students. A dot represents a fixed tap, and an arrow represents an adjustable connection.

RESISTORS SUMMARIZED

Resistors are used in electrical circuits to control current or voltage. In performing their task, resistors generate heat. Resistors have two ratings, their ohmic, or resistance, rating and their power, or wattage, rating.

Although most resistors have fixed values, some are variable and still others may be tapped. Also, a rheostat controls current, while a potentiometer controls voltage.

2·6 CAPACITOR RATING

CAPACITANCE

A capacitor has the ability to maintain a *displacement* of electrons for a limited time. The displaced electrons are electrons removed from one plate of a capacitor and stored on the second plate of the same capacitor. The amount of electrons that can thus be displaced determines the capacitance of a capacitor. Capacitance is measured in *farad* units. However, since a farad is a very large unit, only small fractions of a farad can be utilized, for example, *microfarad* units.

BREAKDOWN VOLTAGE

Electrons removed from one plate of a capacitor and stored on the second plate cause a voltage to be developed *across* the capacitor. The plate losing electrons assumes a positive charge, and the plate gaining electrons becomes negatively charged. This voltage puts a stress on the capacitor *dielectric*, which is placed between these oppositely charged plates. The limit of the electrical *pressure* the dielectric can withstand is the limit of the voltage that can be developed across the capacitor. This limitation is known as the *breakdown voltage* rating of a capacitor.

In summary, capacitors have two ratings, the first, capacitance, measured in farads, and the second, the breakdown voltage.

2·7 CAPACITOR MARKING

The capacitance and breakdown voltage ratings of capacitors are often stamped on the capacitor casing. Color codes are also used on some types of capacitors. Examples of capacitor markings are shown in Figs. 2·15 to 2·17.

2·8 TYPES OF CAPACITORS

Capacitors are classified in several ways, according to shape, size, case, dielectric, and intended application. Two or more classifications are often employed in identifying the type of capacitor involved.

For a simplified description, this text assigns capacitors to one of four classifications: general-purpose, hermetically sealed, electrolytic, and variable.

GENERAL-PURPOSE CAPACITORS

Capacitors illustrated in Fig. 2·15 represent the majority of capacitors found in electronics. None

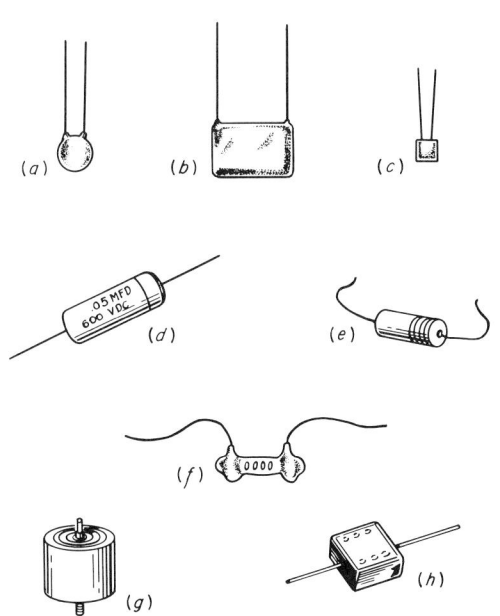

FIG. 2·15 An assortment of general-purpose capacitors: (*a*) Ceramic disk capacitor. (*b*) Ceramic plate capacitor. (*c*) Microminiature ceramic capacitor. (*d*) Molded tubular capacitor. (*e*) Molded tubular capacitor. (*f*) Tubular ceramic capacitor. (*g*) High-voltage ceramic capacitor. (*h*) Molded mica capacitor.

of these capacitors will be over 1½ inches long. Most will be well under 1 inch in length from lead to lead. Unfortunately, identifying marks on these capacitors are not standard. Job 2·9 at the end of this chapter gives instructions on reading the capacitor color code, which is used on capacitors *e, f,* and *h* of Fig. 2·15.

HERMETICALLY SEALED CAPACITORS

Capacitors are sealed in metal containers when an insulating oil or an electrolyte is used. They are also sealed to keep moisture away from the capacitor. The sealing process must be nearly perfect so that no liquid or moisture can get in or out. Two typical hermetically sealed capacitors are illustrated in Fig. 2·16.

ELECTROLYTIC CAPACITORS

The dielectric in an electrolytic capacitor is developed by chemical means. The chemical action is initiated and maintained by a small electric current. This current must flow through the capacitor in *one direction only.* For this reason, when installing electrolytic capacitors, consideration must be given to polarity. Failure to do so will destroy the capacitor and possibly damage other components being used.

An assortment of electrolytic capacitors is illustrated in Fig. 2·17. Notice that in each case the polarity of the leads is indicated in some way. The capacitance rating of the electrolytic capacitors is typically high. Notice also that the capacitors shown are not all single capacitors. Those in Fig. 2·17*a, b,* and *e* are actually two capacitors in one case.

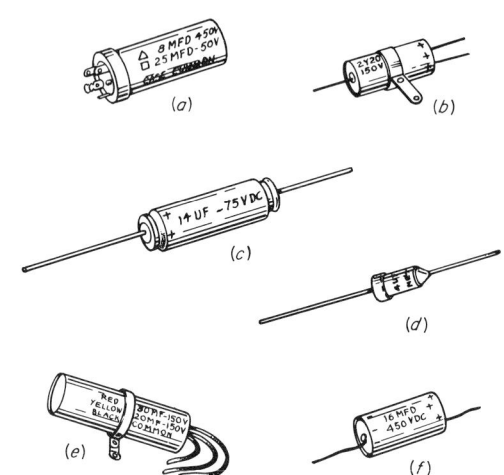

FIG. 2·17 Electrolytic capacitors: (a) Twist-prong-can electrolytic capacitor. (b) Insulated-can dual electrolytic capacitor. (c) Foil-type tantalum electrolytic capacitor. (c) Slug-type tantalum electrolytic capacitor. (e) Wax-filled cardboard electrolytic capacitor. (f) Insulated metal tubular electrolytic capacitor.

VARIABLE CAPACITORS

Capacitance in electrical circuits often has to be adjusted to meet a change in circuit conditions. To simplify the change in capacitance, some capacitors are made variable. Typical variable capacitors are illustrated in Fig. 2·18.

The two-ganged variable capacitor in reality is two separate capacitors adjusted in unison by means of a common tuning shaft. The trimmer, or padder, capacitor is a very small adjustable capacitor used in circuits where infrequent adjustments are necessary. Its identity as a trimmer or padder is determined by its function in the circuit using it.

2·9 SCHEMATIC SYMBOLS FOR CAPACITORS

Symbols used in schematic diagrams to represent fixed and variable capacitors are shown in Fig. 2·19. Notice that only the electrolytic capacitor has polarized leads.

The arrows associated with the variable capacitor symbols indicate that the corresponding capacitors are adjustable; they do not indicate current flow. The broken line joining the two-ganged variable capacitor indicates that both capacitors are adjusted in unison.

FIG. 2·16 Hermetically sealed oil or paper capacitors: (a) Hermetically sealed high-voltage rectangular capacitor. (b) Hermetically sealed bathtub capacitor.

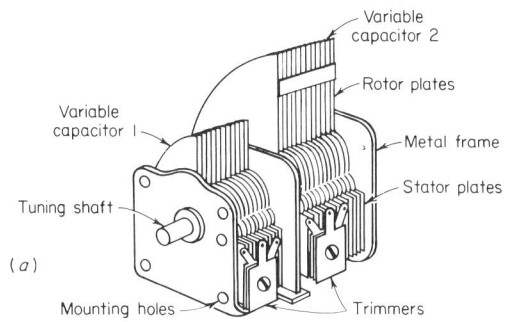

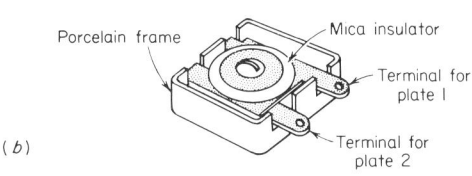

FIG. 2·18 Variable capacitors: (a) Two-gang variable capacitor. (b) Trimmer or padder capacitor.

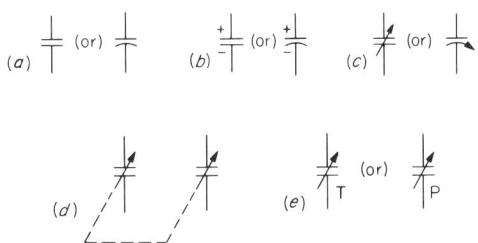

FIG. 2·19 Capacitor symbols: (a) Basic capacitor symbol. (b) Electrolytic capacitor. (c) Variable capacitor. (d) Two-gang variable capacitor. (e) Small trimmer or padder capacitor.

CAPACITORS SUMMARIZED

Capacitors have the ability to maintain an *electrostatic* charge for a limited time. This electrostatic charge is a voltage which exerts a pressure upon the capacitor dielectric. The capacity to build up the electrostatic charge is rated in farads. The ability of the dielectric to withstand the charge is rated in terms of the breakdown voltage.

Capacitors are manufactured in many sizes and forms. Their rating is either stamped on the casing or color-coded. Electrolytic capacitors require strict observance of their lead polarity.

2·10 INDUCTOR RATING

INDUCTANCE

An inductor has the ability to create a magnetic field when an electric current flows through its coil of wire. This magnetic field can be used to cause electric motors to rotate, to operate relays, and to cause the movement of electrons in other nearby coils. It can also set up an opposition to its own current. The ability of an inductor to do all of this is known as *inductance*. Inductance is measured in units called *henrys*.

RESISTANCE

Since inductors are made of wire, the resistance this wire offers to an electric current is important. This resistance is measured in *ohm* units.

IMPEDANCE

The opposition to current created by the magnetic field in an inductor and the resistance the wire in the coil offers to the movement of electrons combine to create a total opposition to electron movement. This total opposition is known as *impedance* and is evaluated in ohm units also. An impedance rating is not found on all inductors.

In summary, inductors are rated according to their ability to set up a magnetic field and to oppose current flow.

2·11 INDUCTOR MARKING

LABELS

Inductance, resistance, and impedance ratings of inductors are stamped on the case housing the inductor or given on attached paper labels. Tubular insulated chokes (see Fig. 2·20d) utilize a color code.

LEADS

If several leads or terminals are required for an inductor, they are identified by a number, letter, or color code. These identifying marks are explained on a sheet of paper enclosed in the packing box. A schematic diagram of the inductor is often included as well.

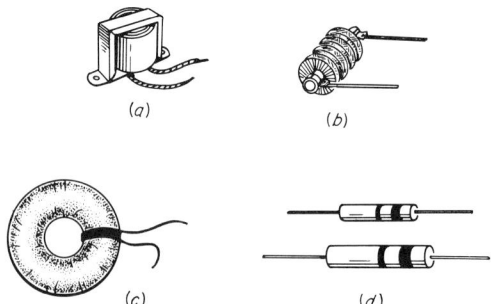

FIG. 2·20 An assortment of chokes: (a) Low-frequency choke. (b) Pi-wound r-f choke. (c) Toroidal choke. (d) Tubular insulated chokes.

2·12 TYPES OF INDUCTORS

Inductors are classified according to the current that will normally flow through them. Some types are designed to operate with d-c electricity, while others are designed to operate with alternating currents. The type of electricity employed determines the type of material used in the core of the coil.

In addition, inductors are identified according to their use. Some are called *chokes*, because their main objective is to restrict current changes. An assortment of chokes is illustrated in Fig. 2·20. Other inductors are constructed with several coils and are called *transformers*. A transformer is used to transfer electrical energy from one coil of wire to another. This transfer is accomplished by utilizing the magnetic field that surrounds an inductor. Other inductors find application in electric motors, relay coils, and antennas of small radio receivers.

D-C INDUCTORS

Inductors that have been designed to operate with direct currents create a steady magnetic field. This magnetic field can in turn cause mechanical movement. Direct-current motors and relays function as a result of this steady magnetic field. Typical electrical motors, as used in electronics, are illustrated in Fig. 2·21. A relay is illustrated in Fig. 2·22.

A-C INDUCTORS

The majority of inductors are designed to operate with alternating currents or with fluctuating or pulsating direct currents. These changes in current

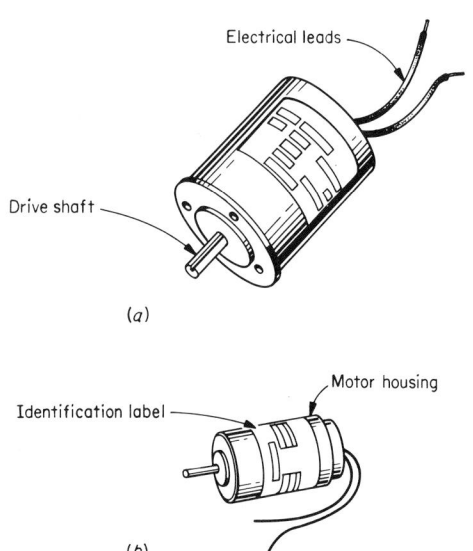

FIG. 2·21 Electric motors: (a) The leads receive an electrical input, while the drive shaft delivers a mechanical output. (b) Miniature type.

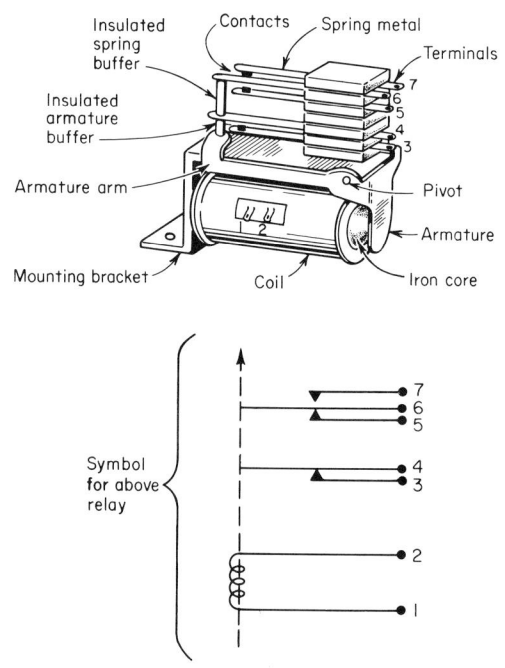

FIG. 2·22 A complete relay with the corresponding symbol.

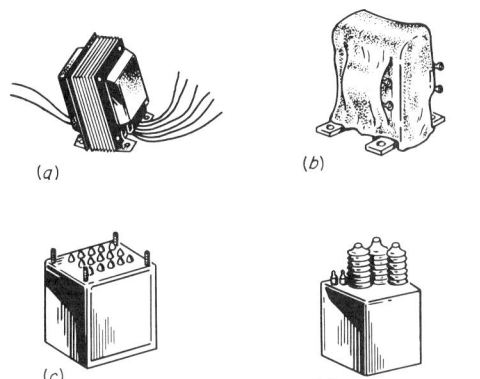

FIG. 2·23 Low-frequency transformers: (a) Power transformer. (b) Encapsulated transformer. (c) Hermetically sealed transformer. (d) High-voltage transformer, cased.

occur periodically, some at a fast rate and others at slower rates. Because of this an inductor to be used with a regularly changing current is further classified according to the *frequency* of current change. Some low-frequency transformers are illustrated in Fig. 2·23. A few high-frequency inductors are illustrated in Fig. 2·24. A word of caution: *never operate a high-frequency inductor at a lower frequency*, and *never operate with direct current an inductor designed to operate with alternating current*. Violation of this caution will cause severe damage to the coil.

INDUCTOR CORES
Direct-current or low-frequency alternating current inductors require soft iron to fill their core. By contrast, high-frequency inductors operate more efficiently with an air core. However, if the frequency is not too high, a powdered-iron core may be required in place of the air core.

2·13 COVERINGS FOR INDUCTORS
Insulation is a minimum covering for inductors. However, it is often necessary to protect an inductor from moisture or from physical damage. A coil can be encapsulated in wax or plastic to protect it from moisture, but a metal case is best to protect it from physical damage. Often the metal case is hermetically sealed to keep out moisture as well.

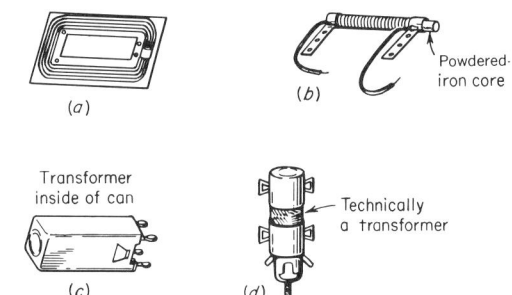

FIG. 2·24 High frequency inductors: (a) Loop antenna. (b) Rod antenna. (c) Intermediate-frequency transformer. (d) Oscillator coil.

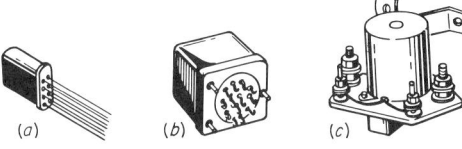

FIG. 2·25 Relay cases: (a) Hermetically sealed miniature relay. (b) Relay case with hook-type terminals. (c) Heavy-duty power relay.

Metallic containers have the additional feature of keeping the magnetic field confined or of keeping a disrupting magnetic field away from the inductor. Quite often this will be the sole purpose of the container. The metallic cases are not always uniform, as evidenced by the assortment of relay cases illustrated in Fig. 2·25.

2·14 SCHEMATIC SYMBOLS FOR INDUCTORS
Symbols used in schematic diagrams to represent inductors are introduced in Fig. 2·26. Notice that the type of core employed will cause the greatest change in the basic symbol. A *laminated-iron* core is made of layers of metal. A powdered-iron core is made of powdered iron that has been molded into shape.

The broken line in the relay-coil symbol does not represent the iron core, but shows that this coil will be magnetically linked to the armature of the relay, which in turn operates the contacts by pushing the buffers (see Fig. 2·22).

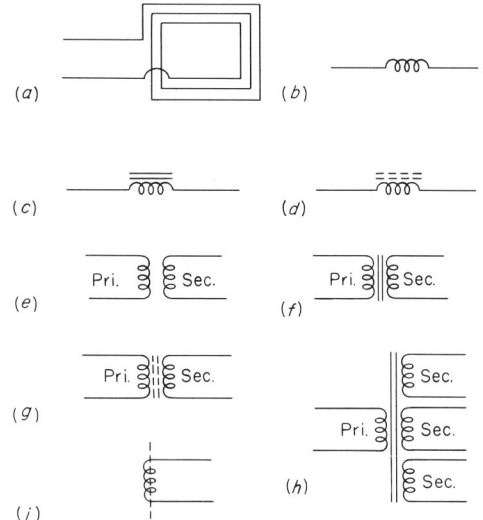

FIG. 2·26 Symbols for inductors. (a) Loop antenna. (b) Air-core choke. (c) Laminated-iron-core choke. (d) Powdered-iron-core choke. (e) Air-core transformer. (f) Laminated-iron-core transformer. (g) Powdered-iron-core transformer. (h) Transformer with three secondary coils. (i) Relay coil.

INDUCTORS SUMMARIZED

Inductors have the ability to create a magnetic field when a current flows through them. This magnetic field makes possible the electric motor and the relay. Inductors are also used to oppose current changes and for transferring electrical energy from one coil to another.

Inductors employ either an air core or a soft-iron core, depending on the type of current that flows in the coil. Inductors are rated either by their resistance, inductance, and impedance or by combinations of these three ratings.

2·15 ACTIVE COMPONENT TYPES

Resistors, capacitors, and inductors act upon an electric current or voltage in a rather "routine" manner. A term often given these parts to describe this trait is *passive components*. By contrast, electronic components which act to change an electrical current or voltage (often violently) by either altering their *level, timing,* or *form* are known as *active components*. Transistors and vacuum tubes fall in this category. Identification of active components becomes a more complicated task than that of identifying passive components because their identifying marks, related symbols and descriptive names are tied closely to their intended function.

Other "general" terms describe certain categories within which these active components fall. One such term is *semiconductors*. Transistors belong to the semiconductor family. Another term is *electron tubes*. Vacuum tubes belong to this family. Although present-day electronics makes extensive use of semiconductors, electronics as a whole had its birth with the advent of the vacuum tube. Semiconductors were later developed as a partial replacement for vacuum tubes, which have certain limitations. Because of this evolution, the student of electronics often finds comparison between these two active components. This text uses this approach in an attempt to acquaint the reader with both parts since there is merit to this method of instruction. However, in keeping with the state-of-the-art, emphasis in the book is on semiconductors.

2·16 SEMICONDUCTOR CLASSIFICATIONS

There are four classifications of semiconductors: (1) *diodes*, (2) *power rectifiers*, (3) *transistors*, and (4) *integrated circuits*. In the diode classification there are semiconductors used for extraction of signal information. The power-rectifier classification includes heavy-duty diodes used in power supplies (battery eliminators). A rectifier is used to change a-c electricity to d-c electricity. Semiconductors used in signal amplification or current control belong to the transistor classification. The integrated-circuit classification includes a mixture of diodes and transistors, as well as some passive components. Figure 2·27 illustrates these four classifications of semiconductors.

DIODES

Diodes are manufactured in several forms and are quite small. Some diodes are seen in Figs. 2·27 to 2·30. For technical and physical reasons, great care must be exercised when working with semiconductors. One of these precautions concerns heat.

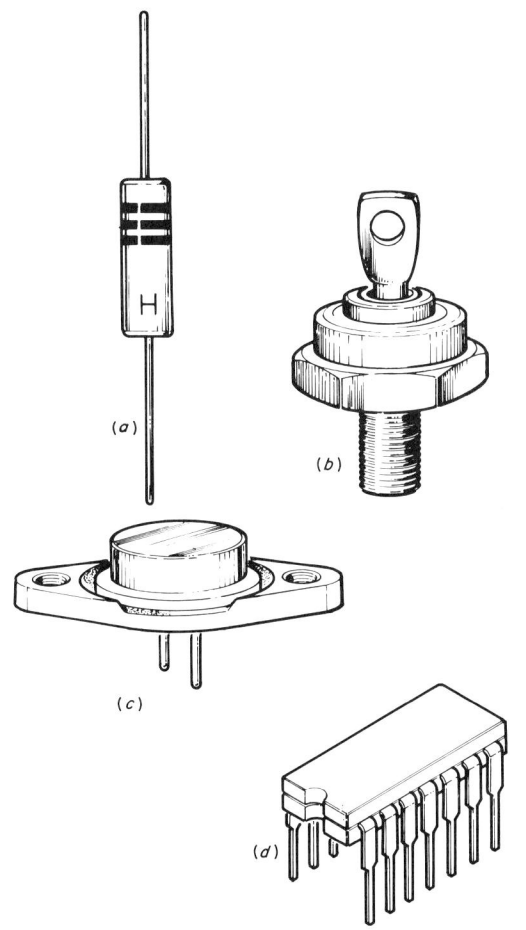

FIG. 2·27 The four general classes of semiconductors: (a) Diode. (b) Power rectifier. (c) Transistor. (d) Integrated circuit.

Excessive heat will damage a diode, rectifier, transistor, or any miniature electronic component. When soldering any semiconductor into a circuit, or attempting to remove one from a circuit, a *heat sink* should always be used. See Fig. 6·14. Also, when wiring a semiconductor into a circuit, instructions relating to polarity should be rigidly followed.

POWER RECTIFIERS

Semiconductor diodes used primarily for power are also manufactured in many forms. Shape variation is dependent upon the heat generated by the rectifiers while in use. Some power rectifiers are quite small, others comparatively large, but all can be held in the palm of the hand. Power rectifiers are shown in Figs. 2·27, 2·28, 2·30, and 2·31. A stud provided for mounting purposes also serves to conduct the generated heat to surrounding metal surfaces. This is done to prevent damage to the power rectifier. A power rectifier having a stud should not be operated off the chassis or the heat generated will be confined to the rectifier, and this will damage it.

TRANSISTORS

Transistors are also found in a variety of sizes and shapes. A few of the transistors commonly encountered are seen in Figs. 2·6, 2·27, and 2·32. Some transistors make use of sockets for installation in the same way vacuum tubes use sockets. Whether sockets are used or the leads are soldered, care

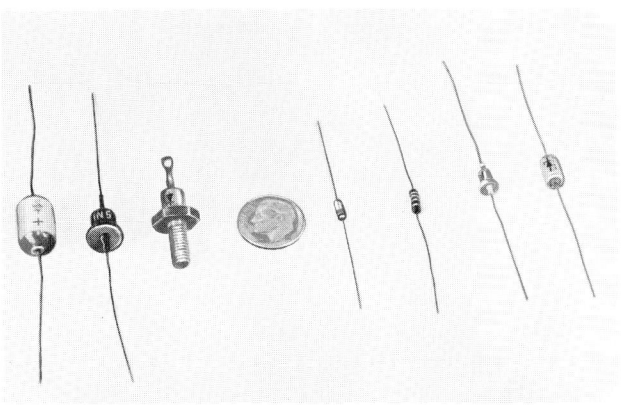

FIG. 2·28 An assortment of semiconductor diodes.

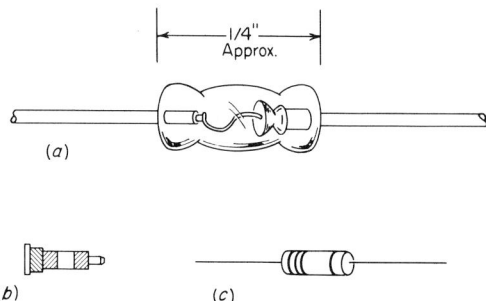

FIG. 2·29 Diodes used to rectify signal voltages. (a) Clear-glass, point-contact diode. (b) Microwave diode. (c) General-service diode with opaque glass case.

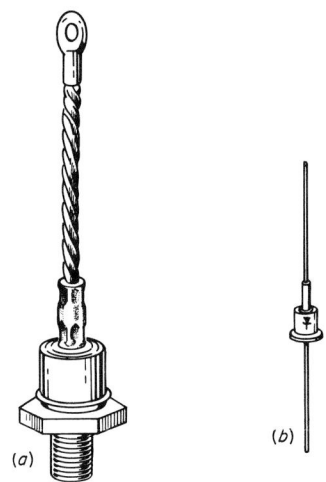

FIG. 2·31 Semiconductors for rectification in power supplies: (a) Heavy-duty power rectifier with stud mounting. (b) Top-hat power rectifier.

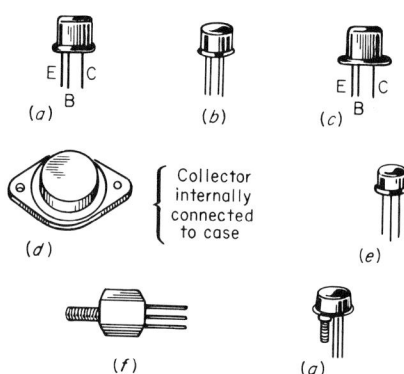

FIG. 2·32 Transistors appear in a variety of shapes. (a) Triode transistor. (b) Tetrode transistor. (c) Triode transistor. (d) Power transistor. (e) Triode transistor. (f) Power transistor with stud mounting. (g) Power transistor with stud mounting.

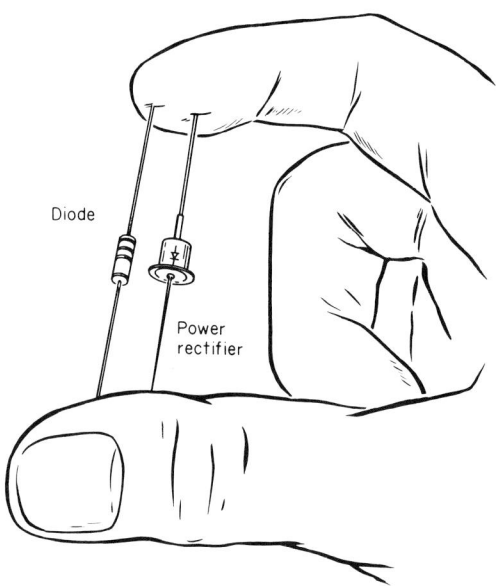

FIG. 2·30 Semiconductors have contributed to the miniaturization of electronics.

should be taken to make correct circuit connections.

POWER TRANSISTORS

Within the family of transistors will be found some that are designed for heavy duty. These handle the heavy currents necessary for power applications. Heavy-duty transistors are known as *power transistors*. The heavy currents involved generate a great deal of heat within the transistor. This heat is sufficient to damage the power transistor if it is not conducted away to surrounding metal.

Power transistors and rectifiers are protected from their own heat generation by the use of permanent heat sinks. A permanent type of heat sink is shown in Fig. 2·33. Such heat sinks are made of heavy aluminum that will spread the heat away from the power transistor or rectifier and allow for faster and easier cooling by convection.

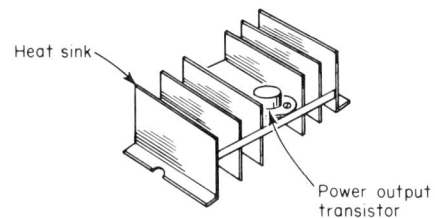

FIG. 2·33 A permanent heat sink.

DRY-METAL RECTIFIERS

Semiconductors previously discussed are made of germanium or silicon. Another semiconductor in popular use is the selenium dry-metal rectifier, shown in Fig. 2·34. This rectifier is also a diode. The selenium rectifier is strictly a power rectifier and is never used to rectify a-c signals for information purposes. The cathode terminal will always be marked with a plus sign, which refers to the *rectified voltage polarity,* normally identified as B+ in power supplies.

INTEGRATED CIRCUITS

An *integrated circuit* (IC) is a device wherein several active semiconductor components are interconnected to several passive components within a single *very small* package in such a way that together they can perform a given electronic task; like building-up (amplifying), modifying (rectifying), or generating (oscillating) an electrical signal. However, integrated circuits are also available in a way that the user can select only a portion of the circuit to use as he may see fit. An IC is normally available in one of three packaged forms: (1) Flat Pack, (2) "TO-5 can" type, and (3) Dual-in-line. These packages are illustrated in Fig. 2·35, and as can be seen are quite small. Actually, the complete circuit or circuits, composed of many active and passive com-

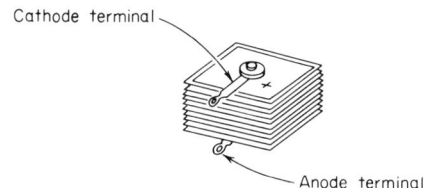

FIG. 2·34 Selenium dry-metal rectifier.

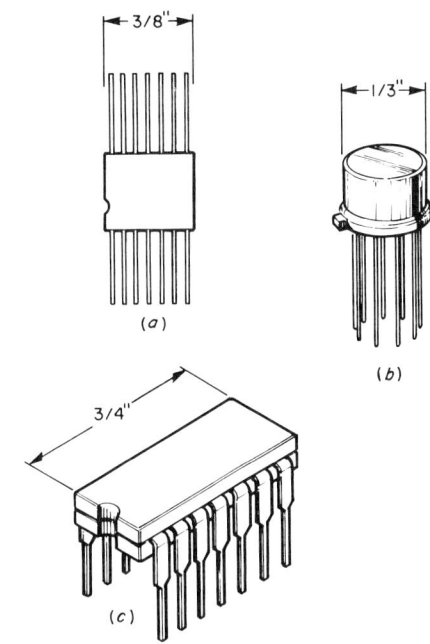

FIG. 2·35 Integrated circuit packages. (*a*) Flat-pack. (*b*) "TO-5" can. (*c*) Dual-in-line. (*Note:* Dimensions are approximate.)

ponents, which are inside these packages are much smaller yet. They are so small that they can only be seen through a microscope. This whole area is therefore referred to as *microelectronics.* Figure 2·36 shows how the integrated circuit appears within a TO-5 package. Chapter 14 covers the subject of integrated circuits in more detail.

IC CLASSIFICATIONS

Integrated circuits are classified in a manner that reflects how they are manufactured. There are basically two such classifications: *monolithic* and *hybrid.* The monolithic IC is one where all components, both active and passive, are fabricated as a unit, *including their interconnections.* The hybrid IC is one where circuit components are not all fabricated as a unit, but *must be interconnected* with wire thinner than human hair. Of course, this work requires highly skilled operators who use binocular microscopes and precision tools and equipment. In fact, the whole process of integrated-circuit fabrication is an exacting task. For now, it is sufficient

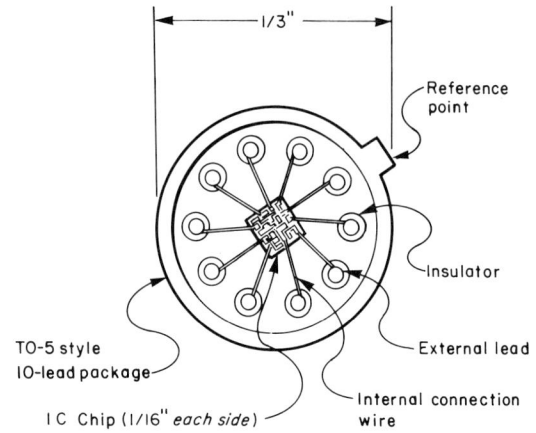

FIG. 2·36 Integrated circuit within a ten-lead TO-5 package.

for the student to be able to identify these devices. Actually, it is not necessary for the electronics assembler to know *how* an IC is manufactured. All he has to know is how to handle them properly during fabrication of electronic devices that use integrated circuits. Some of the summary jobs at the end of this chapter are designed to help a student get better acquainted with integrated circuits of various types.

2·17 ELECTRON-TUBE CLASSIFICATIONS

Physical identification of electron tubes is much simpler than identification of semiconductors. The variation is not as great. However, electron tubes are capable of performing all of the same functions accomplished by semiconductors. The difference is in the way these two devices perform their task. This chapter has not been designed to cover these differences. It should be pointed out, however, that semiconductors operate with much lower electrical power requirements, most generate much less heat during operation, are much smaller, and can withstand higher vibration levels. It is for these reasons that semiconductors find wider application in present-day electronics.

CLASSIFICATIONS

Electron tubes may be classified in two general classifications. One classification is *vacuum tube*, while the other is *gaseous tube*. These classifications stem from the fact that electron-tube electrodes may operate within a vacuum or gaseous atmosphere. Inert gas is used. A vacuum or gaseous atmosphere is possible because the electrodes are housed within airtight chambers. In addition, quite often reference is made to the type of base or socket used when identifying a certain tube.

Electron tubes that have two useful electrodes within their envelope are known as *diodes;* those that have three useful electrodes are known as *triodes;* those with four electrodes *tetrodes;* and those with five electrodes, *pentodes.* Sometimes multiple sets of electrodes are used. A tube with two sets of diodes would be known as a *dual-diode* tube, while one with two sets of triodes would be known as a *dual-triode* tube. Electrodes within an electron tube are identified as *heater (filament); cathode (K); plate (P); control grid (G_1); screen grid (G_2);* and *suppressor grid (G_3).* See Fig. 2·39. The heater is not considered a "useful" electrode. Circuit connections to all electrodes is made via pins in the base of the tube which insert into appropriate tube sockets. These sockets are shown in Fig. 2·37.

The physical appearance of an electron tube is normally like those that appear in Fig. 2·7. Special-purpose tubes often take a different form, such as the picture tube in a television receiver. The reader may find it to his advantage to survey the topic of electron tubes in more detail. Appendix I in this book gives a brief résumé of this subject.

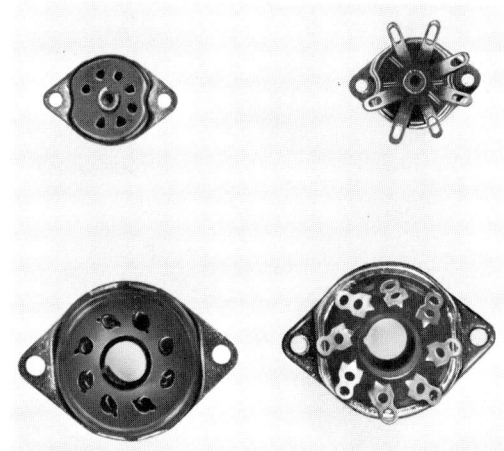

FIG. 2·37 Octal and seven-pin miniature tube sockets.

2·18 SYMBOLS FOR SEMICONDUCTORS AND VACUUM TUBES

Symbols that represent diodes, power rectifiers, transistors, and integrated circuits are shown in Fig. 2·38. The following discussion relates to these symbols by comparing them with vacuum-tube symbols that appear in Fig. 2·39.

DIODES AND RECTIFIERS

The two symbols shown in Fig. 2·38a can be used to represent semiconductor diodes and rectifiers. After all, these two devices are diodes, and they both rectify. When the *plus sign* is used, it represents the rectified output voltage, such as B+. (The plus sign does not represent the applied voltage that can cause the conduction of electrons.) When the letter K is used, it represents the cathode end. In any case, electron flow is from cathode to anode within these diode rectifiers.

Compare these two symbols with the symbols of the vacuum-tube diodes in Fig. 2·39a. In vacuum-tube symbols the letter K is also used to represent the cathode. However, the plus sign is not used for this identification.

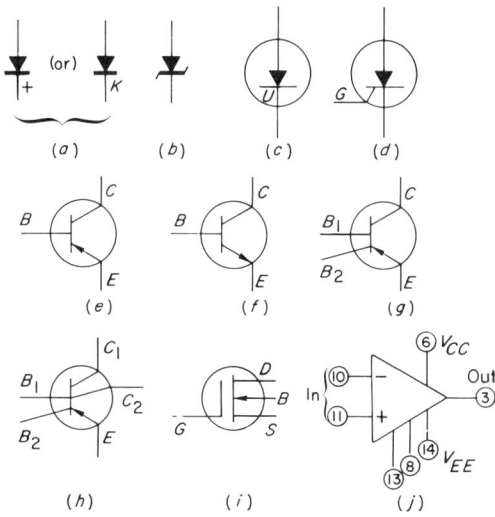

FIG. 2·38 Symbols for semiconductors: (a) Diode (rectifier). (b) Zener diode. (c) Tunnel diode. (d) Silicon control rectifier. (e) PNP transistor. (f) NPN transistor. (g) Tetrode transistor. (h) Pentode transistor. (i) MOS–field-effect transistor. (j) Integrated circuit.

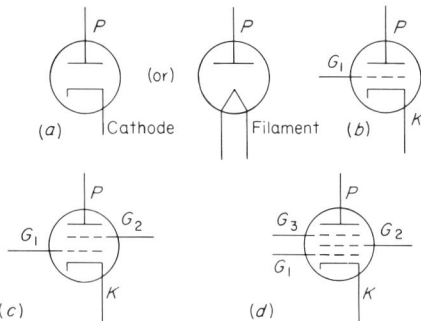

FIG. 2·39 Symbols for vacuum tubes: (a) Both are vacuum-tube diodes. (b) Vacuum-tube triode. (c) Vacuum-tube tetrode. (d) Vacuum-tube pentode.

The Zener and Tunnel diodes shown in Fig. 2·38b, c are special application diodes with which the reader should become acquainted. Although these diodes are not used in great quantities, nevertheless an electronics assembler should become familiar with the symbols and identification aspects of these devices. Their application is discussed in Chap. 14.

TRIODES

The triode symbol for a vacuum tube is fairly simple to comprehend (see Fig. 2·39b). The letter K stands for cathode, G_1 for the control grid, and P for the plate. The triode transistor has three electrodes also, but with two variations. Triode transistors are designated either PNP or NPN, depending on their construction. In Fig. 2·38f we find an NPN transistor as a triode. The letter E represents the emitter. The emitter will always contain the arrow, and it loosely corresponds to the cathode of the vacuum tube. In this case, however, we must remember that *electrons flow against the arrow*, as in the diode. The letter C represents the collector. The collector loosely corresponds to the plate of the vacuum tube. Finally, the control grid (G_1) in a vacuum tube has control over the electrons that travel from the cathode to the plate, and similarly, the base (B) has control over the current flow within the transistor.

Note that, in the NPN transistor shown in Fig. 2·38f, the arrow points away from the base, while in a PNP transistor the arrow points into the base, as shown in Fig. 2·38e.

The *metal-oxide semiconductor* (MOS) *field-effect transistor* (FET) closely compares to the vacuum-tube triode in circuit application. Although this application is discussed in Chap. 14, it should be stated that the *source* (S) electrode has a similar function as the cathode in the tube; the *gate* (G) electrode compares in function to the control grid; and the *drain* (D) electrode compares to the plate. The active *bulk* electrode (B) supports the above-mentioned electrodes, including the actual conducting material found within this type transistor.

TETRODES

In Fig. 2·39c an extra grid G_2 was added to the basic triode. This made the vacuum tube a tetrode. Similarly, with the addition of B_2, a triode transistor becomes a tetrode transistor, as the symbol of Fig. 2·38g shows. A tetrode transistor is illustrated in Fig. 2·32b. Notice the extra lead.

PENTODES

By the addition of another grid to the vacuum-tube tetrode, we have a pentode. This third grid, designated G_3 is seen in Fig. 2·39d. Here again, a tetrode transistor can be made into a pentode transistor by the addition of another electrode, C_2, shown in Fig. 2·38h.

2·19 DESIGNATIONS FOR SEMICONDUCTORS

DIODES AND RECTIFIERS

Most diodes and rectifiers are designated with a 1N followed by other numbers, such as 1N191 or 1N253. However, this 1N prefix is not exclusive; other prefixes will be found for special diodes and rectifiers.

The above designations can either be stamped on the diodes or color-coded in a series of bands. Also marked on the diode (or rectifier) will be some indication as to which end is the cathode and which end the anode. More information on these markings will be given in the job summaries concluding this chapter.

In schematic diagrams, the letters *CR* are often assigned to the symbols of diodes and rectifiers in the same way that *R* is assigned to resistors and *C* is assigned to capacitors.

TRANSISTORS

The identification of transistors is stamped on the casing, not color-coded. The prefix 2N is used for transistor identification, such as 2N109 or 2N356. As for the lead or terminal identification, it is best to obtain it from the technical data available, such as a semiconductor manual or the manufacturer's data sheets, since various lead arrangements are employed. Some of the most popular lead arrangements are covered in Job 2·20 at the end of this chapter. For schematic diagrams, the letter *Q* is often employed as a transistor symbol.

In the case of the field-effect transistor, the 3N prefix is often used; for example, 3N128. However, as in all prefix designations, MOS-FET devices also employ other numbering systems.

INTEGRATED CIRCUITS

For integrated circuits, the prefix CA is used when the basic intent for the circuit is of a general nature, such as with amplifiers. One example is CA3007, which is an integrated circuit designed as an audio-frequency amplifier. This is what is known as a *linear* type of circuit. When the basic intent of the integrated circuit is to function in a *digital* type of electronic device, such as in a computer, the CD prefix is used. An example is the CD2202 which is a multiple digital logic circuit. Again, as stated before, an assembler needs only to concern himself with the physical identification of these devices. However, this includes a need to recognize and "read" appropriate markings and related symbols. Integrated circuits are so diversified that the electronics assembler will do right by only being able to "read" the IC symbol such as shown in Fig. 2·38j. Let the engineers and technicians worry about *what is inside*.

2·20 DESIGNATIONS FOR ELECTRON TUBES

Electron tubes are designated according to five general considerations: (1) the necessary heater voltage,

(2) the intended application, (3) number of electrodes, (4) socket requirement, and (5) type of envelop. Some of this information is contained in the basic tube designating number, such as 6GK6. In this case, the first 6 indicates that approximately 6 volts are needed by the heater element of this tube; the letters GK relate to the intended application for this tube; and the last 6 indicates how many electrodes are connected to a pin in the base, singularly or in combination. Of all this information, the heater voltage rating is the most important to the assembler and test technician. Incidentally, this same tube is available with a heater designed to operate with approximately 10 volts. It is known as the 10GK6 tube. Another version of the same tube is the 16GK6 tube. It is supposed to work with approximately 16 volts at its heater. All other factors remain the same. The tube is basically designed to operate as an audio amplifier. Socket requirements and type of envelop employed are visually noticeable. However, this information is contained in a tube manual.

SEMICONDUCTORS AND ELECTRON TUBES SUMMARIZED

In summary, semiconductors and electron tubes are *active components*. They are used as basic units in amplifiers, rectifiers, and oscillators. As individual components they must either be soldered into a circuit, or they must use a socket for interconnection with passive components. However, in the case of semiconductors, active components and passive components might be fabricated as a microelectronics device known as an *integrated circuit*. In general, semiconductors are much smaller than electron tubes, require less power, generate less heat while in operation, and can withstand vibration better.

2·21 SUMMARY: COMPONENT IDENTIFICATION

A summary for this chapter must stress the need for the ability to recognize electronic parts by all those who work in the design, manufacturing, or testing of electronic products. In fact, this is also recommended for sales and administrative personnel connected with the electronics industry. Identification of electronic components requires recognition of visual, symbolic, and related markings. A basic knowledge of electrical principles is essential for a thorough understanding of some symbols and marks used on components. The following summary jobs have been designed to better acquaint the student with component identification.

EXERCISES

JOB 2·1 How to identify resistors

OBJECTIVES
1. To become acquainted with the various types of resistors
2. To become acquainted with trade magazines and catalogs

MATERIALS REQUIRED

Supplies
Kit of components used in Job 1·1
Old electronics magazines and electronics catalogs
8½ × 11-inch notebook paper and pencil
Scissors
Rubber cement

PROCEDURE
1. Separate all fixed and variable resistors from the kit of components supplied.
2. Look in the magazines and catalogs supplied for pictures of similar fixed and variable resistors.

3. Cut these pictures out and paste them on sheets of notebook paper with rubber cement.
4. Under each picture write a summary regarding the purpose, values, and specifications relating to that specific resistor.
5. For extra credit, cut out pictures of resistors which were not supplied in the kit; paste them on notebook paper and write a summary on each.
6. Assemble all of these resistor identification sheets into a notebook.

QUESTIONS
1. What is the basic purpose of a resistor?
2. Describe a potentiometer.
3. Describe a rheostat.
4. How are resistors rated?
5. Describe the various methods used to mark fixed and variable resistors.

JOB 2·2 How to read the resistor color code

OBJECTIVES
1. To appreciate rated values
2. To learn to interpret the color markings on resistors

MATERIALS REQUIRED

Supplies:
Resistor, 25,000 ohms, 10 percent tolerance, utilizing the color code illustrated in Fig. 2·41

PROCEDURE
1. Note that the resistor supplied has four colored bands.
2. Note that the color bands are closer to one end, as illustrated in Fig. 2·40.
3. The color bands must be read in the proper sequence. Figure 2·40 gives the correct sequence, indicated by numbers.
4. Each color represents a specific number:

Black	= 0	Blue	= 6
Brown	= 1	Violet	= 7
Red	= 2	Gray	= 8
Orange	= 3	White	= 9
Yellow	= 4	Silver	= 10% tolerance
Green	= 5	Gold	= 5% tolerance

5. Using the table of step 4, compare the 25,000-ohm resistor supplied with the illustration of Fig. 2·41.
6. Note that the number 25,000 is composed of the first digit 2, the second digit 5, and *three* 0s. From the table in step 4, the first digit corre-

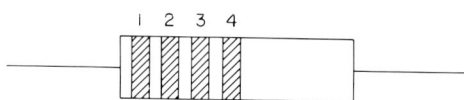

FIG. 2·40 The order for reading the color bands on a resistor.

sponds to red, the second digit corresponds to green, and the number of zeros corresponds to orange since there are three.
7. The third color band actually represents the multiplier of the first two digits, in this case 25 × 1000. However, 1000 contains *three* zeros; hence the use of orange in this resistor.
8. The fourth color band allows for a small variation in the resistance of a resistor. The tolerable resistance variation in the resistor supplied is 10 percent. The resistor can be considered good if its resistance is within the limits of 22,500 and 27,500 ohms.

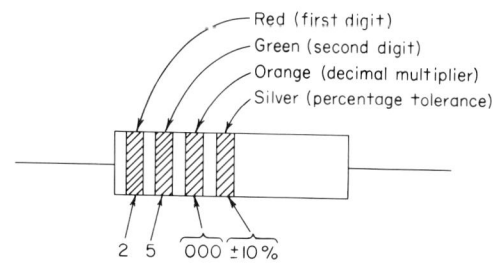

FIG. 2·41 An example for reading the color code on resistors.

9. The color bands do not reveal the wattage rating.
10. Some resistors also use a *fifth* color band. This band indicates the failure rate level, also known as the *reliability level*. When a fifth color band is used, the following table applies:

    | Brown | 1.0 % |
    | Red | 0.1 % |
    | Orange | 0.01 % |
    | Yellow | 0.001% |

    Use of a fifth color band rates the resistor as to the percentage of failures to be expected out of a group of identical resistors operated for 1000 hours, assuming that the resistors are operated at 50 percent of the wattage rating.

QUESTIONS
1. Would the resistor supplied offer 25,000 ohms of resistance in an operating circuit?
2. What is the difference between a rated value and an actual value? Give an example.
3. In step 8 of this job a minimum and a maximum value of resistance were given. Show how these values were obtained.
4. Give your definition of resistance.
5. Explain the difference between resistance in a resistor and in a length of wire.
6. What do resistors resist?
7. What is indicated by a *fifth*, red-color band on a resistor?

JOB 2·3 How to use an ohmmeter

OBJECTIVES
1. To learn about resistance measurements
2. To learn about multiple-unit instruments
3. To learn how to use a VTVM for resistance measurement

MATERIALS REQUIRED

Equipment
Vacuum-tube voltmeter with test leads

Supplies
Resistor, 25,000-ohm, 10 percent tolerance
Copper wire, 1 foot long

PROCEDURE
1. Note that the resistor to be used is the same resistor used in Job 2·2.
2. Study the front panel of the VTVM (see Fig. 19·4).
3. Study the meter and its scales, mounted on the VTVM.
4. Note that, on the meter, the top scale has its zero at the left end. The opposite end of the scale has an *infinity* mark. This top scale will be used in measuring resistance, or ohms.
5. Also note that the resistance scale is nonlinear.
6. This instrument is a multimeter that can be used to measure d-c and a-c voltage as well as resistance. However, voltage and resistance measurements can be made only one at a time.
7. In order to measure resistance the *function switch* must be set at the OHMS position. This switch appears at the right of the panel.
8. The *range switch* must also be set at the $R \times 1000$ position. The range switch appears on the left side of the panel.
9. Do not attach the test leads to the VTVM.
10. Plug in the a-c line cord into a convenient outlet.
11. After a short warmup period, note that the meter *pointer* deflects to the infinity mark. Adjust the pointer to be exactly on the infinity mark with the small knob identified as OHMS.
12. Attach the test probe to the meter.
13. Slide the switch on the probe to A-C OHMS.
14. Clip together the exposed ends of the *ground* lead and the test probe.
15. The pointer should deflect to zero on the resistance scale.
16. If the pointer does not indicate exactly zero, rotate the small knob identified as ZERO until the pointer is exactly at zero.
17. Calibration of the VTVM for correct resistance measurement is accomplished by the OHMS and ZERO knobs.

18. Check and repeat the adjustments for zero and infinity as necessary.
19. Separate the exposed ends of the test leads.
20. Insert the length of copper wire supplied between the exposed test leads.
21. The pointer should indicate zero ohms since a short length of copper wire offers negligible resistance.
22. Remove the copper wire from between the test leads and replace it with the 25,000-ohm resistor supplied.
23. The pointer should indicate approximately 25. However, since the range switch is at the $R \times 1000$ position, 25 must be multiplied by 1000.
24. Compare the results with the tolerance computed in Job 2·2. If the resistor offers resistance beyond the allowable limits, the resistor is no good.

QUESTIONS

1. Is a VTVM an ohmmeter? Why?
2. Explain any difference between the color-code value of resistance and the measured value.
3. What does infinity mean?
4. If you do not know the color-coded rating of a resistor, will the measured value indicate if the resistor is good or bad?
5. Explain why the meter pointer deflected to infinity and then to zero on the meter scale when the test leads were alternately *opened* (separated) and *shorted*.

JOB 2·4 How to determine the wattage rating of a resistor

OBJECTIVES

1. To learn to identify resistors by their power-consumption capabilities
2. To appreciate the physical size of a resistor

MATERIALS REQUIRED

Supplies:
Color-coded resistors, 4700 ohms each, 10 percent tolerance, wattage ratings as follows: ¼ watt, ½ watt, 1 watt, 2 watts
Power resistors, 1000 ohms each, wattage ratings as follows: 10 watts, 25 watts, 100 watts

PROCEDURE

1. Note that the color coding used on the four smaller resistors is identical. This will indicate that each resistor will oppose an electric current by 4700 ohms of resistance, within a tolerance of 10 percent.
2. Note that none of these resistors are of identical physical size.
3. Compare the color-coded resistors with those illustrated in Fig. 2·9.
4. With the aid of Fig. 2·9, which is drawn to exact scale, determine the wattage rating of the color-coded resistors supplied.
5. It should be obvious that a 2-watt resistor can dissipate more heat than the smaller ¼-watt resistor. Yet each resistor is rated to present the same opposition to an electric current.
6. Note that the color code does not reveal the wattage rating of a resistor.
7. To determine the wattage rating of a resistor, it may be compared with one of known wattage rating, provided the resistors are of the same type.
8. Using the knowledge now gained, determine which of the three power resistors supplied has the highest wattage rating.
9. Note that no color code is employed. Since these resistors are designed to dissipate a high degree of heat, paints would burn off or disappear quite rapidly.
10. When a power resistor is new, the wattage rating and resistance rating are stamped on the surface of the resistor, illustrated in Fig. 2·10.
11. Since all painted marks on power resistors burn off quickly, attempt to memorize the wattage rating of resistors whose wattage rating is known.

12. Exact physical dimensions for each resistor of a given wattage rating are provided by resistor manufacturers in their product catalog or other related trade catalogs. It is obviously impractical to consult a catalog every time there is need to identify the wattage rating of a resistor. However, in the interest of training, consult such a catalog if one is available.

QUESTIONS
1. Identify each resistor by its wattage rating to your instructor.
2. Why does a high-wattage resistor get hot?
3. Why can a larger resistor dissipate more heat?
4. Distinguish between the resistive rating and the power rating of a resistor.

JOB 2•5 How to test a variable resistor

OBJECTIVES
1. To become acquainted with variable resistors
2. To recognize a 2-watt potentiometer
3. To learn to test a variable resistor

MATERIALS REQUIRED

Equipment
Ohmmeter

Supplies
10-kilohm 2-watt potentiometer

PROCEDURE
1. Note the physical size of the potentiometer supplied. Remember, in future applications, that this potentiometer has a 2-watt rating, since not all potentiometers have the wattage rating stamped on their case.
2. Note that a potentiometer has three terminals.
3. Since a potentiometer is a type of variable resistor, the total resistance will always appear between two of the terminals.
4. Compare the physical potentiometer with the schematic symbol of a potentiometer shown in Fig. 2•14d.
5. Set the ohmmeter at the $R \times 1000$ position.
6. Measure the total resistance of the potentiometer between the two end terminals and leave the ohmmeter connected to these terminals. The meter should indicate 10.
7. With the ohmmeter still connected, rotate the shaft of the potentiometer. No change in resistance should be noted.
8. Move one of the meter test leads from one end terminal to the center terminal of the potentiometer. Maintain the other test lead on the other end terminal as before.
9. Again rotate the shaft of the potentiometer. The meter reading should now vary between 0 and 10, depending on the rotation of the shaft. Note how the resistance changes with clockwise rotation of the shaft.
10. Maintain the lead to the center terminal, but change the test lead from one end terminal to the other.
11. Repeat step 9.
12. The change of resistance should be opposite with clockwise rotation.
13. Infinite resistance, or a resistance indication outside the tolerance limits of the rated 10,000 ohms during step 6, indicates a faulty potentiometer.
14. If the resistance variation noted in step 9 is erratic, the potentiometer should also be considered faulty.

QUESTIONS
1. What basis is used to determine the wattage rating of a potentiometer?
2. Describe a potentiometer.
3. What terminals should be used to obtain an increase of resistance with clockwise rotation of the potentiometer shaft?

4. Briefly list three applications of a potentiometer.

JOB 2•6 How to test a tapped resistor

OBJECTIVES
1. To become acquainted with tapped resistors
2. To learn to test tapped resistors

MATERIALS REQUIRED

Equipment
Ohmmeter

Supplies
Fixed resistor, tapped, unmarked

PROCEDURE
1. Determine the wattage rating of the tapped resistor supplied by comparing it with other similar resistors of known wattage rating. Do not hesitate to ask your instructor for assistance since the wattage rating of the resistance between terminals will be different.
2. Measure the total resistance of the resistor by measuring between the two extreme end terminals.
3. Measure the resistance between each pair of terminals. You can refer to Fig. 2•14b and c on page 18.
4. If necessary, refer to Job 2•5.

QUESTIONS
1. Why is one wattage rating incomplete for a tapped resistor?
2. What was the total resistance of the resistor supplied?
3. Draw a schematic symbol of the tapped resistor supplied and indicate on the diagram the various resistances involved.
4. How does this job differ from Job 2•5?

JOB 2•7 How to identify capacitors

OBJECTIVES
1. To become acquainted with the various types of capacitors
2. To become acquainted with identification assigned capacitors

MATERIALS REQUIRED

Supplies
Kit of components used in Job 1•1
Old electronics magazines and electronics catalogs
8½ × 11-inch notebook paper and pencil
Scissors
Rubber cement

PROCEDURE
1. Separate all fixed and variable capacitors from the kit of components supplied.
2. Look in the magazines and catalogs supplied for pictures of similar fixed and variable capacitors.
3. Cut these pictures out and paste them on sheets of notebook paper with rubber cement.
4. Under each picture write a summary regarding the purpose, values, and specifications relating to that specific capacitor.
5. For extra credit, cut out pictures of capacitors which were not supplied in the kit; paste them on notebook paper and write a summary on each.
6. Assemble all of these capacitor identification sheets into a notebook.

QUESTIONS
1. What is the basic purpose of a capacitor?
2. Describe a molded capacitor.
3. Describe an electrolytic capacitor.
4. Describe a ceramic capacitor.
5. Describe a variable capacitor.

6. Describe the various methods used to mark fixed and variable capacitors.

JOB 2·8 How to interpret capacitor ratings

OBJECTIVES
1. To become aware of markings on capacitors
2. To learn to recognize capacitors
3. To learn to differentiate between capacity and breakdown voltage ratings

MATERIALS REQUIRED

Supplies
An assortment of capacitors, including electrolytics

PROCEDURE
1. Identify the capacitors at hand with the aid of the illustrations in Figs. 2·15 to 2·18.
2. Carefully examine all markings on the capacitors.
3. Note that some capacitors have numbers stamped on them while others use color codes. (This job will not explain the color code used on capacitors. This will be done in Job 2·9.)
4. Note that capacitors have a *microfarad* rating, indicated by mfd. This rating is the ability of a capacitor to hold an electric charge.
5. Also note that capacitors have a d-c voltage rating. This rating indicates the maximum voltage that a capacitor can withstand while charged.
6. Identify those capacitors that have the word POSITIVE printed on one end (see Fig. 2·42). Some capacitors use *plus signs* instead of the printed word. Capacitors that indicate polarity are *electrolytic* capacitors.
7. Some electrolytic capacitors use colored leads to identify polarity. A black lead is negative, and a red lead is positive. Locate a capacitor of this type.
8. Separate the electrolytic capacitors and the non-electrolytic capacitors. Note that the electrolytic capacitors have a larger microfarad rating.

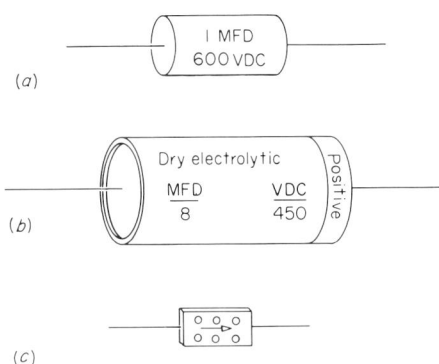

FIG. 2·42 Markings on capacitors.

QUESTIONS
1. What is the meaning of capacitance as applied to capacitors?
2. What is the meaning of breakdown voltage?
3. What does the word electrolytic mean?
4. Why do electrolytic capacitors have a high capacity rating?
5. Which capacitor has the greater ability to hold a displacement of electrons: a 0.1-mfd, 150-volt capacitor or a 0.005-mfd, 600-volt capacitor?

JOB 2·9 How to learn to read the capacitor color code

OBJECTIVES
1. To recognize capacitors that are color-coded
2. To become acquainted with capacitor color-code charts
3. To appreciate personal experience of trained technicians

MATERIALS REQUIRED

Equipment
Capacitor color-code chart

Supplies
Molded tubular capacitor, color-coded
Tubular ceramic capacitor, color-coded
Molded mica capacitor, color-coded

PROCEDURE
1. Identify each of the three capacitors supplied according to the designations in Fig. 2·15.

2. Examine the capacitor color-code chart supplied.
3. Compare the capacitors in your possession with the illustrations on the color-code chart.
4. Note that a knowledge of the color-code table supplied in Job 2·2, step 4 is helpful in interpreting the capacitor color code.
5. Determine from the capacitor color-code chart the end of the capacitor from which to start *reading* the colors.
6. Identify the purpose of each color used. Refer to all the technical data supplied on the chart. Do not hesitate to ask your instructor or any other experienced electronics technician for assistance.
7. Notice the variations encountered in interpreting the ratings of the capacitors supplied. For this reason it is absolutely essential to use an appropriate capacitor color-code chart as well as the help of an experienced electronics technician.
8. Make a list of the capacitor ratings to be taken into account when interpreting the color code of a capacitor.
9. State the purpose of each rating listed in step 8.

QUESTIONS
1. What is a capacitor?
2. Explain the meaning of microfarad.
3. What limitation does the breakdown voltage rating place on a capacitor?
4. What is *temperature coefficient*?
5. Why do capacitors employ a color code?
6. List all the problems to be encountered when reading the capacitor color code.

JOB 2·10 How to investigate capacitors

OBJECTIVES
1. To become acquainted with capacitor construction
2. To appreciate the meaning of capacitance and breakdown voltage ratings through observation of capacitor construction

MATERIALS REQUIRED

Equipment
Small hand tools

Supplies
Molded tubular capacitor
Molded mica capacitor
Electrolytic capacitor
Variable capacitor, air-dielectric

PROCEDURE
1. Carefully dismantle each capacitor supplied, except the variable capacitor.
2. Unroll the tubular and electrolytic capacitors as much as possible.
3. Separate the mica capacitor until each part is clearly visible.
4. Compare the variable capacitor with the dismantled capacitors.
5. Identify the dielectric material on each capacitor.
6. Identify the two opposing plates of each capacitor.
7. Note how the external leads are attached to the capacitor plates.
8. Make a list of the possible places where a short circuit could appear in a *shorted* capacitor.
9. Locate the parts of a capacitor that can cause a change in capacitance and explain the reason for capacitance change.
10. Compare the dismantled capacitors with the schematic symbols given in Fig. 2·19.

QUESTIONS
1. Describe a capacitor in physical terms.
2. What happens during voltage breakdown?
3. What causes a change in capacitance of a capacitor?
4. What is the most probable reason why a capacitor short-circuits?
5. Which capacitor supplied could withstand the highest breakdown voltage? Why?
6. What could cause an *open circuit* in a capacitor?

JOB 2·11 How to identify inductors

OBJECTIVES
1. To become acquainted with the various types of inductors
2. To become acquainted with inductor cores

MATERIALS REQUIRED

Supplies
Kit of components used in Job 1·1
Old electronics magazines and electronics catalogs
8½ × 11-inch notebook paper and pencil
Scissors
Rubber cement

PROCEDURE

1. Separate all fixed and variable inductors from the kit of components supplied. Some of these may be in metal cases.
2. Look in the magazines and catalogs supplied for pictures of similar fixed and variable inductors.
3. Cut these pictures out and paste them on sheets of notebook paper with rubber cement.
4. Under each picture write a summary regarding the purpose, values, and specifications relating to that specific inductor.
5. For extra credit, cut out pictures of inductors which were not supplied in the kit, paste them on notebook paper, and write a summary on each.
6. Assemble all of these inductor identification sheets into a notebook.

QUESTIONS

1. What is the basic purpose of an inductor?
2. Describe a transformer.
3. Describe a low-frequency choke.
4. Describe a high-frequency (r-f) transformer.
5. Describe a laminated-iron core.
6. Describe an air core.
7. Describe a powdered-iron core.
8. Why are transformers often placed within a metal can?
9. How are transformers rated?
10. How are inductors marked?

JOB 2·12 How to investigate transformer and choke construction

OBJECTIVES

1. To appreciate the problems that may arise within an inductor
2. To increase understanding of inductors by means of certain tests

MATERIALS REQUIRED

Equipment
Hand tools, including hack saw
Bench vise

Supplies
Power transformer, defective
R-F choke, defective

PROCEDURE

1. Compare the power transformer with the schematic diagram in Fig. 2·26h.
2. Compare the r-f choke supplied with the schematic diagram in Fig. 2·26b.
3. Place the power transformer securely in the bench vise in preparation for cutting the transformer in half.
4. Cut the power transformer in half in such a way as to reveal the internal construction of all coils and the iron core. Use the hack saw.
5. Note how the iron core envelops the coils of wire.
6. Note how the iron core itself is assembled.
7. Note how the copper wire is wound in layers and how one layer is insulated from the other.
8. Note how the various coils (windings) are wound on top of each other and check the method employed to separate the coils.
9. Carefully dismantle the top surface of the transformer where the leads are connected and investigate the methods employed to attach external leads to the coils.
10. Note the various sizes of wire used in the construction of the different windings.
11. Proceed to dismantle completely the two halves of the transformer, paying close attention to all assembly techniques used in the manufacturing.
12. Carefully disassemble the r-f choke to reveal its construction. Do not cut the choke in half, as was done with the power transformer.
13. Check the way the coil is wound.
14. Check the method employed to attach external leads to the coil.

15. Make a list of the possible problems that could arise within an inductor.

QUESTIONS
1. Define a *transformer winding*.
2. Where is it possible for an inductor coil to become *open*?
3. Is it possible for an inductor coil to have shorted *turns* of wire? Explain.
4. Is it possible for a transformer to have one coil shorted to another? Explain.
5. Explain the difference between a transformer and a choke.
6. What type of core is used in an r-f choke? Why?
7. How is it possible for a coil to become *grounded* to the iron core?

JOB 2•13   How to investigate relay operation

OBJECTIVES
1. To become familiar with relays
2. To appreciate remote control
3. To recognize the relay coil as an inductor
4. To appreciate electromagnetism

MATERIALS REQUIRED

Power source
28-volt direct current

Equipment
Ohmmeter

Supplies
28-volt relay, similar to Fig. 2•22

PROCEDURE
1. Compare the relay supplied with the relay in Fig. 2•22.
2. Locate on your relay the various parts of a relay as designated in Fig. 2•22.
3. Relate the physical parts of the relay in Fig. 2•22 to the various sections of the schematic symbol shown.
4. On the relay supplied locate a set of contacts similar to those shown connected to terminals 3 and 4 in the schematic symbol of Fig. 2•22.
5. Connect the ohmmeter to the corresponding terminals. If no continuity exists, with a finger move the armature arm to cause the contacts to close. Continuity should now be in evidence. Under these conditions the contacts are said to be *normally open*.
6. Maintain the ohmmeter connected to the contact terminals.
7. Apply 28 volts to the relay coil. Notice that the armature arm will move and close the contacts. This can be further proved by observing the continuity effects on the ohmmeter.
8. Alternately apply and remove the 28 volts to the coil and observe the action on the ohmmeter.
9. Notice that while 28 volts is applied to the coil of the relay, the contacts handle only the current supplied by the ohmmeter. A relay can therefore be used to remotely control another circuit.
10. Notice also that when the 28 volts is applied to the coil, the armature of the relay is attracted to the core of the coil. This is caused by the magnetic field that is created by the coil of wire when an electric current flows through it. The magnetic field attracts the *iron* armature. We therefore create a mechanical motion with electrical energy.

QUESTIONS
1. In what respects is the coil of the relay similar to a choke?
2. Make a sketch of the relay supplied and label all parts as in Fig. 2•22.
3. Explain electromagnetism.
4. Draw a schematic diagram illustrating how a lighting system can be remotely controlled with a relay.
5. Explain why the ohmmeter did not have the 28 volts applied to it in step 7.

JOB 2•14   How to associate schematic symbols of inductors with the actual part

OBJECTIVES
1. To learn to interpret schematic symbols
2. To appreciate the design of a schematic symbol

MATERIALS REQUIRED

Supplies

An assortment of inductors to match the symbols in Fig. 2•26

PROCEDURE

1. Note that the loop antenna supplied is similar in construction to the schematic symbol in Fig. 2•26*a*.
2. Note that the air-core choke is made of a single coil of wire with nothing but air for a core, similar to the schematic symbol illustrated in Fig. 2•26*b*.
3. Note that the filter choke is a coil of wire with a core of laminated iron. In Fig. 2•26*c* the schematic symbol for an iron-core filter choke indicates a single coil with laminations of iron adjacent to it.
4. The rod antenna is composed of a single coil of wire with a powdered-iron core. The symbol for a rod antenna is shown in Fig. 2•26*d*. The broken lines indicate that the iron core is composed of small bits of iron, molded to hold form.
5. The schematic symbols of Fig. 2•26*e* to *h* indicate that more than one coil surround a common core. This reveals that these symbols are for transformers, since a transformer is composed of two or more coils.
6. The single broken line through the coil in Fig. 2•26*i* is used to indicate that a mechanical device is directly related to the coil. In the case of the relay, the coil was responsible for creating the magnetic field that attracted the iron armature. Figure 2•26*i* therefore applies to the relay coil.
7. Note that, in all cases illustrated above, the schematic symbol definitely presented a *symbolic picture* of the actual part.

QUESTIONS

1. Does the same principle of symbolic pictures apply to schematic symbols of resistors and capacitors? Explain your answer.
2. Describe what continuity tests could be made on the inductors shown schematically in Fig. 2•26.
3. What parts of an inductor are not included in the schematic-diagram symbols of Fig. 2•26?

JOB 2•15 How to identify semiconductors

OBJECTIVES

1. To become acquainted with the various types of semiconductors
2. To become acquainted with semiconductor designations

MATERIALS REQUIRED

Supplies

Kit of components used in Job 1•1
Old electronics magazines and electronics catalogs
8½ × 11-inch notebook paper and pencil
Scissors
Rubber cement

PROCEDURE

1. Separate all semiconductors from the kit of components supplied.
2. Look in the magazines and catalogs supplied for pictures of similar semiconductors.
3. Cut these pictures out and paste them on sheets of notebook paper with rubber cement.
4. Under each picture write a summary regarding the purpose, designations, and *package* (case) identification relating to that specific semiconductor.
5. For extra credit, cut out pictures of semiconductors which were not supplied in the kit; paste them on notebook paper and write a summary on each.
6. Assemble all of these semiconductor identification sheets into a notebook.

QUESTIONS

1. What is the basic purpose of any semiconductor?
2. Describe a semiconductor diode.
3. Describe a transistor.
4. Describe an integrated circuit.
5. Identify the various semiconductor *packages* you encountered in the kit of components supplied to you. Make this identification orally to your instructor.

JOB 2·16 How to identify semiconductor diodes

OBJECTIVES

1. To learn to recognize a semiconductor diode
2. To learn to identify leads of semiconductor diodes
3. To learn to appreciate polarity markings on semiconductor diodes

MATERIALS REQUIRED

Supplies
An assortment of semiconductor diodes

PROCEDURE

1. Compare the diodes supplied with those shown in Fig. 2·28.
2. Make a comparison of the vacuum-tube diode and the semiconductor diode schematic symbols illustrated in Fig. 14·1.
3. Check the diodes supplied to see if any of them have the symbol illustrated in Fig. 14·1a stamped on their surface. If the symbol is stamped, it will appear in the proper relation to the cathode and anode leads.
4. Check the diodes supplied to see if any of them have a positive mark on one end. This plus mark indicates the cathode end.
5. Check the diodes supplied to see if any of them utilize a series of color bands around one end. The color bands appear at the cathode end.
6. The shape of the diode housing can also indicate which end contains the cathode lead. Compare any identification of the cathode lead as outlined above with the shape of the diode housing, for further clues.
7. If there are any semiconductor diodes supplied which you are not able to identify, consult your instructor.
8. Ask your instructor to explain the role of the cathode and the anode.

QUESTIONS

1. If one lead of a semiconductor diode is identified as the cathode lead, to what is the second lead connected?
2. Why are "heater" leads not included in a semiconductor diode? (Vacuum-tube diodes use heaters.)
3. How many diodes supplied were you not able to identify? Why?
4. To the best of your ability explain why the cathode of a semiconductor diode is assigned a plus mark. (*Hint:* Refer to Chaps. 15 and 16.)

JOB 2·17 How to identify color-coded diodes

OBJECTIVES

1. To learn to recognize a semiconductor diode
2. To learn to interpret the color code used on diodes

MATERIALS REQUIRED

Equipment
Semiconductor diode manual

Supplies
1N264 diode

PROCEDURE

1. Compare the diode supplied with Fig. 2·43.
2. This diode will automatically have the prefix 1N, followed by a series of numbers which are determined from the color bands used.
3. Review the color-code table used in Job 2·2, step 4.
4. Interpret the color bands as numbers, using the table in Job 2·2, step 4.
5. The three color bands grouped close together are the only color bands used in this phase of identification. The first color band to be considered is the band nearest the end of the diode.
6. Since red represents number 2, blue number 6, and yellow number 4, the color bands reveal a 264 number series.
7. Combine the 1N prefix with the 264 obtained from the color bands, and we have a 1N264 diode.
8. The final red band on the diode illustrated in Fig. 2·43 is used to identify the manufacturer of this diode. Some manufacturers will use an alphabetical letter to identify themselves instead of a color band, as illustrated in Fig. 2·27a.
9. For technical information concerning this diode refer to the semiconductor diode manual sup-

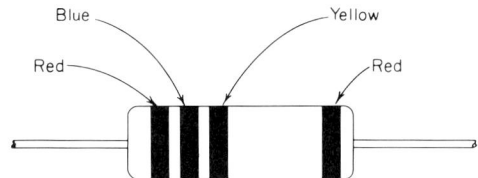

FIG. 2·43 A color-coded semiconductor diode.

plied. Use the 1N264 designation to locate the data required.

10. The lead nearest the three color bands is connected to the cathode. The opposite lead is connected to the anode.
11. The manufacturer's color band is not important when seeking technical information in a diode manual. It is important only when it becomes necessary to know the manufacturer of the diode in question. An electronics parts dealer can supply the information relating to the manufacturer of the diode when only the manufacturer's color band is known.

QUESTIONS

1. What is a semiconductor diode?
2. Explain the meaning of cathode in a semiconductor diode.
3. Sketch a semiconductor diode having the designation 1N034. Label the necessary color bands.
4. Who manufactures diodes with a *red* color band near the anode lead?

JOB 2·18 How to interpret power-rectifier markings

OBJECTIVES

1. To learn to recognize semiconductor power rectifiers
2. To learn to interpret markings on power rectifiers
3. To become acquainted with selenium rectifiers

MATERIALS REQUIRED

Supplies
Top-hat power rectifier
Selenium rectifier

PROCEDURE

1. Compare the *top-hat power rectifier* supplied with the one illustrated in Fig. 2·31b.
2. Note that the physical outline of the rectifier resembles a gentleman's top hat.
3. Note that the diode symbol painted on the surface is the same symbol shown in Fig. 2·38a. This symbol is painted in the same relationship as the position of the cathode and anode, which are inside the *hat*.
4. The wide rim of the hat is therefore at the cathode end of the diode, while the anode and its lead appear at the opposite end.
5. Locate other power rectifiers, using catalogs or trade magazines if necessary. Identify them to your instructor.
6. Inspect the selenium rectifier supplied.
7. Note that at one end plate a *plus* mark appears.
8. The terminal at the end where the plus mark appears is the cathode terminal of the rectifier. The opposite terminal is attached to the anode of the rectifier.

QUESTIONS

1. Where is a power rectifier used?
2. Why does a power rectifier have its cathode lead identified?
3. Describe a selenium rectifier.
4. What does the word power imply in connection with rectifiers?
5. Make a list of all the power rectifiers you were able to locate, using the proper technical description for each.

JOB 2·19 How to identify transistors

OBJECTIVES

1. To learn to recognize a transistor
2. To learn to use a transistor manual
3. To become acquainted with transistor elements

MATERIALS REQUIRED

Supplies
An assortment of transistors
Various transistor manuals

PROCEDURE

1. Compare the transistors supplied with those shown in Fig. 2·6.
2. Notice the transistor identifying number, such as 2N104, that appears printed on all transistors.
3. Notice the various lead arrangements used. (Job 2·20 teaches the identification of the individual leads.)
4. Look over the various transistor manuals supplied.
5. Select the transistor manual which you can best understand and use it to identify each transistor in your possession. Use the identifying number, such as 2N104, to locate the transistor information in the manual.
6. Inspect the transistor schematic symbols used in the manual.
7. Get acquainted with the technical specifications for transistors given in the manual.
8. Refer to the schematic symbol of each transistor in your possession and identify the internal elements by name.

QUESTIONS

1. Approximately how much smaller is the average transistor than vacuum tubes found in radio receivers?
2. Approximately how much less does a transistor weigh than an average vacuum tube?
3. Do transistors use a color code? If so, how?
4. Did any of the transistor manuals contain basic transistor theory?
5. Did any of the transistor manuals contain typical circuit application of the transistors?
6. Was any mention made of hole current in the technical specifications given for an individual transistor?
7. How can one tell which is the base element in a transistor symbol?
8. How can one tell which is the collector element in a transistor symbol?
9. How can one tell which is the emitter element in a transistor symbol?
10. Is the collector element of a transistor ever soldered internally to the case of the transistor, thereby dispensing with the need for a collector lead?

JOB 2·20 How to identify transistor leads and sockets

OBJECTIVES

1. To learn to recognize transistors
2. To learn to identify the leads of transistors
3. To learn to recognize transistor sockets

MATERIALS REQUIRED

Equipment
Transistor manual

Supplies
An assortment of triode transistors with various lead arrangements
An assortment of transistor sockets

PROCEDURE

1. Study the triode transistor symbols in Fig. 2·38e and f. Note that the emitter is represented with an arrow. The arrow points away from the base in an NPN transistor and toward the base in a PNP transistor. The base and the collector remain the same.
2. Compare the lead arrangement of the transistors in your possession with the illustrations of Fig. 2·44.
3. Identify the leads of the transistors supplied, determining the emitter, base, and collector leads. Use Fig. 2·44 as a guide.
4. Verify your answers by consulting the transistor manual.
5. Repeat step 3 to identify the terminals of the transistor sockets supplied.

QUESTIONS

1. Is there a *key* used to identify transistor leads? Explain your answer.
2. Compare the lead designation of transistors with the lead designation of vacuum tubes.
3. Identify the leads of at least one transistor to your instructor.
4. Why is a color dot used in a transistor?

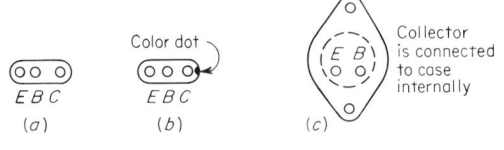

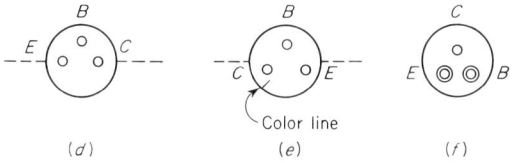

FIG. 2·44 Various lead arrangements used on transistors.

5. List the advantages and disadvantages of using sockets with transistors.
6. Why do some transistors have only two wire leads extending outward?
7. Draw a diagram showing FET lead arrangement. (Use a 3N128.)

JOB 2·21 How to identify transistor cases

OBJECTIVES

1. To become acquainted with the various transistor cases
2. To appreciate manufacturer's data manuals

MATERIALS REQUIRED

Equipment
Transistor manual
"Guide to RCA Solid-State Products (SPG-201E)"

Supplies
An assortment of transistors having different type cases

PROCEDURE

1. Study the OUTLINES section of the transistor manual. This section normally appears toward the back pages of the manual.
2. Notice that this section illustrates the silhouette outline of transistor and diode cases, giving the various applicable physical dimensions and lead arrangements.
3. You will notice such transistor outlines as TO-3, and TO-5. These outlines have been defined by the *Joint Electron Device Engineering Council* (JEDEC) and are therefore properly designated as the *JEDEC TO-3* and *JEDEC TO-5* transistor cases. However, the abbreviated terms TO-3 and TO-5 are well accepted.
4. For a photographic view of these transistor cases (packages), refer to the OUTLINES section of the "Guide to RCA Solid-State Products" pamphlet supplied. Notice the large variation in transistor cases available. Also notice the various diode outlines, such as the DO-1 and the DO-5.
5. With the aid of the transistor manual and the Guide mentioned above, proceed to identify the cases of the transistors supplied to you. Identify the cases *to your instructor* in terms of the TO-number designation.

QUESTIONS

1. Can you tell from the transistor symbols that appear in Fig. 2·38 the type of cases being used?
2. Identify the case of the power rectifier (diode) in Fig. 2·27b.
3. Identify the case of the transistor in Fig. 2·27c.
4. Identify the case of the transistor in Fig. 2·32e.
5. Identify the case of the transistor in Fig. 2·44c.
6. Field-effect transistors (FET) are often found in cases such as that illustrated in Fig. 2·32b. What is the TO-number of this case?

JOB 2·22 How to identify integrated-circuit packages

OBJECTIVES

1. To become acquainted with various integrated-circuit packages
2. To appreciate manufacturers' data pamphlets and manuals

MATERIALS REQUIRED

Equipment
Integrated-circuits manual
"RCA Integrated Circuits Product Guide (CDL-820B)"
"Guide to RCA Solid-State Products (SPG-201E)"

Supplies

An assortment of integrated circuits in different packages

PROCEDURE

1. Study the OUTLINES section of the integrated-circuits manual. This section normally appears toward the back pages of the manual.
2. Notice that this section illustrates the silhouette outline of integrated-circuits packages, giving the various applicable physical dimensions and lead orientation.
3. Also study the DIMENSIONAL OUTLINES section of the "RCA Integrated Circuits Product Guide." Notice that some integrated-circuits packages *borrow* the transistor outline identification for their case numbering and call it the *TO-5 style* package. The reason is that the two cases are practically identical, the main difference being that only three leads are used in a transistor, but integrated circuits use either 8 leads, 10 leads, or 12 leads when encased in TO-5 style packages. Notice, however, that integrated circuits come in flat packages and dual in-line packages as well.
4. For a photographic view of integrated-circuit packages, refer to the OUTLINES section of the "Guide to RCA Solid-State Products" pamphlet supplied.
5. With the aid of the integrated-circuits manual and the pamphlets supplied, proceed to identify the integrated-circuits packages made available to you. Identify the packages to your instructor, orally.

QUESTIONS

1. Is there any relation between the integrated-circuit symbol shown in Fig. 2•38*j* and the physical appearance of any integrated-circuit package seen in the above manuals?
2. Identify the integrated-circuit package illustrated in Fig. 2•27*d*.
3. Identify the integrated-circuit package illustrated in Fig. 2•35*a*.
4. Identify the integrated-circuit package illustrated in Fig. 2•35*b*.
5. Identify the integrated-circuit package illustrated in Fig. 2•35*c*.

JOB 2•23   How to identify electron tubes

OBJECTIVES

1. To become acquainted with electron tubes
2. To survey electron-tube theory
3. To become acquainted with electron-tube nomenclature

MATERIALS REQUIRED

Supplies
Kit of components used in Job 1•1
8½ × 11-inch notebook paper and pencil

PROCEDURE

1. Separate all vacuum tubes from the kit of components supplied. (Gaseous tubes are not used in radio and television receivers.)
2. Note that by just viewing the physical outlines of any tube, including the pins at the base, it is impossible to tell the type and basic intent for the tube. One would have to refer to a tube manual for this information. However, in order to understand the tube manual, the basic concepts of electron-tube theory are necessary.
3. Look over the questions at the end of this job. If you can answer them, do so. If not, continue with step 4 below.
4. Read and study Appendix I at the end of this book. The basic concepts of electron-tube theory are covered in this section of the text.
5. After studying Appendix I, work Jobs 2•24 and 2•25 before proceeding. These jobs deal with tube sockets and tube manuals, respectively.
6. Having gained some basic knowledge regarding electron tubes by working steps 4 and 5 above, proceed to identify the tubes found in your kit of components. Record the identification by listing each tube and writing a brief summary about each tube.

QUESTIONS

1. Describe a vacuum-tube *diode*.
2. What is the purpose of the *heater* in a vacuum tube?
3. What is the purpose of the *plate* in a vacuum tube?

4. Describe a *filament*.
5. Describe a vacuum-tube *triode*.
6. What is the purpose of the *control grid* in a vacuum tube?
7. What are the battery requirements of a vacuum tube?
8. What is the *envelope* of a vacuum tube?
9. How are vacuum tubes marked?
10. How can you tell which pin in the base of the tube connects to the plate electrode?

JOB 2·24 How to identify vacuum-tube sockets

OBJECTIVES
1. To become acquainted with electron-tube sockets
2. To appreciate the need for a tube socket
3. To learn to work with sockets

MATERIALS REQUIRED

Supplies
Octal-tube socket
Miniature socket, 7-pin
6J5 vacuum tube

PROCEDURE
1. Compare the sockets in your possession with those in Fig. 2·37 and 2·45.
2. Note the key slot in the octal socket.
3. Note the wide space between two pin holes in the miniature socket. This space is the *key*.
4. Viewing the octal socket from the terminal side, locate the terminal for pin 2. Count clockwise from the key.
5. Note that, when locating the same terminal from the opposite side of the socket, one must count counterclockwise from the key.
6. Carefully install the 6J5 vacuum tube in the octal socket. Be sure that the key of the vacuum tube slides into the key slot on the socket.
7. Note that any leads to be connected to the tube may be soldered to the terminals of the socket.

QUESTIONS
1. What does the word octal mean?
2. List the advantages of a tube socket.
3. List the disadvantages of a tube socket.

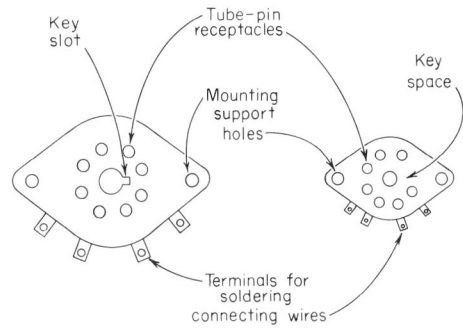

FIG. 2·45 Identification of electron-tube sockets.

JOB 2·25 How to use a vacuum-tube manual

OBJECTIVES
1. To become acquainted with electron-tube manuals
2. To learn to extract information from a tube manual

MATERIALS REQUIRED

Supplies
An assortment of tube manuals

PROCEDURE
1. Look over the various tube manuals in your possession.
2. Select the tube manual that best conveys information to you.
3. Note that all tube manuals list electron tubes by code numbers and letters, such as 6J5.
4. Also note that the schematic diagram of the internal elements of the electron tube is given.
5. Note that, on the schematic diagram, the numbers of the pin terminals are indicated.
6. Electrical specifications about each tube are also given. Among these specifications are found the heater voltage and current.
7. Note also that the purpose of each tube is given.

QUESTIONS
1. What is the heater voltage of a 6J5 vacuum tube?
2. What is a 6J5 vacuum tube used for?
3. What are the heater pin terminals of a 6SN7 tube?
4. Is a 6J5 a diode or a triode?

JOB 2·26 How to investigate electron-tube construction

OBJECTIVES
1. To acquire a better understanding of the internal elements of an electron tube
2. To gain better insight into possible flaws in an electron tube

MATERIALS REQUIRED

Equipment
Small hand tools, including a hammer

Supplies
An assortment of discarded electron tubes utilizing glass envelopes

Miscellaneous
An old handkerchief or similar piece of cloth

PROCEDURE
1. Place an electron tube on the workbench.
2. Cover the tube with the piece of cloth.
3. Lightly hit the electron tube with the hammer to break the glass envelope. DO NOT REMOVE THE CLOTH. The cloth serves to prevent glass from flying out.
4. Carefully remove the cloth.
5. With the aid of hand tools, slowly dismantle the electron tube. Study each part dismantled.
6. Repeat the preceding five steps with another electron tube. However, dismantle the tube just enough to expose all the elements as shown in Fig. 2·46.
7. Identify each element.
8. Point out to your instructor the heater, cathode, control grid, and plate.

QUESTIONS
1. Is it possible for the cathode to chip off?
2. Is it possible to have a *short circuit* between the heater and the cathode?
3. If the control grid of a vacuum tube became loose, what would most likely happen?
4. Where would an *open circuit* of the plate be most likely to occur?
5. What happens when the envelope of a vacuum tube cracks?

FIG. 2·46 A vacuum tube opened for inspection.

JOB 2·27 How to test vacuum tubes

OBJECTIVES
1. To learn about the various tests connected with vacuum tubes
2. To learn to identify vacuum-tube terminal pins
3. To learn to make continuity tests on vacuum tubes
4. To appreciate a vacuum-tube tester

MATERIALS REQUIRED

Equipment
Ohmmeter

Supplies
6J5 vacuum tube

PROCEDURE
1. Hold the vacuum tube with the base pointing toward you.
2. Notice the ridge on the plastic center post. This is the *key*.

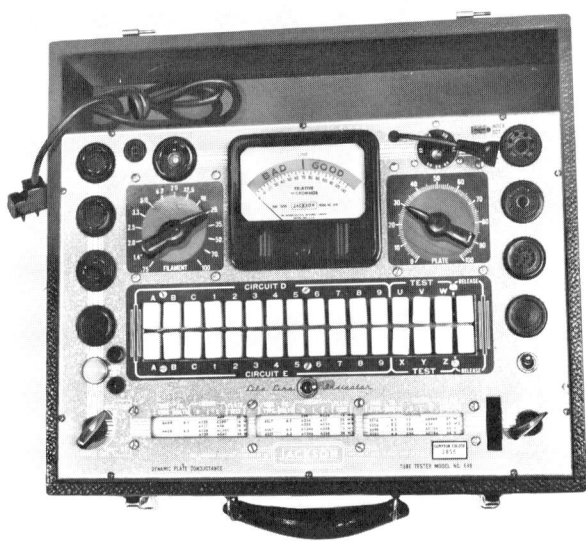

FIG. 2·47 The electron tube tester.

3. Count the metal pins clockwise from the key and locate pins 2 and 7. These pins are the terminals of the heater leads.
4. Connect the ohmmeter leads to pins 2 and 7 and proceed to make a continuity test of the heater. A good heater will show continuity.
5. Check to see if a short circuit exists between the heater and cathode. This is accomplished by making a continuity test between pin 2 and pin 8, the cathode pin. There should be no continuity (see Fig. A1·9).
6. The emission quality of the cathode cannot be tested with an ohmmeter.
7. To test the overall conductance of the vacuum tube, a complete testing circuit is required.
8. Study the tube tester shown in Fig. 2·47. With a tube tester, all the tests mentioned above can be made, giving a complete indication of the quality of an electron tube.

QUESTIONS

1. What does the continuity test of the heater in a vacuum tube indicate?
2. Why can a continuity test reveal a short circuit between elements in a vacuum tube?
3. Can a continuity test reveal an intermittent short?
4. Why is the key important in a vacuum tube?
5. What is thermal emission?
6. What is conductance?
7. Is a tube tester necessary to test the heater of a vacuum tube?

3

ELECTRONICS PACKAGING

Progress of electronics to the present-day state-of-the-art has been made possible by several factors. Among these has been the development of new electrical and electronic components along with improvements on many older units. For examples, there are the development of transistors and the miniaturization of components. Just as important has been the evolvement of methods to interconnect these components so that they could be packaged in compact, useful arrays. This chapter deals primarily with the latter development—the *packaging* of electronic devices.

Changes in packaging techniques have been closely related to the search for methods to miniaturize whole electronic units. So successful has been this search, that the field of *microelectronics* evolved. We can classify this field into three groups according to circuit wiring technique. The first group is that of *wired circuits*; the second, *printed circuits*; and the third, *integrated circuits*. The following sections describe these three group classifications and give a survey of manufacturing and assembly techniques used with each.

3·1 WIRED CIRCUITS

A *wired circuit* is identified as an interconnection of *discrete components* (individual separate parts) by means of wires, which are attached one at a time. Discrete components include resistors, capacitors, inductors, transistors, and vacuum tubes. These components are normally held in place by their leads which are secured to a terminal board, or to socket terminals of some other component. A wired circuit can be seen in Fig. 4·4. Routing of wires is normally rather haphazard. Methods for wire-wrapping of leads, and the soldering of these leads, center around the type of terminal used at *tie points*. See Chaps. 6 and 8.

Electronics assemblers often use a *wiring diagram* in order to guide them during fabrication of electrical products. Use of these diagrams ensures that those products that utilize *wired circuits* are identical. Further, a wiring diagram is of great help to the electronics technician who must later locate a certain component or wire during troubleshooting procedures.

3-D PACKAGED CIRCUITS

In an attempt to reduce space taken by a wired circuit, three-dimensional (3-D) packaged circuits were developed. A *3-D packaged circuit* (also known as *cordwood module*) can be described as a compact wired circuit where discrete components have been bundled to form a cube or some similar physical configuration in order to save space. This type of circuit may be made of passive components only, or can include semiconductors as well. After completely interconnecting and soldering all component leads, the *cube form* is normally encased in a container and sealed. Only *circuit terminal leads* extend outside this box. This circuit is therefore said to be *potted*. See Fig. 3·1.

Leads which protrude out of the box are those which are necessary for the application of operating voltages, input signals, and the leads needed to extract an output signal. This type of circuit assembly is quite compact and can withstand wide variations in shock, moisture, and temperature changes. However, the package is normally a permanent assembly that does not allow for circuit repairs in the event of faulty operation.

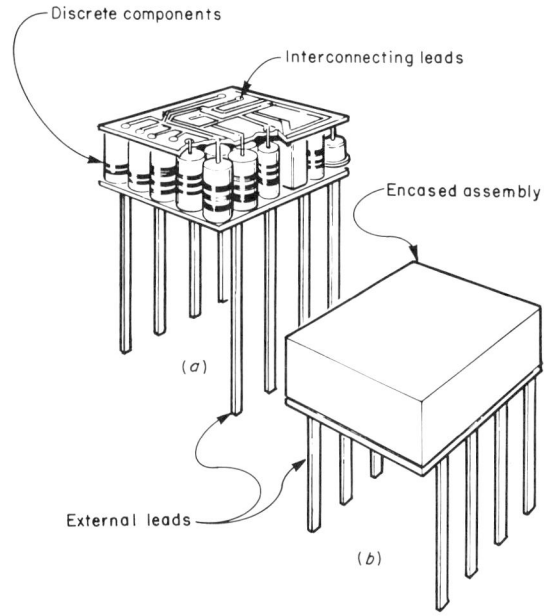

FIG. 3·1 3-D packaged circuit (cordwood module). (a) Discrete components wired in cube form. (b) Cube of components encased.

3·2 PRINTED CIRCUITS

A *printed circuit* is described as an interconnection of discrete components by means of *thin ribbon conductors* which are permanently bonded to a board made of laminated plastic. These boards have conductors prewired into circuit arrangements prior to having *components mounted on the boards*. A printed-circuit board can be seen in Fig. 3·2. These printed-circuit boards can be easily mass produced. Mass-production of these boards is accomplished by a photographic and etching process; therefore, the designation *printed circuit*. The term *etched circuit* is also used. A comparison of a conventional wired circuit with a printed circuit is seen in Fig. 3·3. A closer view of a printed circuit is seen in Fig. 5·3.

ADVANTAGES OF PRINTED CIRCUITRY

Printed circuits have important advantages by allowing for uniformity in mass production and by practically eliminating assembly wiring errors. Further, both sides of the board can have different

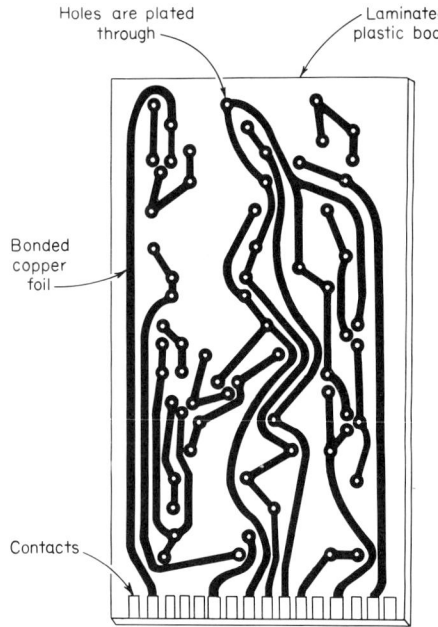

FIG. 3·2 Printed-circuit board.

printed circuitry, allowing for increased compactness.

Components are mounted on the boards in such a way that the leads are inserted through holes in the circuit board and are soldered on the opposite side. See Figs. 3·4 and 3·5. Usually these holes are an extension of the printed wiring and are known as *plated-through holes*. Exact soldering skills are required when soldering leads to printed circuits.

APPLICATIONS

A printed-circuit board with all the necessary component parts mounted on it is called a *module*. A module may contain several amplifiers, oscillators, rectifiers, or combinations of similar circuits. Modules are used in large quantities in computers and other electronic systems that require multiples of a given circuit. A connector is used at the end of a wire harness for engaging a printed-circuit board. One such connector is shown in Fig. 7·34c.

3·3 INTEGRATED CIRCUITS

The *integrated circuit* is the ultimate in miniaturized circuit configurations. The circuit is so

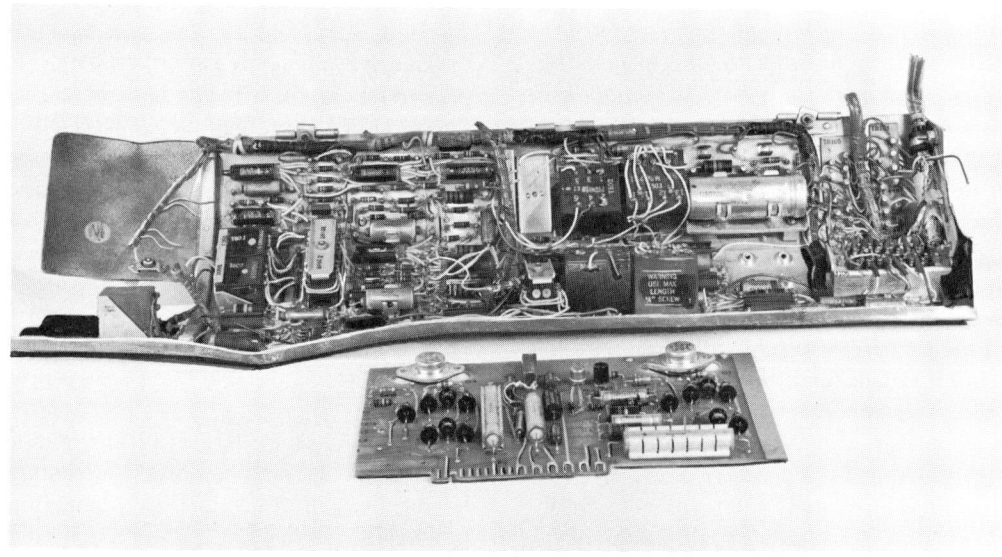

FIG. 3·3 A comparison of a conventional wired circuit with a printed circuit.

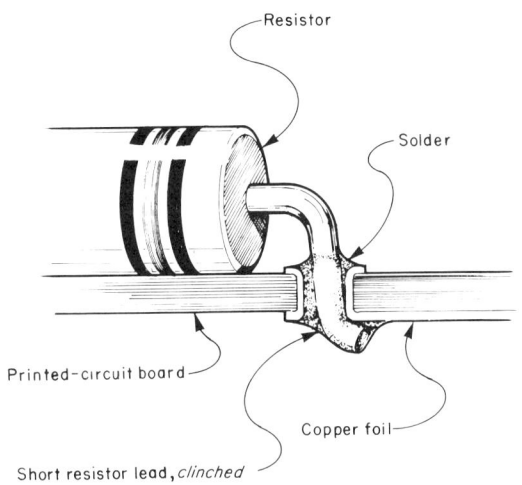

FIG. 3·4 Cross section of a printed-circuit board having leads of components *clinched* and soldered.

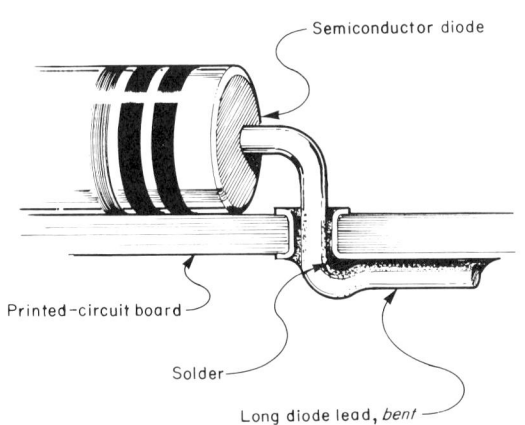

FIG. 3·5 Cross section of a printed-circuit board having leads of components *bent* and soldered.

small that it must be observed under *magnifying optical devices*. What has contributed to this miniaturization is the ability of modern technology to *form* many of the necessary components (such as resistors and capacitors) while at the same time *interconnecting* the leads of these components to each other. Therefore, discrete components are not used within these circuits, and interconnecting of components by manual means is not necessary. Fig. 3·6 shows two integrated circuits compared to discrete components. An integrated circuit is also referred to as an *IC*. Integrated circuitry in effect uses the principles of printed circuitry and 3-D circuitry all wrapped into one. Photography, etching, interconnecting of components to form circuits, and *casing* the whole assembly into a solid unified miniature mass all takes place. Here too, only leads essential to the application of operating voltages and input signals, and for the extraction of the output signal are made available. Whole amplifiers, oscillators, rectifiers, and switching circuits are possible in the IC format. In order to make it convenient to handle these circuits, they are further packaged in larger containers such as those illustrated in Fig. 2·35. It is handling of these *larger* IC packages that this text emphasizes. Detailed knowledge of what is inside an IC and how it was manufactured can be classified as "nice to know," but considered nonessential for the average electronics assembler.

Actually, integrated circuits are manufactured in a very precise fashion using photographic, chemical, vacuum, and microscopic techniques. Although interesting, the process for manufacturing integrated circuits is beyond the scope of this text. However, precise soldering techniques which can be attained by studying this book will help qualify a person for entry into the field of *microelectronics* assembly. Precision soldering can be singled out because this involves the efficient use of small tools, working under optical magnifiers, working with small wires and components, learning to control the application of heat, and in general, close attention to detail. Figure 3·7 shows a magnified illustration of a *hybrid* integrated circuit in a *Flat-Pack* package. Compare the size of the IC to the size of the fingernail. Figure 2·36 shows an illustration of a *monolithic* integrated circuit in a TO-5 package.

3·4 SUMMARY: ELECTRONICS PACKAGING

In summary, progress and application of electronics technology is closely related to methods used to *package* complete circuits in useful, efficient arrangements. Although no one method of packaging can be considered superior over other methods for all applications, certainly methods that have contributed to circuit miniaturization have done much

FIG. 3·6 Integrated circuits compared to discrete components. (a) Electron tube, seven-pin miniature. (b) Resistor. (c) Integrated circuit (TO-5 package). (d) Integrated circuit (microminiature flat-pack).

to advance the use of electronics. For example, the use of portable electronic computers has become practical because the many circuits that a device such as this needs can be packaged into small units. Further, the use of navigational equipment, computers, and communication systems which are necessary in space vehicles are possible due to the *small* and *light* electronic packages that comprise these devices. Be that as it may, well rounded electronics manufacturing personnel have a working knowledge of all three basic methods used for packaging electronic-circuit arrays; namely, *wired circuits*, *printed circuits*, and *integrated circuits*. This is necessary since all three are widely used, and since

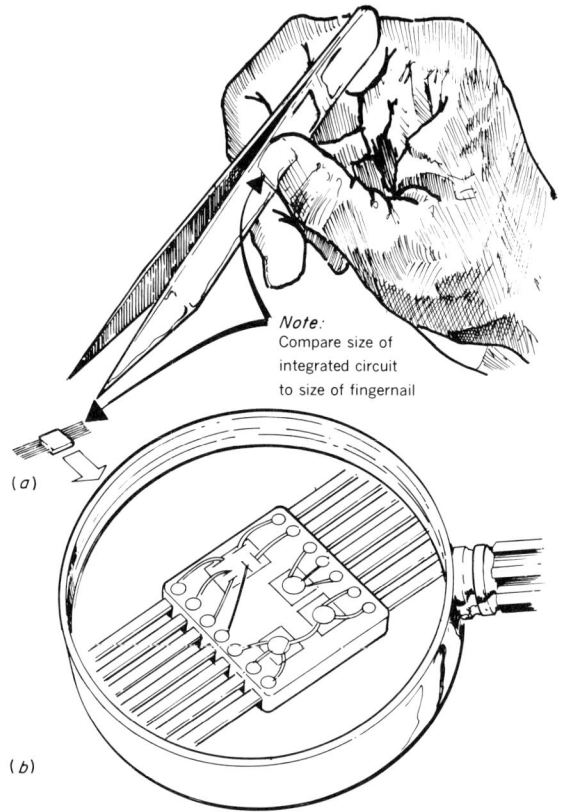

FIG. 3·7 Magnified illustration of a hybrid integrated circuit in a flat-pack package (relative sizes are approximate).

employment circumstances may require knowledge of all three systems.

The following jobs will help the student become better acquainted in the area of electronics packaging.

EXERCISES

JOB 3·1 How to investigate a wired chassis

OBJECTIVES
1. To become acquainted with the method used to interconnect electrical and electronic components with wires
2. To appreciate the quality of good workmanship in electronics fabrication

MATERIALS REQUIRED

Equipment
Radio or television receiver with wired-circuit-type chassis (commercially fabricated, and without vacuum or picture tubes installed) as illustrated in Fig. 4•4.

PROCEDURE
1. Place the chassis supplied upside-down on top of the workbench. Be careful not to damage any of the parts on the top side.
2. Carefully inspect how wires are used to interconnect the various electronic components, and the quality of workmanship involved.
3. Answer the following questions on paper.

QUESTIONS
1. Are the wires bundled into harnesses?
2. How many wire splices did you observe?
3. Were terminals used as tie points for wire leads?
4. Was the soldering smooth and bright, or dull and coarse?
5. Were wire leads clamped in any way to keep them in place?

JOB 3•2 How to investigate a 3-D wired circuit

OBJECTIVES
1. To become acquainted with three-dimensional wired circuits
2. To appreciate a compact, packaged circuit

MATERIALS REQUIRED

Equipment
Hand tools for mechanical assembly
Bench vise
Safety glasses

Supplies
Discarded, defective, epoxy-encapsulated, 3-D wired circuit

PROCEDURE
1. Carefully examine the 3-D wired circuit supplied. (See Fig. 3•1.)
2. Notice that all that is visible is a hard cube with leads.
3. Place the cube within the jaws of a vise and tighten securely.
4. Put on the safety glasses to protect your eyes.
5. Use the hand tools to *gradually* chip away the epoxy covering.
6. As the various components are exposed, use smaller tools to pick away at the epoxy surrounding the part. Pay attention to every detail.

QUESTIONS
1. How many component parts were you able to expose?
2. Is it practicable to attempt repairs on a 3-D packaged circuit?
3. List the advantages and disadvantages of a 3-D wired circuit.

JOB 3•3 How to investigate a printed-circuit board

OBJECTIVES
1. To become acquainted with printed-circuit boards
2. To appreciate pre-wired circuits

MATERIALS REQUIRED

Equipment
Hand tools for mechanical assembly
Bench vise

Supplies
Discarded, defective printed-circuit board

PROCEDURE
1. Carefully examine the printed-circuit board supplied. (See Fig. 3•2.)
2. Notice the ribboned conductors that serve as leads. Also notice the plated-through holes and the solder pads. Finally, notice the external circuit contacts that appear along one edge.

3. Place the printed-circuit board within the jaws of the vise and secure tightly. Do not bother to protect the surfaces.
4. With a hacksaw, cut the printed-circuit board in half.
5. With sharp, pointed instruments lift off from the laminated board the conductor copper foil. Pay attention to every detail.

QUESTIONS
1. Write a statement regarding your observations.
2. List the advantages and disadvantages of a printed-circuit board.

JOB 3•4 How to investigate an integrated circuit

OBJECTIVES
1. To become acquainted with integrated circuits
2. To appreciate microminiature electronics

MATERIALS REQUIRED

Equipment
Hand-type magnifying glass
Industrial microscope

Supplies
Open-case, demonstration integrated circuits (assorted)

PROCEDURE
1. Wash and dry your hands thoroughly.
2. Place the assortment of integrated circuits supplied on top of a piece of a clean, plain, blank notebook paper.
3. Inspect the integrated circuits with the aid of a magnifying glass. (See Fig. 3•7.)
4. View some of the integrated circuits through an industrial microscope. (See Fig. 5•8.)
5. Carefully replace the integrated circuits in their storage container.

QUESTIONS
1. Write a statement regarding your observations.
2. How many *monolithic* integrated circuits did you observe?
3. How many *hybrid* integrated circuits did you observe?
4. Differentiate between a monolithic and hybrid integrated circuit.

4

DIAGRAMS

The art of communication between individuals takes many forms. Yet, regardless of the form, it is the objective of any communication system to convey (and receive) instructions or information, for social, business, or other purposes.

In electronics work two basic methods of conveying instructions or information are employed. One is language. In the United States, English is used extensively, for obvious reasons. In addition, a *language of symbols* is used extensively.

Symbols that represent parts are used in various arrangements to convey intelligence. It is therefore necessary for electronics personnel to read and write the language of symbols.

Individual electronics symbols are comparable to individual English words. As words are arranged together to form sentences, *symbols are arranged to form basic circuits*. In English, groups of sentences make paragraphs, while in electronics, *groups of circuits make stages*. Also, just as the paragraphs make the chapter, the *stages make the units*. Finally, several chapters combine to make a book, and likewise, the units *combine to make a system*. As can be seen, the organization of written information is as important as the individual symbols.

In daily life, one sees very small children picking up words and trying to use these words, with a little help here and there. When the child goes to elementary school, he receives formal teaching to increase his vocabulary and also to learn how to read and write. In the higher grades he is again taught the English language but with greater emphasis on detail. And if he goes to college, he will find that more advanced training in English is required. Moreover, he will probably find that he has to become familiar with a foreign language as well, to increase his proficiency in his major course of study. This is approximately what happens in electronics. The beginning student will pick up a symbol here and there, and maybe use it properly. But as he advances in his electronics training, it becomes necessary to learn more about the details connected with the language of symbols. He will find that, in addition, it is necessary to learn the language of the *mechanical assembler*, since electronics parts are assembled mechanically first and then electrically.

This chapter deals primarily with the *organization of symbols* into the various types of *diagrams* encountered by electronics personnel. The student will find explanations of individual electronics symbols throughout this book.

4·1 SCHEMATIC DIAGRAM

The schematic diagram is the most important diagram used by electronics personnel. To interpret a schematic diagram, a knowledge of the symbols that represent the parts is an absolute necessity. Diagrams used to show the interconnection of component parts in this text are illustrated in schematic form in most cases.

Quite often the novice believes he is reading a schematic diagram, when in reality he is making use of a wiring diagram. A comparison of a schematic diagram and a wiring diagram is illustrated below. Figure 4·1a shows a *wiring diagram* of a simple circuit, and Fig. 4·1b is the *schematic* counterpart of the same circuit. Figures 4·2 and 4·3 are complete schematic diagrams of radio receivers. Figure 4·2 utilizes semiconductors. Figure 4·3 uses vacuum tubes.

4·2 WIRING DIAGRAM

Wiring diagrams are usually easy to understand.

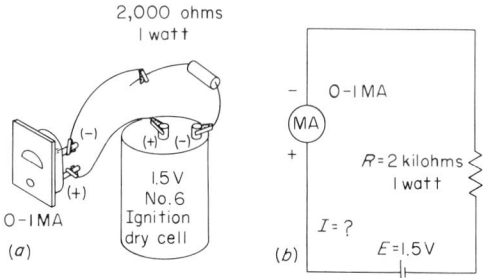

FIG. 4·1 (a) Wiring diagram. (b) Schematic diagram.

For this reason most manufacturing firms employing large numbers of employees with scant training make use of the wiring diagram extensively. Refer to Fig. 4·1a again and notice how easy it is to follow this type of diagram. However, although it has this advantage, the wiring diagram has many limitations. For this reason electronics assemblers desiring promotion and advancement must learn to use the schematic diagram.

There are many variations and interpretations of wiring diagrams. Some wiring diagrams deal with the wiring found within a chassis (see Fig. 4·4); others are *interconnection* wiring diagrams that deal with the interconnection of various units into a system (see Fig. 4·5). Wiring diagrams are subject to the use of pictorial illustrations, schematic symbols, and block diagrams. However, as a rule, the pictorial type of line drawing is used.

It should also be mentioned that the wiring diagram, of whatever variation, is both useful and necessary. The electronics technician must be able to interpret and use all forms of wiring diagrams, in addition to his mastery of schematic diagrams.

4·3 INTERCONNECTION DIAGRAM

As mentioned above, the interconnection diagram is a form of wiring diagram. Its purpose is to show how one electrical unit is electrically connected to another unit of the same system. There are many forms and methods used in making an interconnection diagram. This diagram is widely used in the electrical field, and the automotive industry also uses it, as do home builders, builders of industrial plants, the aircraft industry, the shipbuilding in-

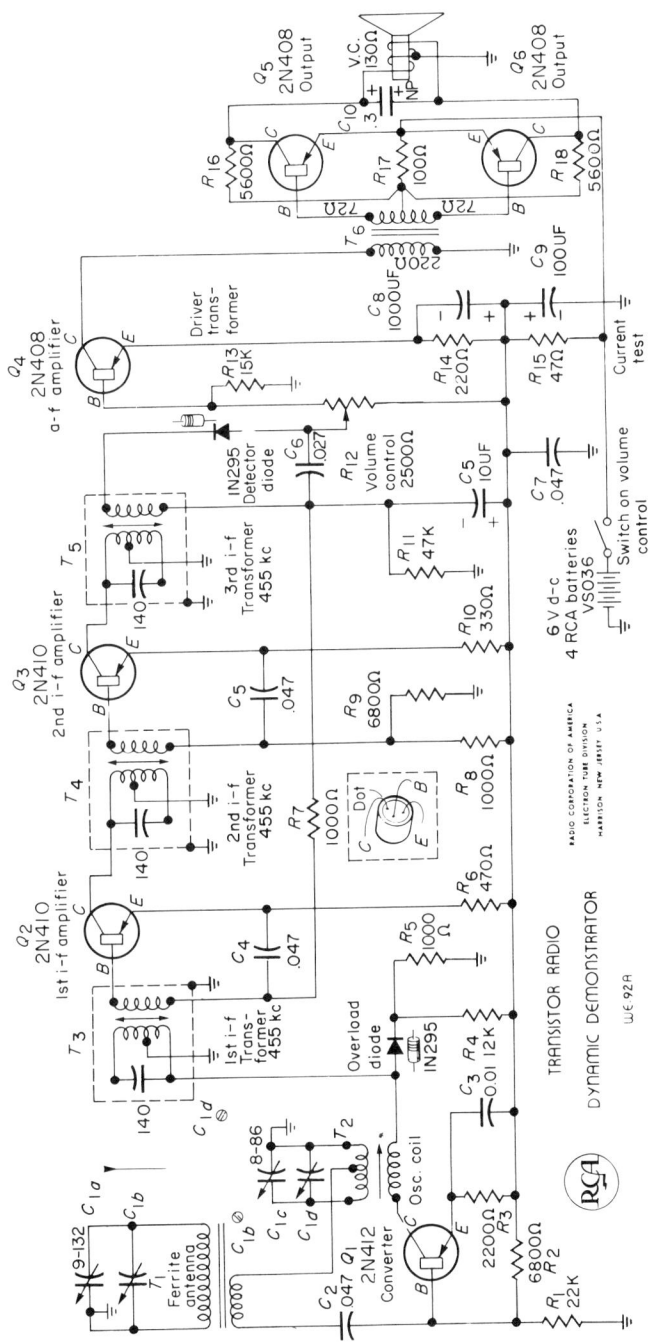

FIG. 4·2 A schematic diagram of a superheterodyne radio receiver using semiconductors. (Radio Corporation of America, Electron Tube Division, Harrison, New Jersey.)

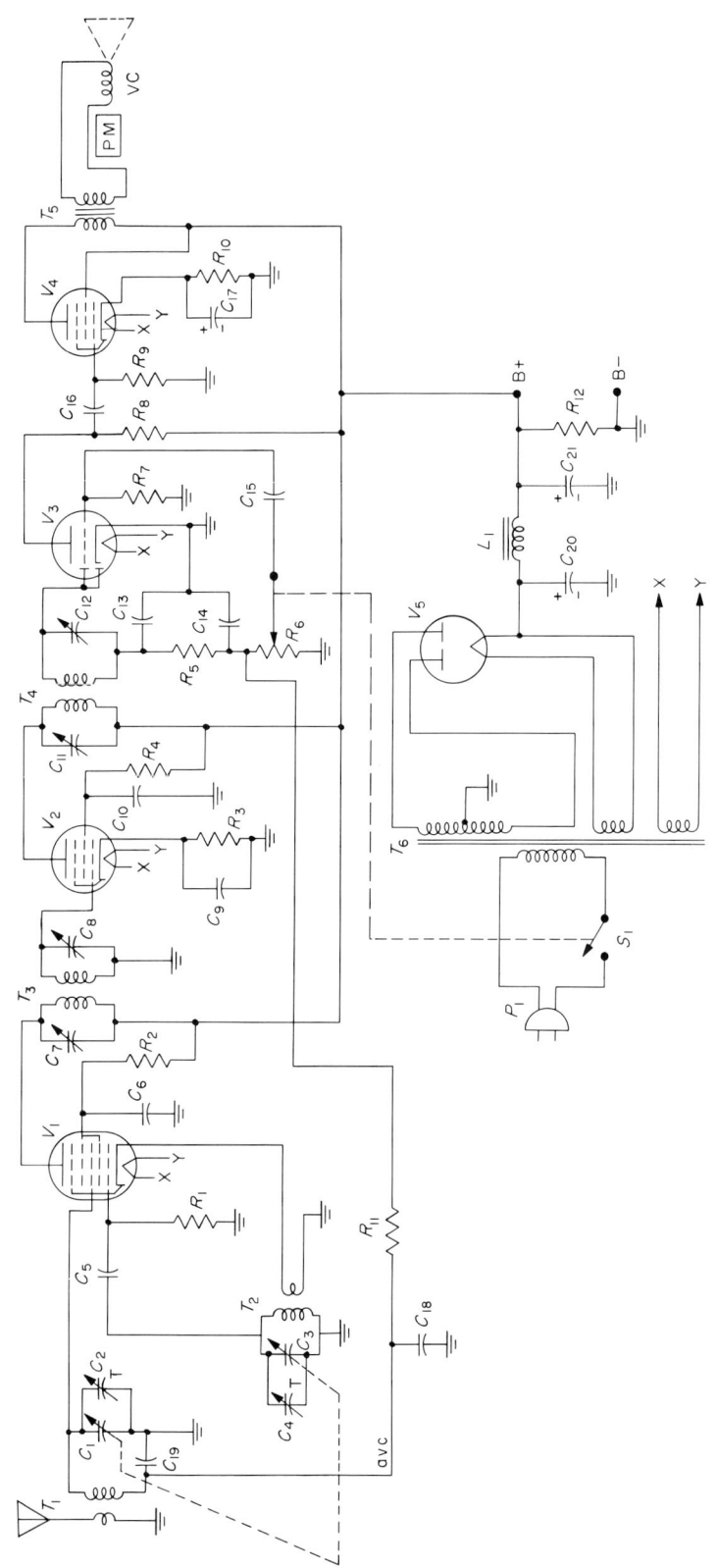

FIG. 4·3 A schematic diagram of a superheterodyne radio receiver using vacuum tubes.

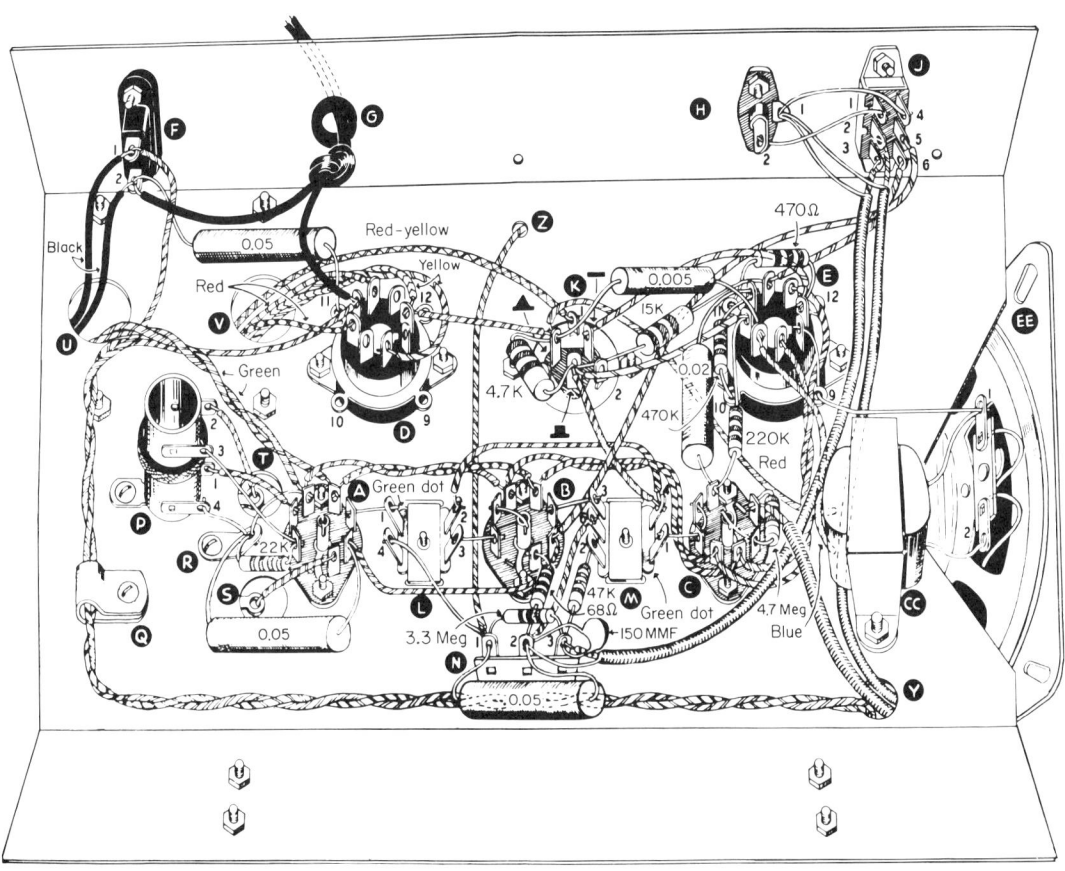

FIG. 4·4 Chassis wiring diagram of a radio receiver. (*The Heath Company, Benton Harbor, Michigan.*)

dustry, and our own electronics industry. To attempt to give examples of all the methods used in making interconnection wiring diagrams is beyond the scope of this book. However, having become acquainted with one such wiring diagram, the student can with practice adapt himself to other, similar forms of wiring diagrams. Figure 4·5 is an example of an interconnection type of wiring diagram. Notice that single wires from one unit join other wires (either from the same unit or from other units) in wire bundles. Because these bundles of wire *deliver* the single wires that have joined it to other units, they are often called a *wire harness*. Symbols are also used to represent wire bundles. One such symbol is seen in Fig. 7·38a.

4·4 BLOCK DIAGRAM

A block diagram is primarily concerned with general information. A block may represent a single stage, or it may represent a unit composed of various stages. Generally, a block diagram illustrates the path the signal takes between stages or between units. At times additional information may be included in a block diagram. An example of a block diagram can be seen in Fig. 4·6. Each block represents a group of circuits within that *stage*.

The electronics technician and engineer make extensive use of block diagrams for communication. However, electronics assemblers have little need for block diagrams in everyday work. Lead men (or

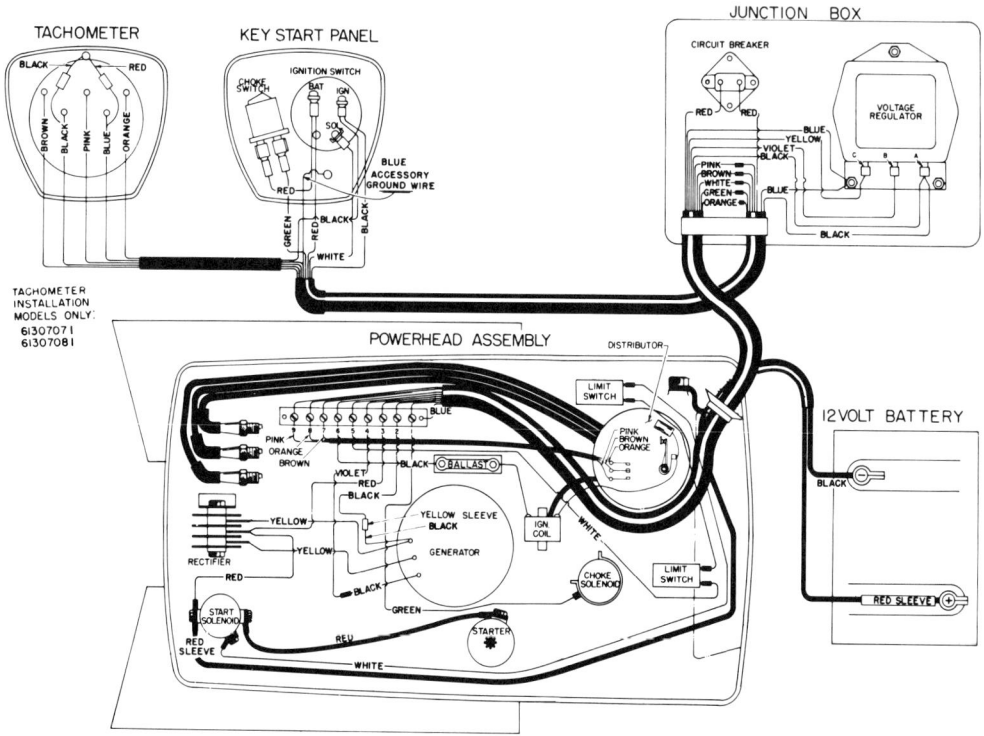

FIG. 4·5 An example of an interconnection type of wiring diagram. (*McCulloch Corporation, Scott Division, Minneapolis, Minnesota.*)

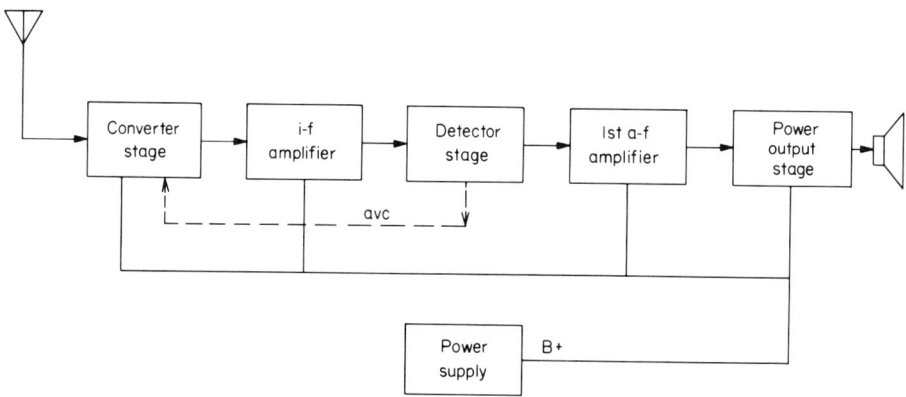

FIG. 4·6 Block diagram of a superheterodyne radio receiver.

women) and other group leaders will of course find the knowledge of block diagrams very helpful in communicating with technicians and engineers.

4·5 SYMBOLIC DIAGRAM

A symbolic diagram is a variation of a block diagram, the main difference being that where the normal block diagram uses a rectangle for the individual block (with the identification of the stage written within the block), the symbolic diagram makes use of *symbols that represent blocks or stages*. Symbolic diagrams find extensive use in the area of computers. A few examples of symbolic notations as used in symbolic diagrams appear in Fig. 4·7. Block-diagram and schematic-diagram symbols often find application in symbolic diagrams.

4·6 LAYOUT DIAGRAM

A layout diagram gives the physical location of parts mounted on a chassis. It can also give the location of a hole in the chassis or any other physical identifying mark. Layout diagrams may appear in various stages of completion and can be made to reveal the view from either the top, bottom, or sides. Actually, the layout diagram is primarily intended for the manufacturing and assembly of mechanical parts. However, since the electronics assembler and technician have to know the location of parts or holes, the layout diagram becomes a tool for both the mechanical and electrical personnel. Figure 4·8 is an example of a layout diagram showing the location of parts in the underside of a radio chassis. Another example of a layout diagram is given in Fig. 4·9. Here the top view of the same radio receiver is shown.

4·7 PRINTED-CIRCUIT LAYOUT DIAGRAM

Printed-circuit layout diagrams are normally a combination of a wiring diagram and a parts layout diagram. This type of diagram shows a view of the *foil (printed) circuit layout*, and at the same time indicates the *layout of parts* which are mounted on the board itself. See Fig. 4·10. In this diagram the "wiring" is emphasized. When the parts layout is to be emphasized, the part outline will be dark, and the foil circuit will be light. Parts can be mounted on either side of the *laminated* board. However, in most applications parts are mounted on the side of the laminated board that is *opposite* to the side that contains the foil (printed) circuit.

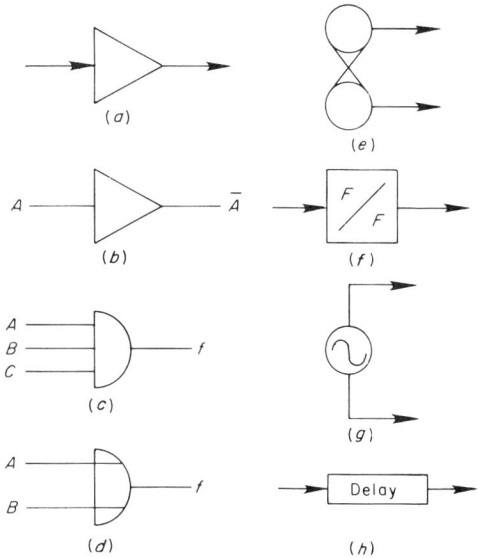

FIG. 4·7 Typical symbolic notations: (*a*) Amplifier stage. (*b*) Inverter stage. (*c*) AND gate, three-input. (*d*) OR gate, two-input. (*e*) Free-running multivibrator. (*f*) Flip-flop circuit. (*g*) Alternating-current signal source. (*h*) Delay circuit.

4·8 PICTORIAL DRAWING

A pictorial drawing is another means of conveying information relative to the location of component parts. This type of illustration is a fairly complete line drawing of a subassembly section, as shown in Fig. 4·11.

4·9 EXPLODED ILLUSTRATION

An exploded illustration serves as a visual guide for the mechanical assembly of small, detailed parts. Either line drawings or photographs of the parts may be used. However, the same principle of visual guide for assembly is used. An exploded illustration developed to serve as a visual guide in the assembly of a variable capacitor, oscillator coil, and the associated hardware on a radio receiver chassis is shown in Fig. 4·12.

4·10 DRAWING DIAGRAMS

All the diagrams discussed above will deliver information or instructions to the electronics assembler

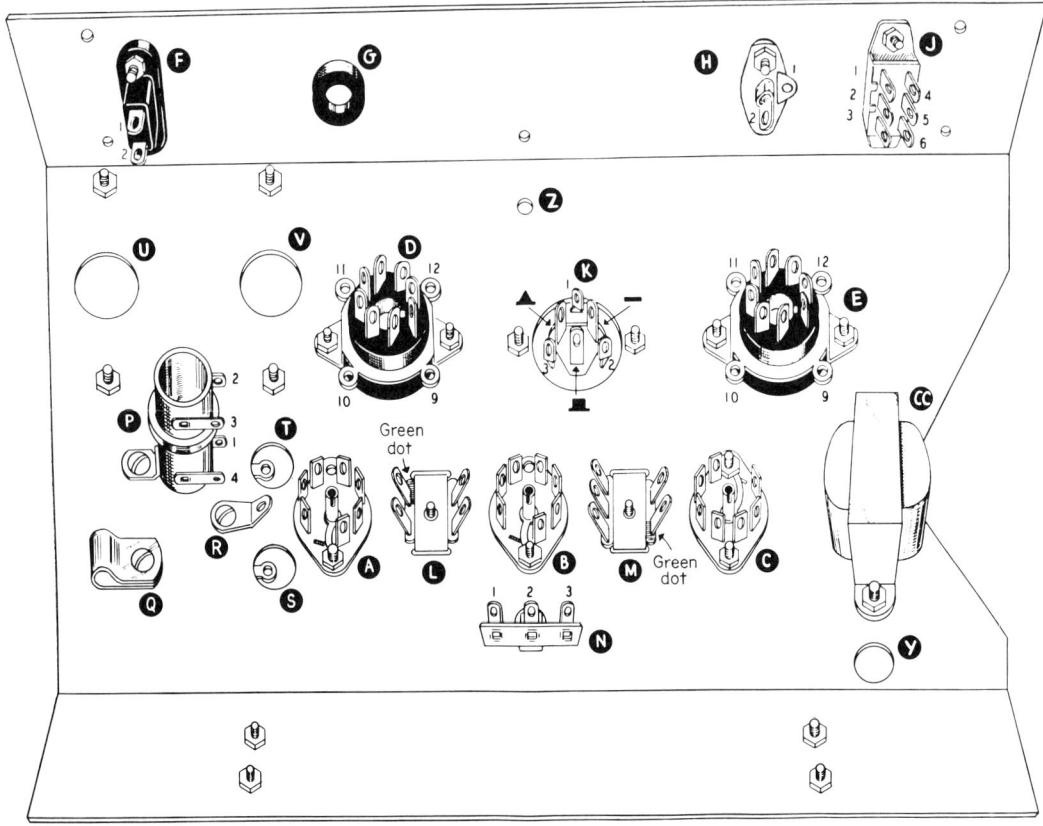

FIG. 4·8 A layout diagram illustrating the bottom view of a radio receiver chassis. (*The Heath Company, Benton Harbor, Michigan.*)

or technician. These diagrams are made by expert draftsmen. However, often *live communication* will take place between two or more people working on the same project. This will call for the immediate drawing of diagrams in order to illustrate a point or to make oneself properly understood. At other times the technician or assembler will have to make written reports that call for certain illustrations. Again, the diagram most widely used for communication by electronics personnel is the schematic diagram. The student should therefore practice making some of these diagrams. This in a way is like learning to *write*, as well as being able to *read*.

4·11 BLUEPRINTS

Various types of diagrams convey information to the electronics assembler or technician. Each type of diagram has a specific objective. These diagrams may be small, or they may be quite large. They are seldom original, single copies. Usually they are a reproduction of an original drawing that is kept in a safe place so that additional copies may be made when needed. Methods employed for reproducing these drawings or diagrams are usually some kind of photographic process.

The reproduced copies are normally referred to as *blueprints*. However, many blueprints are not blue at all; they may appear in white, brown, black,

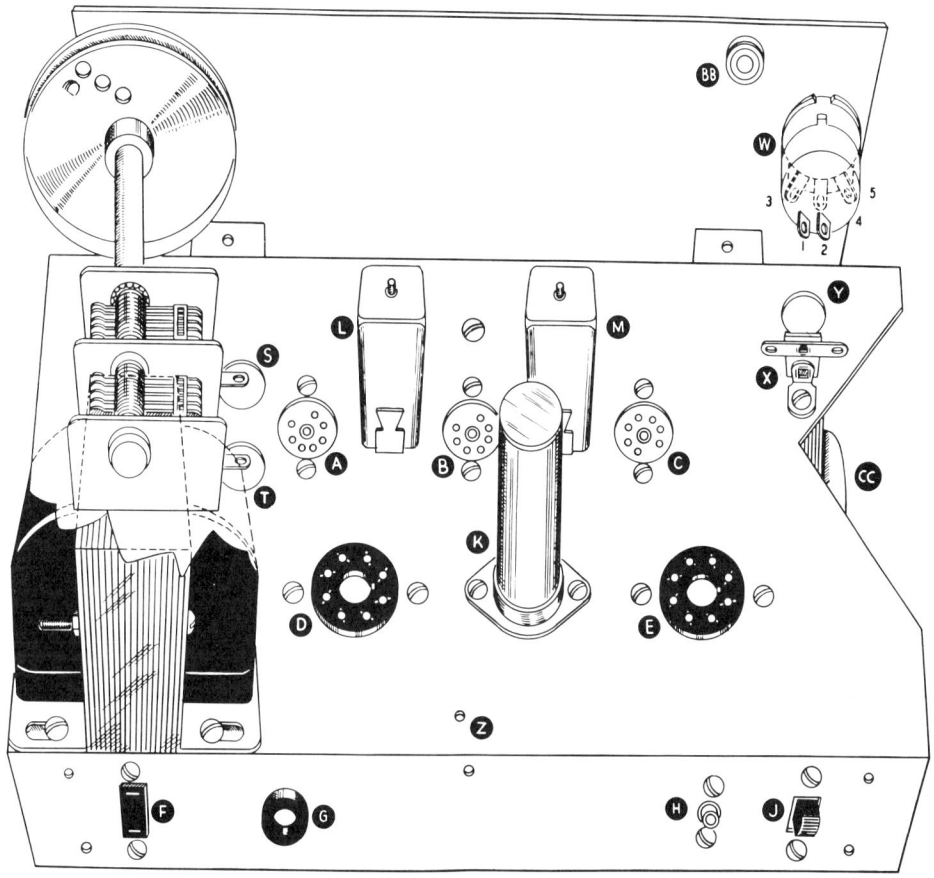

FIG. 4·9 A layout diagram illustrating the top view of the same radio receiver chassis shown in Fig. 4·8. (*The Heath Company, Benton Harbor, Michigan.*)

purple, blue, or other colors. At times the background is colored and the lines are white, or the background can be white with colored lines. Most blueprints fade with prolonged exposure to light, regardless of the color used. Therefore it is best to keep all blueprints covered when not in use.

To the electronics worker the method of reproducing diagrams is not as important as the content of the blueprint. However, the care of blueprints is important.

BLUEPRINT CONTENT

A blueprint may contain a reproduction of a schematic diagram, a wiring diagram, a block diagram, a pictorial drawing, a symbolic diagram, a layout diagram, an exploded illustration, an interconnection diagram, or other diagrams.

Blueprints may appear in sections, or they may have final and complete information. Blueprints containing exact detailed information are called *detail blueprints*. Other prints may be concerned solely with a section of the overall assembly, and are therefore called *subassembly blueprints*. Other blueprints cover a completed unit of an overall product, and are referred to as *unit assembly blueprints*. A blueprint that illustrates how the various

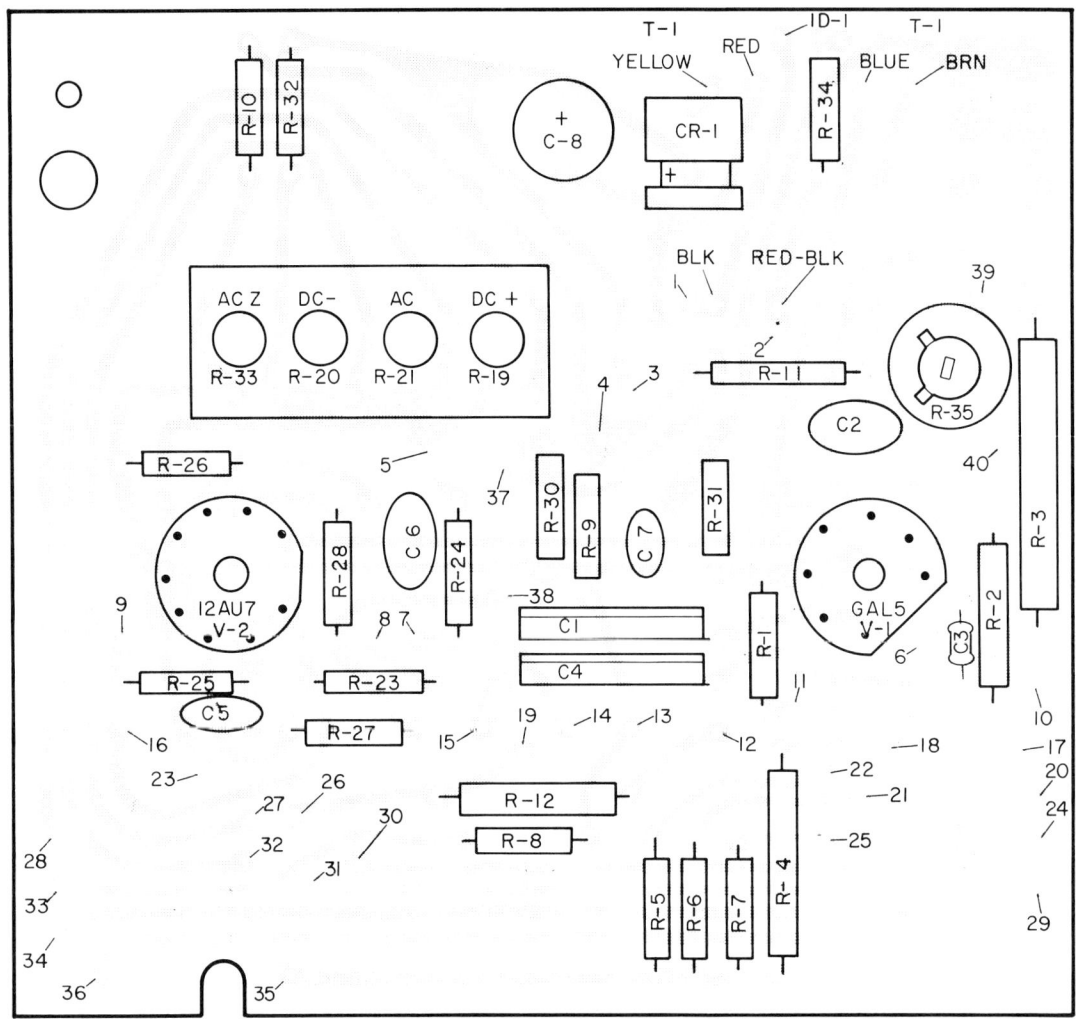

FIG. 4·10 Printed-circuit layout diagram. (*Radio Corporation of America, Electron Tube Division, Harrison, New Jersey.*)

units fit together to make the final product are called *final assembly blueprints*.

CARE OF BLUEPRINTS

Blueprints are designed to be used repeatedly. Because of this, they are marked and identified in a standard manner and in a standard place on the print. This is done so that prints can be stored when not in use and found quickly when needed and with a minimum of handling.

Most prints have to be folded for storage. They should be folded neatly, so that the identifying marks can be seen without unfolding. Usually, previous folds and creases will indicate the proper folding procedure, but the blueprint identification must be visible when the final fold is completed.

Pencil or ink marks should never be made on a blueprint without proper authorization. The hands should be clean and dry when handling blueprints, and blueprints should be kept in a dark dry place when not in use.

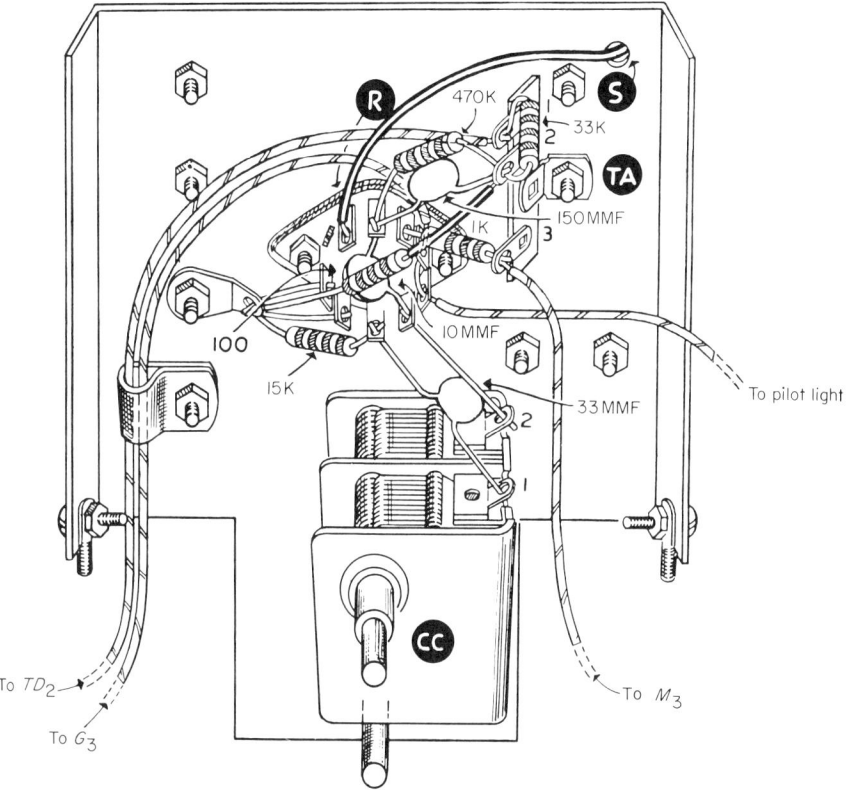

FIG. 4·11 A pictorial drawing of a subassembly unit. (*The Heath Company, Benton Harbor, Michigan.*)

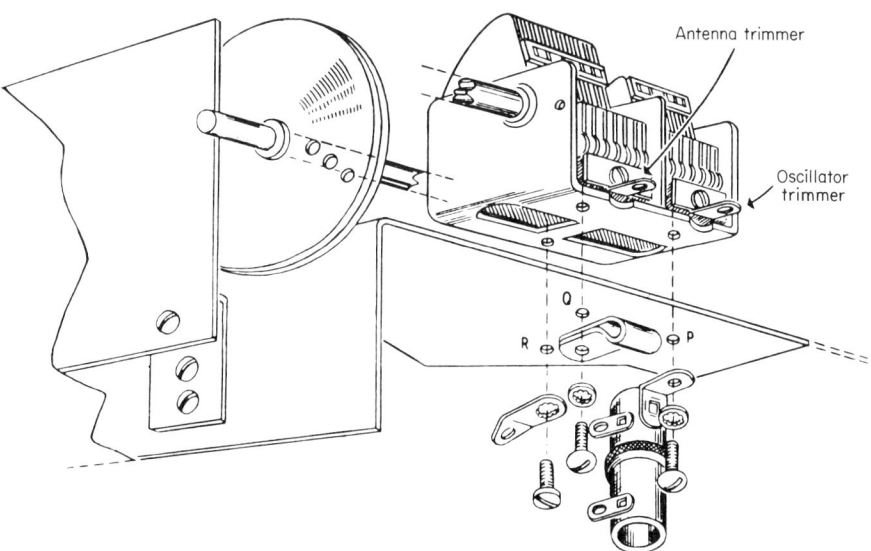

FIG. 4·12 An exploded illustration. (*The Heath Company, Benton Harbor, Michigan.*)

4·12 LINEAR MEASUREMENT

As stated earlier in this chapter, "the objective of any communication system is to convey (and receive) instructions or information." In order to be able to communicate effectively in the manufacturing industry, the worker must therefore be able to interpret not only technical symbols which appear in blueprints, but he must also be able to interpret the length of objects or the distance between points. In order to do this, a *knowledge of numbers* and *what they represent* is essential. The ability to read blueprints can only be developed if the trainee has a minimum knowledge of *linear measurement*. This is due to the fact that in order to make use of *quantitative* information which may be present in a blueprint, he must *bring this information to life*.

ENGLISH SYSTEM

Some blueprints use the English system of linear measurement. This system uses the yard, the foot, and the inch as the basic units of measurement. Fractions of each are also used extensively. For example, the blueprint may call for a wire lead that is 18 inches long; or it may call for it as a wire lead that is 1½ feet long; or it may call for a wire lead that is 1.5 feet long. In all three cases, the length of wire-lead needed is the same. Since information regarding the length of wire lead is the same, and since this information may appear in any of the three forms indicated, it becomes necessary for the assembler to be able to interpret whatever numerical system is used. This requires knowledge pertaining to *whole units*, *common fractions*, and *decimal fractions*.

METRIC SYSTEM

Another system often employed in linear measurement is the metric system. This system uses the *meter*, the *centimeter*, and the *millimeter* as basic units of measurement. Decimal fractions of these units are also employed. For example, a blueprint that utilizes the metric system for measurement may call for two small holes to be drilled 2.54 centimeters apart; or it may call for these same two holes to be 25.4 millimeters apart. Actually, in both cases the holes are one inch apart, since one inch equals 2.54 centimeters or 25.4 millimeters. Therefore, when using the metric system, it is not only necessary to know how to work with multiples of ten, but a person should be able to *interpret* metric measurements in terms of *inches* or the *yard*. For example, when asked to cut a wire one meter long, a person should *know* that the length will be a few inches longer than a yard.

The metric system of measurement is used in most countries of the world. Even in the United States of America, a country in which the English system of measurement is generally employed, most of the tool and machine industry uses the metric system. In November of 1970, the National Aeronautics and Space Administration (NASA) adopted the metric system for all its measurement requirements. Because of this, this text, which refers to NASA specifications often, will drill the student in both the English and metric systems of measurements.

Actually, the metric system is quite easy to work with. For the person acquainted with the English system, two main concepts must be readily kept in mind. The first being that the *meter* is the basic unit of length (1 meter = 39.37 inches), the second concept is that in electronics work most linear measurements will be in *decimal fractions* of the meter. Decimal fractions means in this case that the meter is divided into one hundred parts, or one thousand parts. One one-hundredth (1/100) part of a meter is known as a *centimeter*, and is abbreviated *cm*. One one-thousandth (1/1000) part of a meter is known as a *millimeter*, and is abbreviated *mm*. Therefore, 100 cm equals one full meter; and 1000 mm also equals one full meter. A little reasoning will reveal that *10 mm equals 1 cm*.

4·13 MEASURING DEVICES

The *yardstick*, the *foot rule*, the *meter stick*, and the *metric scale* are all length-measuring devices that should be mastered by persons training for work in any assembly or manufacturing facility. While these are somewhat basic measuring instruments, they are often made in compact, flexible

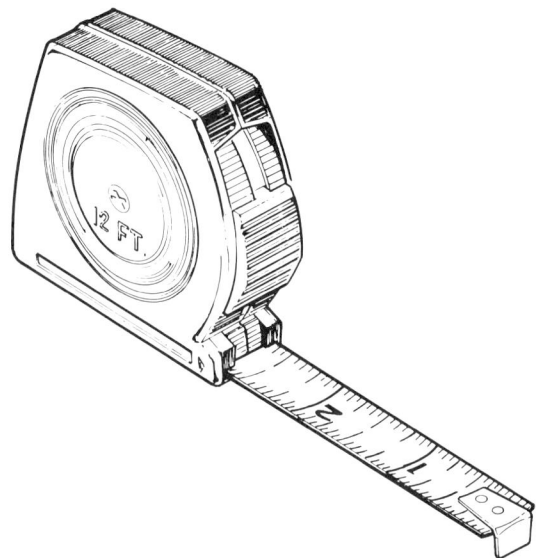

FIG. 4·13 Flexible measuring tape.

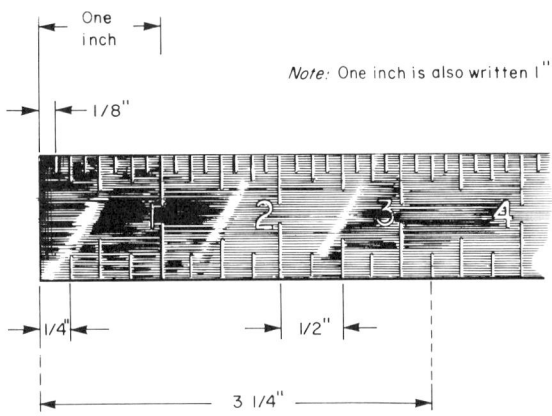

FIG. 4·14 Portion of a foot rule, indicating fractions of an inch. (Not drawn to scale.)

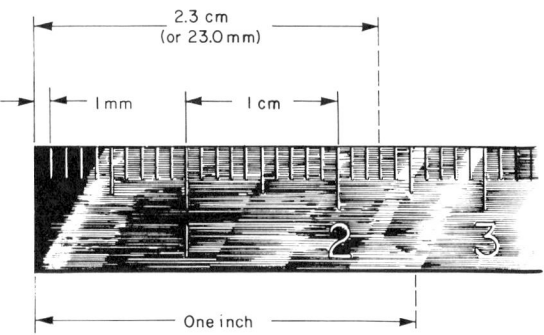

Note: One inch equals 2.54 cm.

FIG. 4·15 Portion of a metric scale, indicating centimeter and millimeter units. (Not drawn to scale.)

units for convenience. (See the *measuring tape* shown in Fig. 4·13). A thorough acquaintance with basic measuring devices is essential for promotion to a position of responsibility. It also lays the foundation for an understanding in the use of *calipers*, *micrometers*, and other precision measuring tools. Figure 4·14 shows an expanded view of a portion of a foot rule. Figure 4·15 shows an expanded view of a portion of a metric scale. Each illustration uses 1 inch as a comparison factor.

4·14 SUMMARY

A student learning to read and write the English language has to practice. This is also true in learning the language of symbols and numbers.

The following job summaries are designed with this precept in mind: *practice*. The diagrams and linear measuring devices most likely to be used by the average electronics assembler or technician have been selected for summary.

EXERCISES

JOB 4·1 How to make schematic diagrams

OBJECTIVES
1. To learn the principles of schematic diagramming
2. To review the schematic-diagram symbols previously learned
3. To become proficient in making schematic diagrams

MATERIALS REQUIRED

Supplies
8½ × 11-inch notebook paper
Pencil
12-inch ruler
Assorted coins (money)

PROCEDURE
1. Inspect the schematic diagrams in Fig. 15·1.
2. Notice that all lines are either horizontal or vertical. There are no slanted lines.
3. Notice that all corners are right angles. There are no rounded corners.
4. Notice that all symbols used are neat and uniform.
5. On your paper draw the two schematic diagrams of Fig. 15·1. Use a coin to draw a circle.
6. Identify each symbol used.
7. On another paper draw the two schematic diagrams that appear in Fig. 15·4.
8. Identify each symbol used.
9. On a third sheet of paper draw the complete schematic diagram of the radio receiver in Fig. 4·3. Turn the notebook paper so that the punched holes appear at the top.
10. The first step in drawing a large diagram is to draw the circles of the vacuum tube or transistor envelopes in the same relative position as they appear in the diagram from which you will copy. If the diagram is an original schematic, then you must estimate how much circuitry will be involved and place the envelope circles accordingly.
11. Notice that, in Fig. 4·3, the antenna appears at the left and the loudspeaker appears at the right. This is because the signal is received by the antenna and delivered by the loudspeaker. In other words, a schematic diagram is drawn so that the signal is allowed to travel from left to right.
12. In addition, power supplies are normally drawn at the bottom of the schematic diagram. In Fig. 4·3 the circuit around tube V_5 is the power supply.
13. When you have completed drawing the schematic diagram of Fig. 4·3, have your instructor check it.
14. Be prepared to identify each symbol used in the radio receiver schematic diagram on Fig. 4·3.

QUESTIONS
1. List *all* symbols used in Fig. 4·3 under one of the following classifications: resistor, capacitor, inductor, vacuum tube, and miscellaneous.
2. What do the broken lines in Fig. 4·3 indicate?
3. Show your instructor where two wires cross but do not touch each other electrically in Fig. 4·3.
4. For extra credit, copy the complete schematic diagram of the transistor radio receiver in Fig. 4·2. Be neat and thorough.

JOB 4·2 How to make a chassis wiring diagram

OBJECTIVES
1. To learn to recognize a chassis wiring diagram
2. To learn to interpret a chassis wiring diagram
3. To appreciate a chassis wiring diagram

MATERIALS REQUIRED

Equipment
Small table-model radio receiver, vacuum-tube type, no printed circuits
Schematic diagram for the radio supplied

Supplies
8½ × 11-inch notebook paper
Pencil
12-inch ruler

PROCEDURE
1. Place the radio receiver supplied so that the underside of the chassis is exposed (see Fig. 4·4).

2. Check to see that no part mounted on the topside of the chassis is under any pressure that may damage it.
3. Sketch a pictorial presentation of the things you see underneath the chassis supplied. Your sketch should resemble the illustration of Fig. 4•4.
4. Notice that a chassis wiring diagram is similar to a photograph of the subject.
5. Locate as many components, terminals, and wires as possible on the chassis wiring diagram, and their corresponding symbol on the schematic diagram supplied.

QUESTIONS

1. Make comparisons between a schematic diagram and a chassis wiring diagram.
2. Is a wiring diagram important? Why?
3. What is shown on a chassis wiring diagram that is not shown on a schematic diagram?

JOB 4•3 How to make a block diagram

OBJECTIVES

1. To learn to use a block diagram
2. To learn to make a block diagram

MATERIALS REQUIRED

Equipment
Radio receiver used in Job 4•2
Schematic diagram for the above receiver

Supplies
8½ × 11-inch notebook paper
Pencil
12-inch ruler

PROCEDURE

1. Compare Figs. 4•3 and 4•6.
2. Note that each *block* represents a complete circuit. Although the detailed circuit is not shown in Fig. 4•6, the signal can be traced as it progresses from the antenna to the loudspeaker. This is indicated by the arrows.
3. Note that B+ voltages are applied to some of the stages, indicated by the lines between the power supply and the individual stages requiring B+.
4. The block diagram of Fig. 4•6 corresponds to the schematic diagram of Fig. 4•3.
5. With the help of your instructor, make a block diagram of the radio receiver supplied. It should resemble the block diagram of Fig. 4•6.

QUESTIONS

1. List the advantages of a block diagram.
2. List the information needed to make a block diagram.
3. Make a block diagram of the transistor radio receiver shown schematically in Fig. 4•2.

JOB 4•4 How to make a symbolic diagram

OBJECTIVES

1. To become acquainted with the principles of symbolic diagrams
2. To be able to distinguish between a block diagram and a symbolic diagram
3. To appreciate symbolic notations

MATERIALS REQUIRED

Supplies
Paper
Pencil
6-inch ruler

PROCEDURE

1. Make a *block* diagram of the following statement: "A flip-flop circuit delivers a signal to an amplifier. The amplifier delivers the signal to a second amplifier through a delay line." Label the blocks with the words flip-flop, amplifier, and delay line as applicable.
2. On the same paper make a *symbolic* diagram of the statement given in step 1. The symbol for a flip-flop is given in Fig. 4•7f. Figure 4•7a represents an amplifier. The delay line is symbolized in Fig. 4•7h.
3. In making the symbolic diagram it is only necessary to replace each block with the appropriate symbol given in Fig. 4•7.

4. Have your instructor compare the diagrams you have made.

QUESTIONS
1. Make comparisons between a block diagram and a symbolic diagram.
2. List the advantages and disadvantages of using symbolic diagrams.

JOB 4·5 How to make a layout diagram

OBJECTIVES
1. To learn to recognize a layout diagram
2. To learn to use a layout diagram
3. To learn to make a layout diagram

MATERIALS REQUIRED

Equipment
Radio receiver used in Job 4·2

Supplies
8½ × 11-inch notebook paper
Pencil
12-inch ruler

PROCEDURE
1. Inspect the top-view layout diagram in Fig. 4·9.
2. Compare the diagram of Fig. 4·9 with the top view of the radio receiver supplied.
3. Make a top-view layout diagram of the radio receiver supplied.
4. Identify on the diagram you make the vacuum-tube identification numbers (example: 12BA6).
5. The diagram you make should be similar to that illustrated in Fig. 4·9.
6. Look at the layout diagram in Fig. 4·8 of the bottom view of the same chassis used in Fig. 4·9.

QUESTIONS
1. List the possible uses for a layout diagram.
2. How can a layout diagram be used to locate the power-supply section?

JOB 4·6 How to read a printed-circuit layout diagram

OBJECTIVES
1. To learn to read a printed-circuit layout diagram
2. To learn to associate a printed-circuit layout diagram with an actual PC board and the related schematic diagram
3. To learn how to look for defects in the *wiring* of a printed-circuit board

MATERIALS REQUIRED

Equipment
Printed-circuit board, complete with parts installed
A printed-circuit board layout diagram for the PC board supplied (see Fig. 4·10)
Schematic diagram of the circuit etched on the PC board supplied
Table lamp, approximately 50 watts

PROCEDURE
1. Inspect thoroughly the printed-circuit board supplied. Look at both sides of the board.
2. Turn the lamp ON.
3. Allow the light from the lamp to shine *through* the partially transparent, laminated printed-circuit board.
4. Notice that the silhouette of the parts which are mounted on the opposite side show through. If the opposite side contains the etched circuitry, then the ribboned conductor *wiring* shows through in silhouette form.
5. Compare what you see with the printed-circuit layout diagram supplied. Notice that the diagram is a pictorial representation of what you see.
6. Attempt to trace the circuitry of an obvious component which is mounted on the PC board, using the silhouette method and the schematic diagram supplied.
7. Since the *wiring* is permanent on a PC board, one must assume that if the unit worked satisfactory at one time, the circuit arrangement must be all right. However, if a break in the etched conductor occurs, the use of test instruments may be necessary, including the use of a magnifying glass.

QUESTIONS
1. Is it possible to have the wrong *wiring* between components mounted on a printed-circuit board? Explain your answer.

2. Is it possible to mount a component on a printed-circuit board wrong? Explain.
3. How can an *open circuit* occur on a printed-circuit board?
4. Is the printed-circuit layout diagram essential? Why?
5. How would you use a magnifying glass to locate a break in an etched conductor?

JOB 4·7 How to read a foot rule

OBJECTIVES

1. To learn to interpret linear-measurement specifications
2. To review conversion of common fractions to decimal fractions
3. To compare various linear measuring devices

MATERIALS REQUIRED

Equipment
Yardstick (wood)
Foot rule (metal)
6-inch scale (plastic)
Steel measuring tape, at least 10 feet long (see Fig. 4·13)
Assorted precut measuring standards (metal) labeled **A**, **B**, **C**, etc.

Supplies
Notebook paper and pencil

PROCEDURE

1. Compare the various measuring devices provided for your use.
2. Study the illustration of Fig. 4·14. Ask your instructor to help you interpret it if necessary.
3. Convert the various common fractions indicated in Fig. 4·14 to decimal fractions. For example, when reading the long dimension indicated, it can be stated as $3\frac{1}{4}$ inches or 3.25 inches.
4. Using every type of measuring device provided, measure the length of the measuring standard which is labeled **A**. Record these measurements using *common fractions*. Also make a note of problems encountered.
5. Repeat step 4, this time measuring the *width* of the standard. For better accuracy, start the measurement at the 1-inch line; then subtract 1 inch from the final reading. This eliminates the problem of a damaged edge of the ruler, which would throw the total reading off a little.
6. Repeat steps 4 and 5 using the measuring standard labeled **B**. However, this time use *decimal fractions*.
7. Repeat steps 4 and 5 using each measuring standard provided by your instructor, starting with the standard labeled **C** and continuing in alphabetical sequence. Record all measurements in *feet, inches, and fractions of inches* as applicable. Use common fractions for half of your measurements and decimal fractions for the other half.

QUESTIONS

1. Measure the length of your notebook paper and record it in inches and decimal fractions. What is the length?
2. Measure the width of this notebook paper and record it in inches and common fractions. What is the width?
3. Measure and record the length and width of the table top on which you are working. What are the dimensions in feet, inches, and fractions of an inch (using both common and decimal fractions)?

JOB 4·8 How to read a metric scale

OBJECTIVES

1. To learn to interpret linear measurements in both English and metric measuring systems
2. To learn to use metric measuring devices
3. To review conversion factors between English and metric units of measurement

MATERIALS REQUIRED

Equipment
Meter stick (wood)
Metric scale (metal)
Combination metric-and-inch, steel measuring tape (approximately 2 meters long)
Assorted precut measuring standards (metal) labeled **A**, **B**, **C**, etc.

Supplies
Notebook paper and pencil

PROCEDURE

1. Compare the various measuring devices provided for your use.
2. Study the illustration of Fig. 4•15. Ask your instructor to help you interpret this illustration if necessary. Notice that common fractions are not used.
3. Measure the length and width of the measuring standard which is labeled **A**. Record these measurements in *centimeters* and decimal fractions of a centimeter. Also make a note of the problems encountered.
4. Again measure the length and width of the measuring standard labeled **A**, but this time record the measurements in *millimeters* and decimal fractions of a millimeter.
5. Compare the results of step 3 and 4.
6. Repeat steps 3, 4, and 5 using the measuring standard labeled **B**.
7. Repeat steps 3, and 4, and 5 using the rest of the measuring standards provided by your instructor, starting with the standard labeled **C** and continuing in alphabetical sequence. Record all measurements in meters, centimeters, and millimeters as applicable. Fractions of millimeters should be expressed in decimal fractions. (An example of a measurement could be: 1 meter, 29 centimeters, and 4.5 millimeters.)
8. Compare all of the measurements taken during this job with the measurements taken during Job 4•7, assuming the measuring standards used were the same. If the measuring standards used were different, repeat steps 3 through 7 using the English system of measurement and then compare the results obtained when using both systems. (Remember that 1 inch equals 2.54 cm; and that 1 meter equals 39.37 inches.)

QUESTIONS

1. In your opinion, which system of measurement is easier to use (English or metric)? Why?
2. Measure the length of the wooden meter stick with the steel measuring tape. Is there any difference in the indicated lengths of the stick? Would you expect a difference? Why?
3. Why is it important for an electronics assembler to be able to measure the length of objects?
4. Which system of linear measurement should the electronics assembler know (English or metric)? Why?
5. Express the example given in step 7 in terms of *feet* and *inches* for the total length.

5

TOOLS

Preceding chapters have dealt with some of the basic aspects connected with electronics. However, while performing their tasks both the electronics assembler and the electronics technician must handle hardware. Skill in the use of hand and power tools and knowledge of their limitations are prerequisites to working with hardware. Tools are necessary because the assembly and disassembly of hardware are impossible tasks with hands alone. This chapter will acquaint the student with a variety of *hand* and *power tools* normally used in electronics work.

Tools used in electronics fall into two groups. One group includes tools needed for *electrical assembly*. The second group includes tools designed primarily for *mechanical assembly*.

5·1 HAND TOOLS FOR ELECTRICAL ASSEMBLY

Electrical assembly is centered around wires and *soldering*. Soldering is the process of uniting two separate metals, such as copper wires, by means of a third metal *alloy* which is in a molten state. Instruction on the art of soldering is given in Chap. 6. However, the tools used in soldering and in preparing wires for soldering are covered below.

SOLDERING IRON

In soldering, the soldering iron itself must be considered the primary tool. A typical soldering iron used in electronics is shown in Fig. 5·1. Soldering irons are manufactured in many shapes and sizes. They are rated according to the amount of *heat* they can exert. However, a soldering iron is measured in

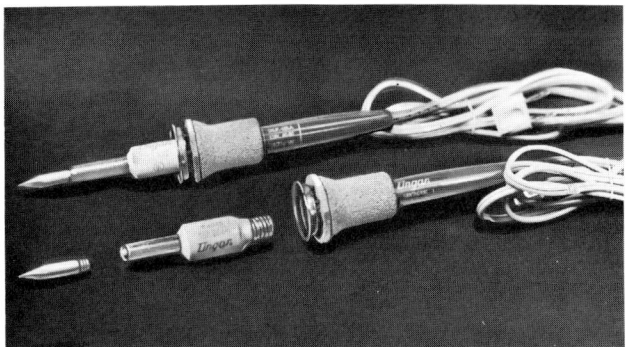

FIG. 5·1 The pencil-type soldering iron and its component parts.

watts rather than in degrees of temperature, the watt being the electrical unit for power. The higher the wattage rating of a soldering iron, the greater amount of heat it will exert.

The soldering iron shown in Fig. 5·1 is of the *pencil type*. The wattage rating of this soldering iron is 47½ watts. Electrical power is obtained from a 110-volt source through the *line cord*.

The average soldering iron will have a wattage rating lower than 100 watts, but higher-wattage soldering irons are available. One should select the soldering iron according to the job at hand. A soldering iron that is too hot can damage the work; a soldering iron not hot enough will not do an adequate job. Soldering irons having a rating of approximately 40 watts are quite satisfactory for most electronics work. Control units are available to regulate the amount of heat generated by a soldering iron. See Fig. 5·2.

SOLDERING AID

The soldering aid is a handy tool that provides for easy manipulation of the wire being soldered. Soldering aids are available in a variety of forms, each type having been designed to help make certain soldering tasks easier. They can be used to bend wire and to clean surfaces that are to be soldered or that have been soldered. They can also be used to help position a wire before soldering. A typical soldering aid is shown in Fig. 5·3. The point at one end is slotted in order to straddle the wire to be handled.

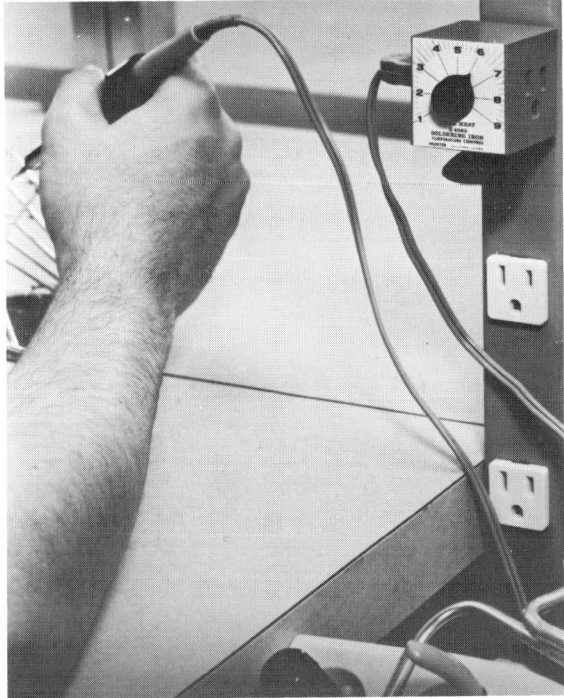

FIG. 5·2 Soldering-iron heat control unit. (*Hunter Tools, Santa Fe Springs, California.*)

Special skill is often required to remove a wire from a joint that has been soldered. This task is made easier with a soldering aid.

Soldering aids should never be used to pry in an electrical circuit when the power is applied. The shaft of a soldering aid is made of metal; therefore it is a conductor of electricity.

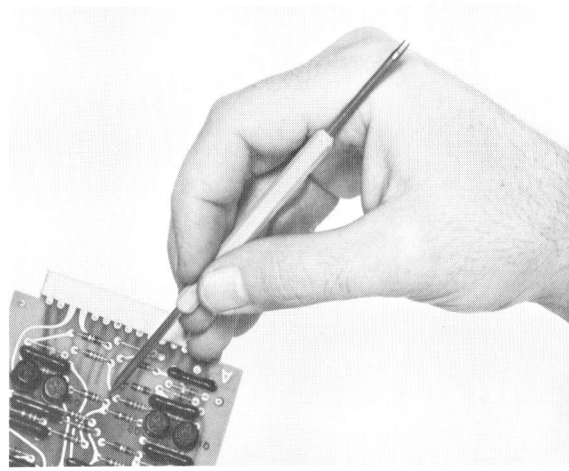

FIG. 5·3 Soldering aid. (*Hunter Tools, Santa Fe Springs, California.*)

LONG-NOSE PLIERS

Long-nose pliers are used for gripping and twisting wires. This tool can be seen in Fig. 5·4. Although long-nose pliers are found in many sizes, a small size with a long, slender nose is best for electronics work. Long-nose pliers of 4½ inches in length are popular. If a longer pair can accomplish a given task better, the longer tool should be used.

DIAGONAL-CUTTING PLIERS

Another tool in the family of pliers is the diagonal-cutting pliers. Two popular terms for these cutters are *diagonals* and *dykes*. Dykes is a nickname for diagonals and is widely used in the industry.

Diagonal-cutting pliers are used to cut wire. The type used in electronics work is designed to cut soft metal wire, and no attempt should be made to cut iron or steel wire. See Fig. 5·5. The size to be used is determined by the diameter of the wire to be cut. Diagonals that are up to 5 inches long are popular in the electronics industry.

Diagonal-cutting pliers should be made of good-quality tool steel. This will ensure a clean, sharp cut. The taper to the tips of the cutters should be rather pointed, to allow the user to reach a wire that is to be cut in close quarters.

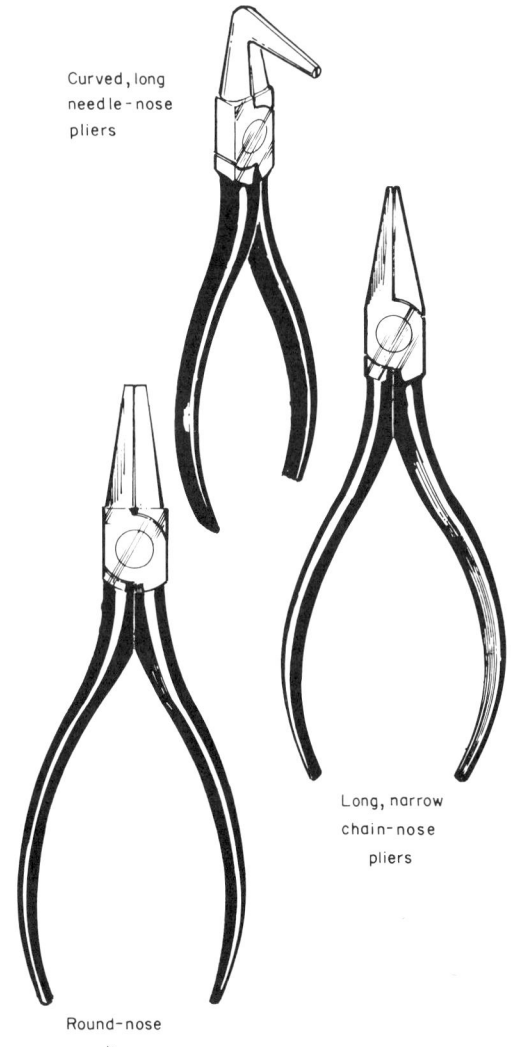

FIG. 5·4 An assortment of long-nose pliers.

SCISSORS

Another useful cutting tool is the scissors. These are used primarily to cut cords or tapes used to tie wire bundles. Figure 5·6 illustrates an industrial type of scissors available for this task.

WIRE STRIPPER

A wire stripper, as seen in Fig. 5·7, is used to remove a plastic or cloth type of insulation from

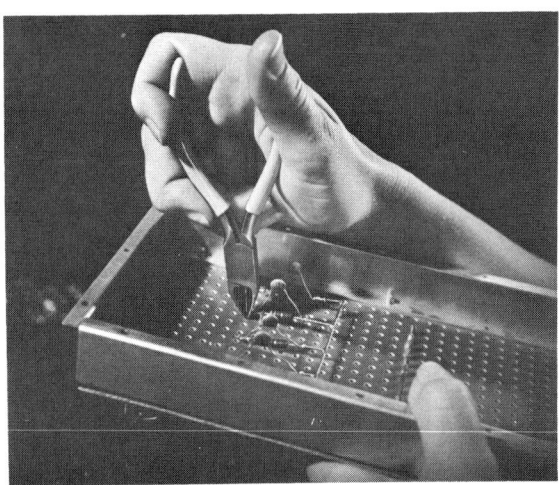

FIG. 5·5 Diagonal-cutting pliers in use. (*Hunter Tools, Santa Fe Springs, California.*)

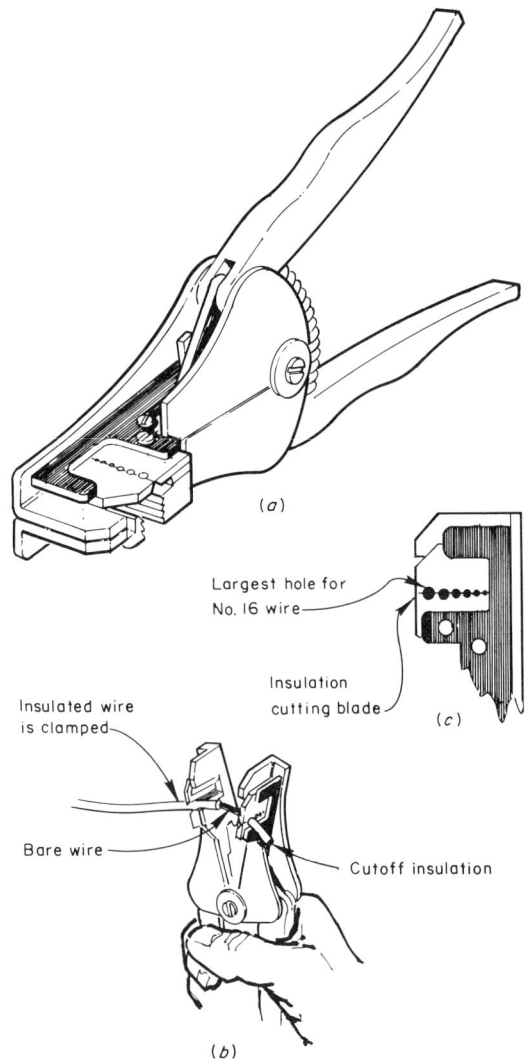

FIG. 5·7 Precision wire stripper.

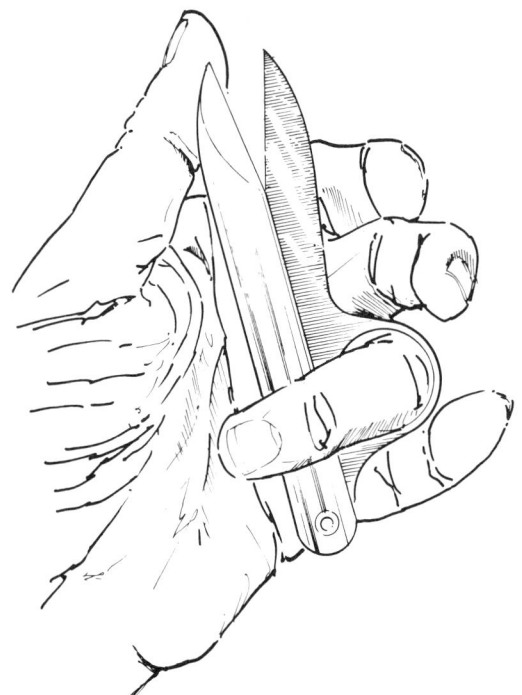

FIG. 5·6 Industrial scissors for electronics assembly work.

around a wire. Insulation is removed from a conductor to expose the wire for soldering.

Various types and sizes of wire strippers are on the market. Some are fully mechanical, while others require an electrically heated element to cut the insulation. This latter type of wire stripper is known as a *thermal wire stripper* (illustrated in Fig. 7·14).

Most mechanical wire strippers are adjustable to cut the insulation to a safe depth, since an overcut may damage the wire conductor. Again, the selection of the type of wire stripper is determined by the job at hand.

INK ERASER

An ordinary ink eraser serves as a wonderful tool to clean a conductor before soldering. Most solid wires and other similar conductors are usually coated with a light film of grease or oxides that has to be removed before a sound soldering job can be accomplished. An ink eraser does a good job of cleaning without causing damage to the conductor, but it has its limitations. For instance, it is not possible to erase the enamel insulation of a conductor, nor is it possible to remove visible grease with an eraser. Other methods or materials must be employed to do heavy cleaning.

5·2 TOOLS FOR MICROELECTRONICS

Components used in microelectronics are so small that special tools and equipment are used to work with these parts. Tools that the electronics assembler may expect to find at his workstation are illustrated in this section. Figure 5·8 illustrates the use of an industrial *microscope*, while Fig. 5·9 illustrates a suction-type pickup tool used to handle small integrated circuits and similar small units of hardware. Gloves, full-length white coats, and shoe covers are often used when handling these tiny components, since lint and dust become a major problem when dealing with such small tolerances.

5·3 HAND TOOLS FOR MECHANICAL ASSEMBLY

Since electronic parts have to be mounted before they can be soldered, or may have to be replaced when faulty, the electronics worker will find it necessary to use a few hand tools designed primarily for mechanical work. Tools for this purpose are so numerous, it is an impossible task to cover every type and size that may be encountered. Therefore only the most typical and representative mechani-

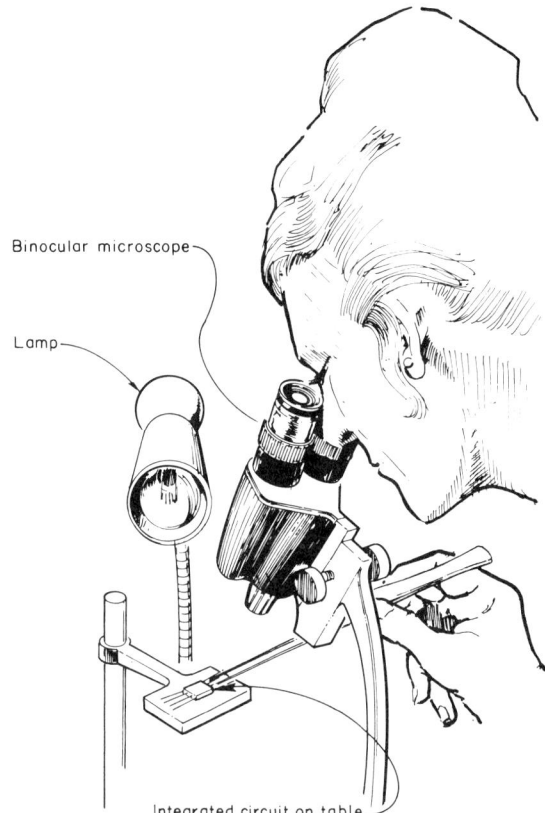

FIG. 5·8 Industrial microscope.

cal-assembly tools used by the average electronics assembler or technician are discussed below.

PHILLIPS SCREWDRIVER

The Phillips screwdriver is used on recessed head screws. These screws have four walls that are driven by this type of screwdriver. Special attention should be given the point of the Phillips screwdriver, shown in Fig. 5·10. Attention is also drawn to Fig. 9·3, showing a Phillips head screw.

The use of the proper size screwdriver is very important. If the point of the screwdriver does not fit properly into the head of the screw, the walls of the screw will be damaged. In electronics work, Phillips screwdrivers having a size identification of No. 1 or No. 2 are frequently used.

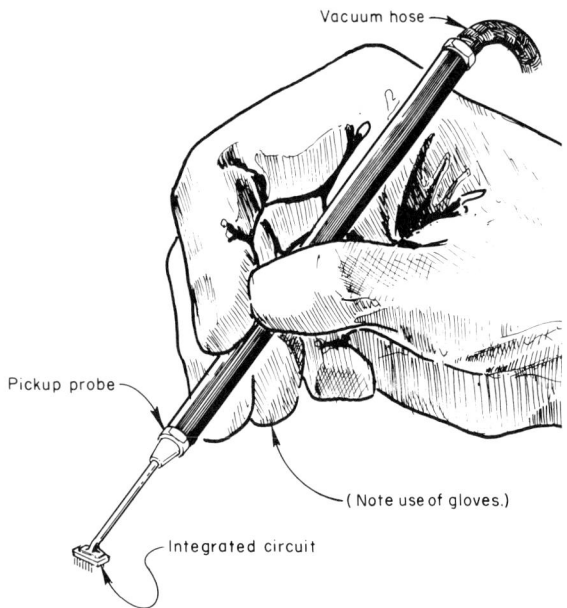

FIG. 5·9 Suction pick-up tool.

BLADE SCREWDRIVER

A popular tool used in many trades is the blade screwdriver. This screwdriver is designed to be used on screw heads that have a slot for driving purposes. These slotted screw heads have two walls on which driving pressure can be exerted. Compare the blade screwdriver with the Phillips screwdriver in Fig. 5·10. Slotted head screws are also shown in Fig. 9·3.

Again, use the proper-size blade screwdriver for a given slotted screw head. Failure to do so will damage the screw head. The blade width of this type of screwdriver seldom exceeds $1/4$ inch when its use is restricted to electronics fabrication. However, if a wider blade is required to fill the slot of a screw head, do not attempt to use the smaller size.

NUT DRIVER

Another frequently used tool in electronics work is the nut driver, shown in Fig. 5·10. This tool is designed as a fixed-socket wrench. The nut driver is used to hold or turn threaded nuts. A threaded nut is illustrated in Fig. 9·2.

Nut drivers are rated according to the size of the nut they can hold. An electronics worker should have available nut drivers as small as $3/16$ inch in size and an assortment of sizes up to $3/8$ inch. For specialized work other sizes may be required. The length of the shaft is determined by the job at hand, but it is found that 4 inches of shaft length is adequate for most applications.

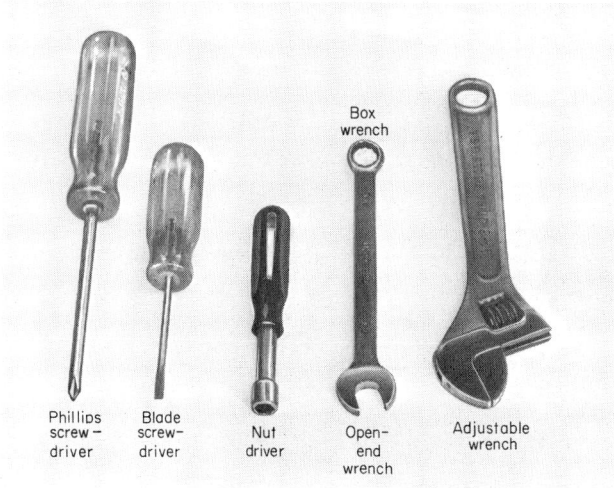

FIG. 5·10 Tools for mechanical assembly.

BOX WRENCH

The wrench appearing next to the nut driver in Fig. 5·10 is of a combination type. At the top end of this combination wrench is a box wrench. A box wrench is a type of socket wrench, having the same application as the nut driver. The information given above for the nut drivers, with relation to socket sizes, also applies here.

OPEN-END WRENCH

An open-end wrench has the same function as the nut driver and the box wrench. An open-end wrench is shown in Fig. 5·10, at the bottom end of the combination wrench. The information previously given about socket sizes also applies to the open-end wrench. Selection of the nut driver, the box wrench, or the open-end wrench depends upon how much space is available. Proper selection will make the job at hand easier.

ADJUSTABLE WRENCH

An adjustable wrench may be employed if used with caution when the proper nut driver, socket set, box wrench, or open-end wrench is not available. It is poor practice to use pliers to secure or loosen threaded nuts. An adjustable wrench is shown in Fig. 5·10.

SOCKET SET

Although nut drivers, box wrenches, and open-end wrenches will normally take care of most nut-turning needs in electronics fabrication, at times it may be necessary to use a tool with more *leverage, reach,* and *flexibility*. Such a tool is a set of sockets with appropriate handles, extensions, and universal joints. The basic elements of a socket wrench set are shown in Fig. 5·11. It is possible to obtain socket wrench sets in various sizes. The size of the set is determined by the size of nuts to be turned and by the size of drive that couples the handle to the socket. Sockets come with either 6-point or 12-point grips. Most socket sets are available with at least seven different size sockets.

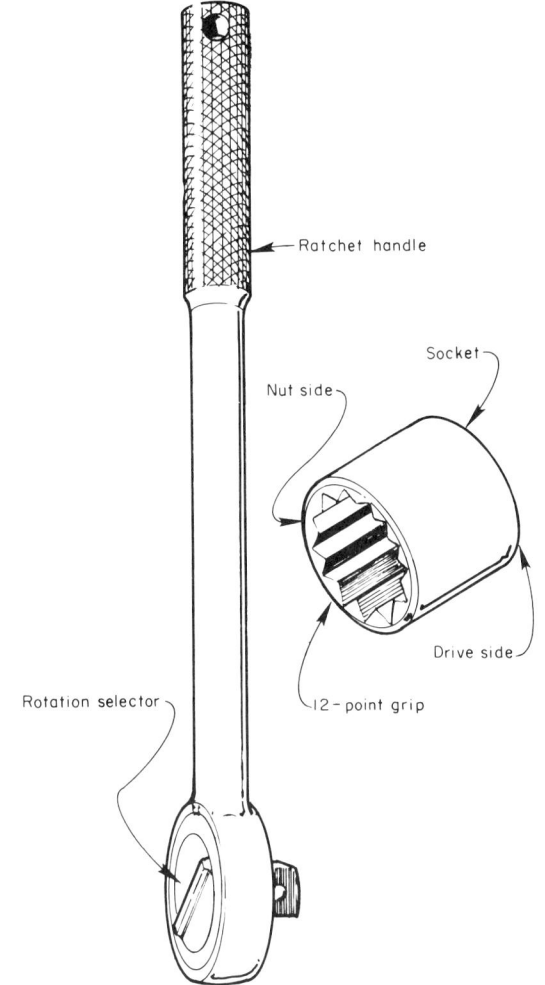

FIG. 5·11 Socket wrench set.

MEASURING AND LAYOUT TOOLS

Besides the measuring tape, foot rule, and metric scale discussed in Chap. 4 under *linear measurements*, Sec. 4·12, a person working in fabrication of electronic products may have need to install an unexpected bracket or similar mechanical device. As such, he will have need for special tools to help him with the necessary measuring and layout. In most instances, a person does not merely point to a spot and say "I'll drill a hole there." He has to *plan* his work, no matter how simple it is. This planning may partially be done in his head, but the actual *practice* will be done on the chassis he may be working on. Two tools that will help with this layout will be the *divider* (Fig. 5·12), and the *combination square* (Fig. 5·13).

If the job at hand requires precision measuring tools, it may be necessary to use a *vernier caliper* (Fig. 5·14a), or a *micrometer* (Figure 5·14b). These tools are used to measure the *thickness* of objects.

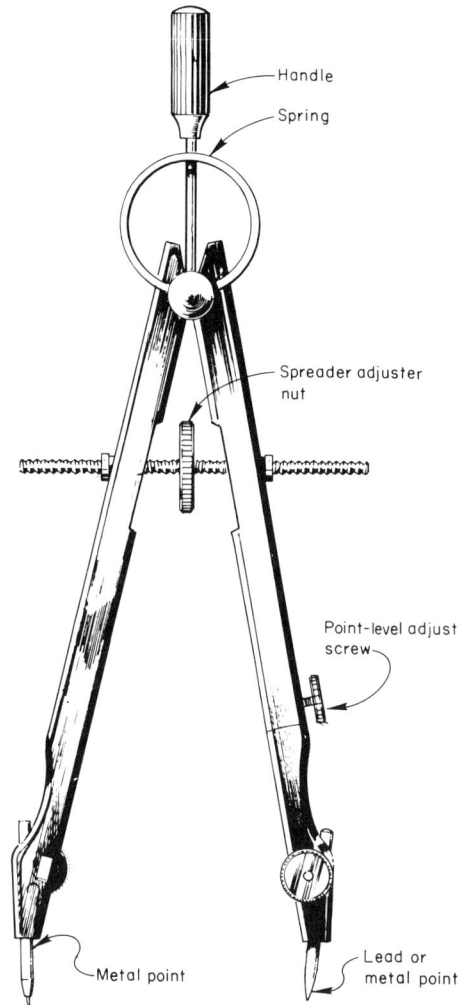

FIG. 5·12 The divider.

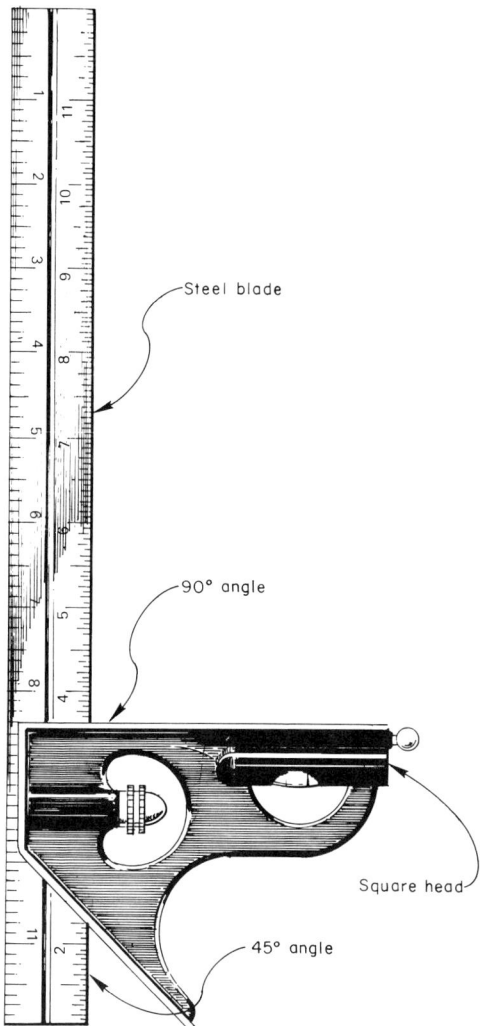

FIG. 5·13 Combination square.

MARKING TOOLS

In order to mark the spot where a hole is to be drilled or a cut made, a lead pencil may be used. However, for more exacting accuracy, especially on metal, it is best to *scratch* a mark with a pointed instrument. The *scratch awl* (Fig. 5·15) serves nicely for this. If, after making the scratch to mark the spot where a hole is to be drilled, the supervisor gives approval to proceed with the drilling, it will

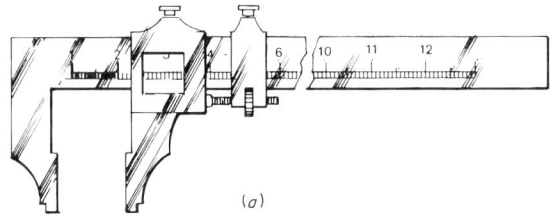

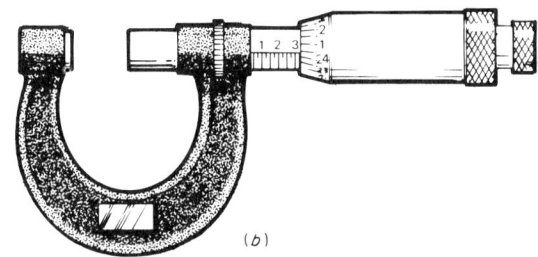

FIG. 5·14 Precision measuring tools. (*a*) Vernier caliper. (*b*) Micrometer.

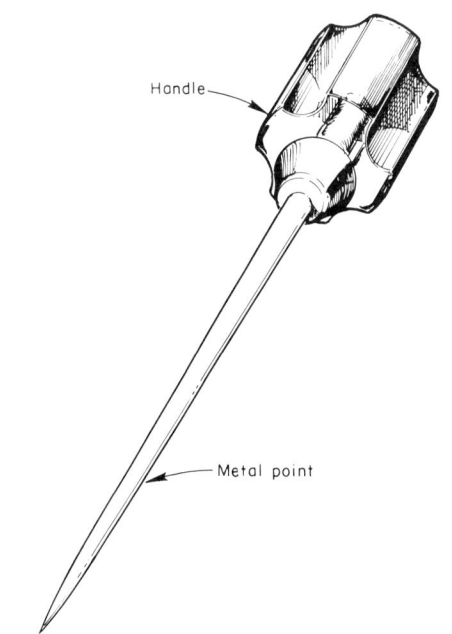

FIG. 5·15 Scratch awl.

be necessary to make a *pronounced* mark with a *center punch* (Fig. 5·16). A center punch will actually leave an indentation in the metal that serves to confine a rotating drill to that exact spot. See Fig. 9·14 that shows a center punch being used. As can be seen, a *ball-peen hammer* is used to hit the center punch. *A scratch awl should never be used as a center punch.*

HOLDING DEVICES

Prior to drilling or cutting any piece of metal, it should be secured firmly as a *safety factor*. However, common sense will tell us that if the metal is secured during drilling or cutting, the job will be neater and more exact since it will not wobble during the operation. It should be stated that although examples given in this section deal mainly with metal parts, *any* material to be drilled or cut should be secured. About the most popular holding device widely used is the *bench vise* (Fig. 5·17). However, if a bench vise is not available or is not practical

FIG. 5·16 Center punch.

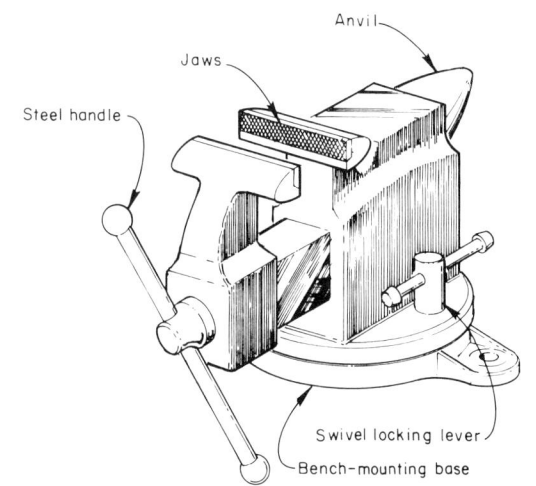

FIG. 5·17 Bench vise.

to use, use a *"C" clamp* (Fig. 5·18a). When using a drill press, it will be necessary to secure the part being drilled on a *drill-press vise* (Fig. 5·18b). Here again, it may be necessary to use a "C" clamp.

Many times the part to be held while work is being accomplished will be fragile and delicate. When this is the case, place a protective covering over the jaws of the holding device, or use special holding devices as shown in Fig. 5·19. Notice that this vise has rubber protective covering over the jaws, and that it is fully adjustable to provide more flexibility. Actually, this vise is designed to hold

FIG. 5·19 Fully adjustable holding vise. (*Hunter Tools, Santa Fe Springs, California.*)

printed circuits or cable connectors during soldering operations.

DRILLING AND CUTTING TOOLS

Drilling of holes will require the use of a *portable hand drill* (Fig. 5·20) or a *drill press* (Fig. 9·18). Although *twist drills, chuck wrench,* and related work-holding devices are essential for hole drilling operations with these *power tools*, an absolute accessory must be a pair of *safety glasses*. See Fig. 9·18.

The familiar *hacksaw* is another cutting tool often used. This tool is used most often to cut shafts of potentiometers and rotary switches. Proper use of this cutting device centers closely with the selection of the proper cutting blade, and the proper tension on the blade. See Fig. 5·21.

FILING AND GRINDING TOOLS

In practically all cases, when a piece of metal is cut, whether it be a hole or a straight cut, rough edges with *burrs* will remain. These rough edges must be made smooth, and the burrs must be removed. To do this an assortment of *files* are available. Figure 5·22 illustrates these files. At times, it may be advisable to use power tools for this pur-

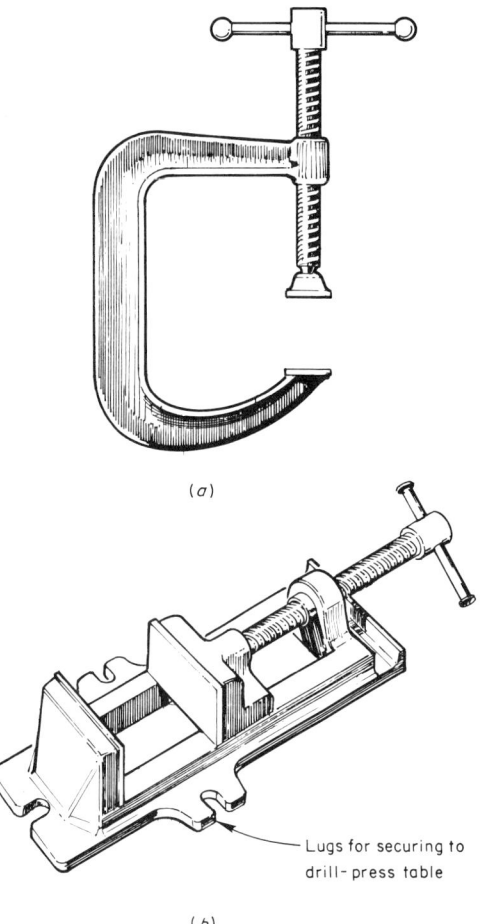

FIG. 5·18 Work-holding devices: (*a*) "C" clamp. (*b*) Drill-press vise.

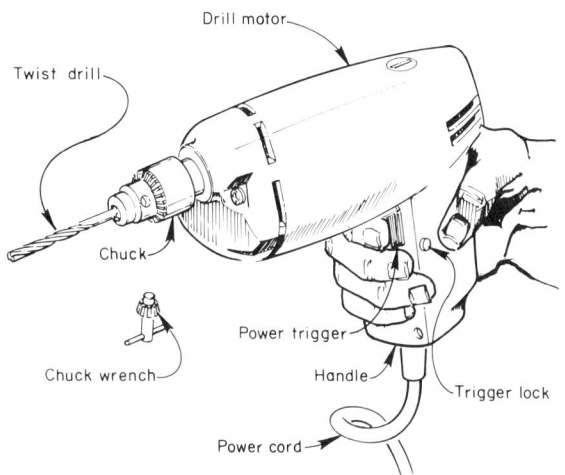

FIG. 5·20 Portable hand drill.

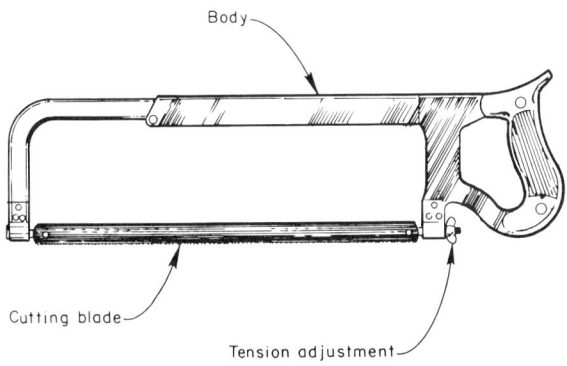

FIG. 5·21 Hacksaw.

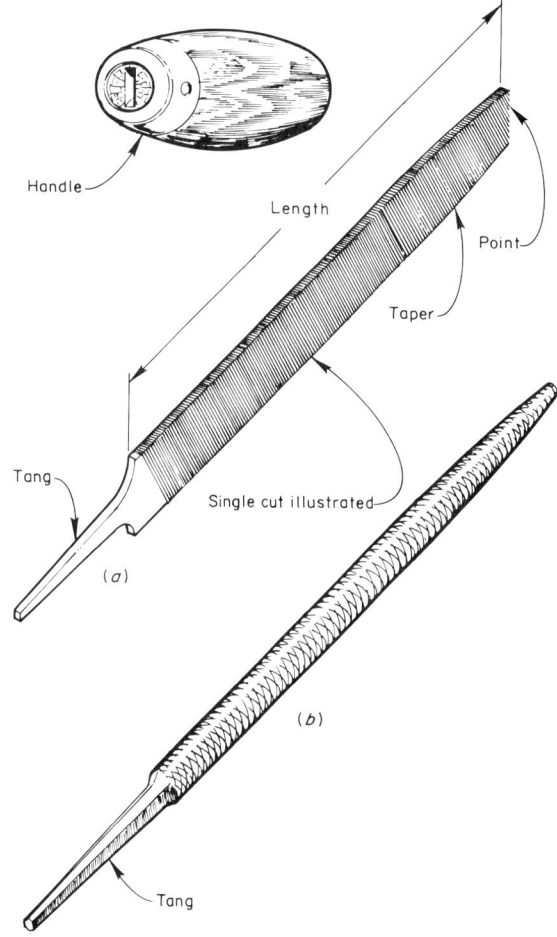

FIG. 5·22 (a) Flat file, single cut. (b) Round file.

pose. The *rotary grinder* (Fig. 5·23) or the *bench grinder* (Fig. 5·24) may be appropriate. The rotary grinder will be useful also when attempting to enlarge a hole. An assortment of rotary cutters are available for this tool. The bench grinder can also be used to sharpen twist drills and shape brackets. As mentioned during the discussion on power drills, *safety glasses are an absolutely necessary accessory to any power tool*, especially the rotary type. Even though the bench grinder has a built-in safety-glass window, personal safety glasses should still be worn by *everyone* in the immediate vicinity of an operating machine.

5·4 SUMMARY ON TOOLS

In summary, tools are used by people working in electronics for the assembly or disassembly of hardware and component parts used in the trade. A variety of hand and power tools are available, therefore making mandatory an intelligent selection of the proper tool for a given job. The proper selection is made when the individual has wide knowledge of tools and knows their application and limitations. Further, proper selection will increase the safety factor, providing safe practices are followed when using the tools.

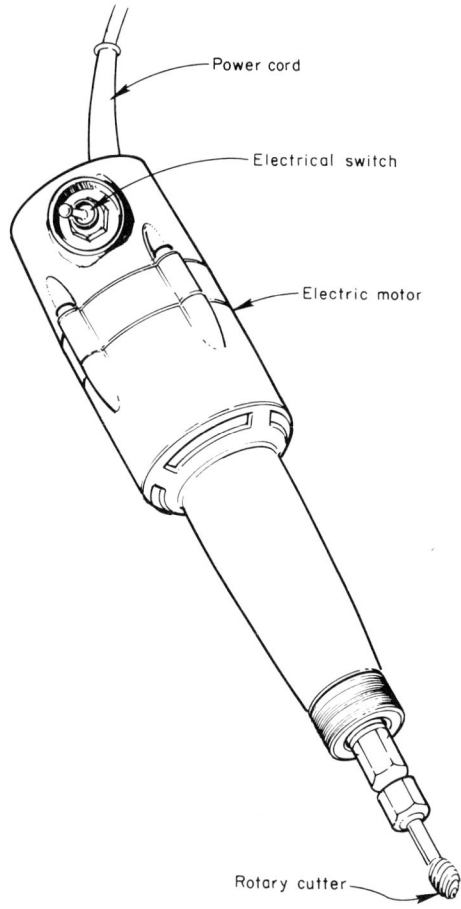

FIG. 5·23 Portable rotary grinder.

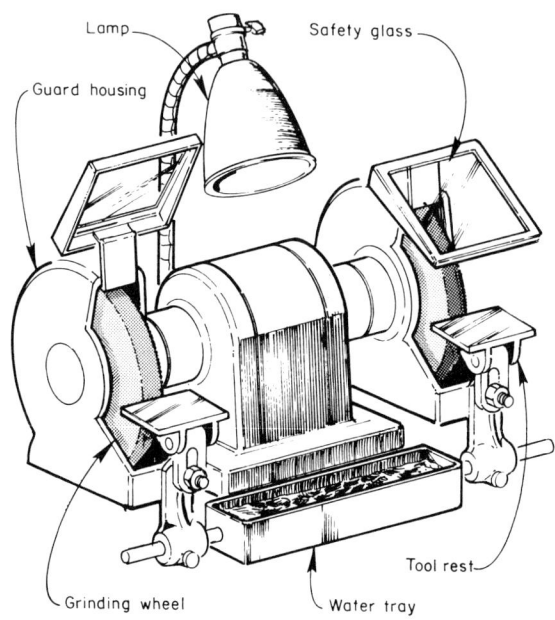

FIG. 5·24 Bench grinder.

The following jobs are intended to further acquaint the student with tools used by electronics assemblers and technicians. Specific uses of some of the tools are covered in succeeding chapters.

EXERCISES

JOB 5·1 How to get acquainted with hand tools used for electrical assembly

OBJECTIVES
1. To learn to recognize hand tools
2. To learn about the difference in tool sizes
3. To become aware of safety measures in connection with hand tools

MATERIALS REQUIRED

Equipment
Assorted hand tools used in electrical assembly

Supplies
Paper and pencil

PROCEDURE

1. Make a sketch of the long-nose pliers supplied.
2. On this illustration, identify every section of this tool.
3. Write a simple statement about the primary purpose of this tool.
4. State the purpose of each section identified on the sketch.
5. List any safety factors involved in connection with the use of this tool.
6. Write a simple statement regarding the factors involved when the manufacturer assigns a size or capacity rating for this tool.
7. List the probable difficulties that may be encountered if a person using this tool does not select one of proper size when working on a given application.
8. Repeat steps 1 to 7 for each tool supplied.

QUESTIONS

1. Which of the tools supplied are designed primarily for electrical-assembly work?
2. Which of the tools supplied are designed primarily for mechanical-assembly work?
3. Which of the tools supplied are the most dangerous to human safety? Why?
4. What precautions should be taken to avoid injury when using hand tools?
5. What factors must be considered when selecting a screwdriver?
6. What factors must be considered when selecting a pair of diagonal-cutting pliers?
7. What factors must be considered when selecting an open-end wrench?
8. Make a list of any additional tools you believe should be included in the toolbox of any electronics assembler.

JOB 5·2 How to get acquainted with tools and equipment used in microelectronics assembly

OBJECTIVES

1. To become aware of special tools and equipment used in the electronics industry
2. To get an insight on methods that can be used to keep abreast of progress in industry

MATERIALS REQUIRED

Supplies
Assortment of trade journals and catalogs
Notebook paper and pencil

PROCEDURE

1. Look through the trade journals and catalogs supplied, paying special attention to pictures or sketches of tools and equipment used in electronics fabrication. Notice that most of this type information which appears in trade journals is found in the sales promotion pages catering primarily to advertisement. Catalogs normally have a special section for tools and equipment. Some catalogs are in pamphlet form and specialize in certain items, such as in tools for electronics assembly. Since trade journals are normally monthly publications, they tend to have the latest information regarding new electronics-fabricating techniques and of the tools and equipment associated with such techniques. In the area of microelectronics, this is quite important since the industry constantly attempts to improve methods and processes associated with fabricating, handling, and assembling the extremely small components and circuit arrays. Therefore, the *outsider* has to develop ways and means to keep up with the latest developments. The use of advertising pages in trade journals as you are now doing is an excellent way to attempt to do this.
2. On notebook paper, sketch and give a brief summary regarding any special tool or piece of equipment designed primarily for use in the fabricating, handling, or assembly of components used in the field of microelectronics, or in related miniature component fabrications.

QUESTIONS

1. Were you able to find out about a tool or equipment used in electronics assembly which you were

not aware of? (If you had no luck in one issue, keep trying as new issues come out.)

2. Is descriptive text regarding the use and application of the new tool or equipment better in the journal, catalog, or special pamphlet? Explain, or give an example.

JOB 5·3 How to get acquainted with tools and equipment used in mechanical assembly

OBJECTIVES

1. To become better acquainted with tools and equipment used in mechanical phases of electronics fabrication
2. To learn to properly identify mechanical-assembly hand tools

MATERIALS REQUIRED

Equipment
Toolbox of assorted hand tools as illustrated in this chapter
Power tools as illustrated in this chapter

Supplies
Notebook paper and pencil

PROCEDURE

1. Lay out on the tabletop all tools found in the toolbox.
2. Put back in the toolbox those tools which would *not* be used in some mechanical assembly procedure.
3. Under each tool remaining on the tabletop, place a clean sheet of notebook paper. On this paper write the name of the tool and include a one-sentence summary regarding the possible uses for this tool.
4. Call your instructor and ask him to check your display.
5. Put all tools back in the box.
6. Since power tools are not normally available in great quantity, ask your instructor to direct you to where you can see power tools similar to those illustrated in this chapter. It may be that he might direct you to the machine shop or auto shop.
7. On a notebook sheet of paper, list the power tools that you were able to see. Write a summary statement regarding the possible uses of these power tools in electronics fabrication. Also include comments regarding hazards to health and body, and safety precautions to take when using this equipment.

QUESTIONS

1. List other tools that were not in the toolbox but which you think should have been included. Give your reasons.
2. Write a statement regarding the use of the proper tool for a given job. Include the thought of safety and size consideration.
3. List the hand or power tools that can contribute to an eye injury if not used properly. Give examples.
4. Why is it important for a person working in electronics fabrication to know how to use hand and power tools used in mechanical assembly?

6

SOLDERING PRINCIPLES

Soldering can be traced to very early cultures. Jewelry taken from the oldest Egyptian tombs contain excellent soldered joints. Until comparatively recently, the art of soldering was the prerogative of the skilled, master craftsman. Today, soldering is done regularly by semiskilled laborers in factories throughout the world. Even amateur craftsmen and hobbyists find it necessary to solder.

SOLDERING DEFINED
Soldering may be defined as the process of uniting two clean pieces of metal with a thin layer of a third metal applied in a molten state.

In metalwork, there are three types of soldering: brazing, silver soldering, and soft-soldering. The last, soft-soldering, is practiced in electrical and electronic work. The term *soldering* implies here that soft soldering is taking place. In other words, one solders wires together. The metal that is heated to a molten state is referred to as the *solder*. Solder is an alloy of tin and lead, and is normally manufactured in wire form. Figure 6·1 shows three different wire sizes of solder.

6·1 SOLDERING REQUIREMENTS
In order to solder properly, three requirements must be met: first, everything must be clean; second, a proper solder alloy must be used; and third, sufficient heat must be employed.

FIG. 6·1 Solder in wire form as shown is popular in the following diameter sizes, expressed in inches: 0.015, 0.030 and 0.060.

CLEANLINESS

In preparing a metal surface for soldering, all insulation must be removed. Any grease, oil, scale, oxides, or other foreign matter must also be removed. The removal of insulation and grease is obvious. However, oxidation may invisibly contaminate an apparently clean surface, rendering it useless for good soldering.

A flux is used to remove these oxides, thus ensuring a good union of the metals. Solder that is manufactured in wire form usually will contain one or more cores of flux. Note the flux core in the solder shown in Fig. 6·4. Flux normally employed in electronics work is *resin*. Acid or soldering paste should never be used as flux for soldering electronic joints. In time, these harsh cleaning agents will corrode the delicate wires and terminals.

SOLDER ALLOY

The solder alloy used in electronics work is normally composed of 60 percent tin and 40 percent lead. However, for certain applications, other ratios of tin and lead may be required. The ratio of tin and lead content determines the hardness, strength, and melting point of the solder.

HEAT REQUIREMENTS

In order to melt the solder, heat must be applied. Further, the surfaces to be soldered must be preheated before application of the solder alloy. Heat is normally applied with a soldering iron. Soldering irons are shown in Figs. 6·2 and 5·1. Heat is transferred from the soldering-iron tip to the joint to be soldered, as well as to the solder itself.

6·2 TINNING

Among the first steps in the process of soldering is tinning. Tinning is accomplished by spreading a thin layer of molten solder over a surface that

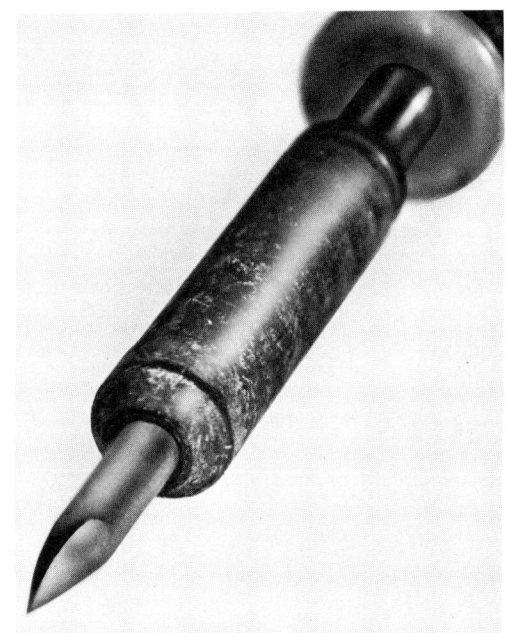

FIG. 6·2 A standard heavy-duty soldering iron.

has previously been cleaned and preheated. Figure 6·3 shows a wire being tinned.

Tinning is done to the soldering-iron tip, as well as to the surface to be soldered.

TINNING FOR HEAT TRANSFER

A soldering-iron tip is tinned to allow for maximum transfer of heat to the joint to be soldered (see Fig. 6·4). By tinning the point of the soldering-iron tip, solder is made to penetrate the surface of the metal tip to a molecular depth, forming a thin cushion of molten solder through which heat can be transferred.

TINNING FOR PENETRATION

Whenever possible a surface to be soldered should be tinned first. By tinning, penetration to molecular depth by the solder is ensured. This penetration is necessary in order for a thorough bond between the solder and the surfaces being soldered to be accomplished. *Wetting* is a term used to designate that solder is penetrating a surface to molecular depth.

6·3 MECHANICAL AND ELECTRICAL CONNECTION

MECHANICAL CONNECTION

When a wire is to be soldered to a terminal, or to another wire, a tight mechanical connection should be accomplished first. Solder alone is not sufficient to support any appreciable weight. However, when the product being manufactured is designed for use in the aerospace industry, NASA requests that mechanical connections of electrical wires be such that visual inspection can readily be made. Through close inspection, a thoroughly reliable soldering job can be assured. To provide this *window* for inspection, NASA specifies that a wire lead connected to a terminal be *wrapped* around that terminal considerably less than a full turn. The exact specifications are called out and illustrated in this text, along with conventional procedures.

A wire lead is *wrapped* around a terminal to accomplish the mechanical connection. When two wires are joined together (without the aid of a tie point), they are said to be *spliced*. Splicing is primarily a mechanical engagement. Several wire-splicing methods are illustrated in Fig. 6·5. (Note: NASA does not allow splicing.)

ELECTRICAL CONNECTION

Joining of two or more electrical conductors requires that no electrical resistance be present between such conductors. For this reason, a me-

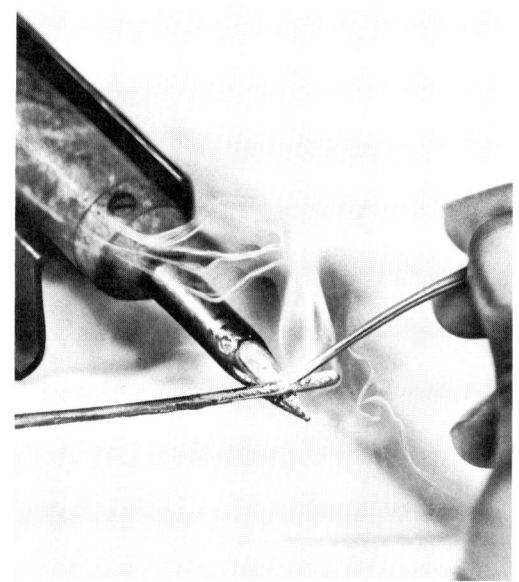

FIG. 6·3 The proper way to tin a wire when using a soldering iron.

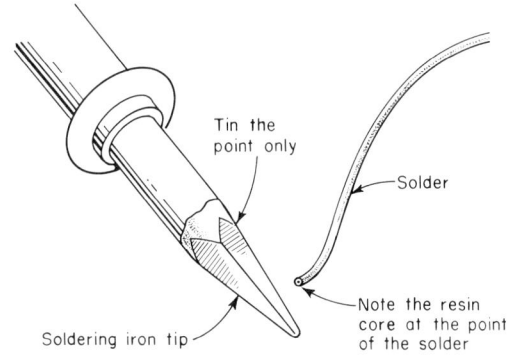

FIG. 6·4 Tinning the soldering-iron tip.

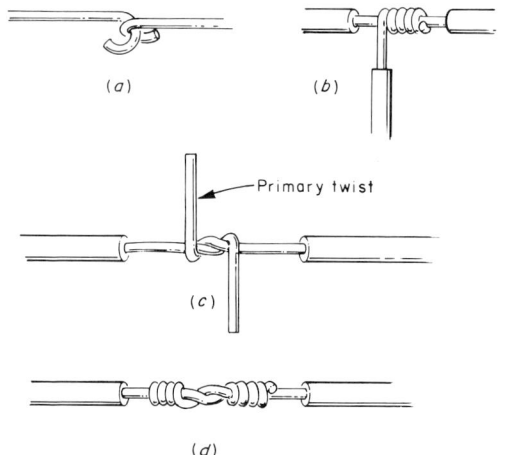

FIG. 6·5 Wire-splicing methods: (*a*) Hook splice. (*b*) Tap splice. (*c*) Western Union splice.

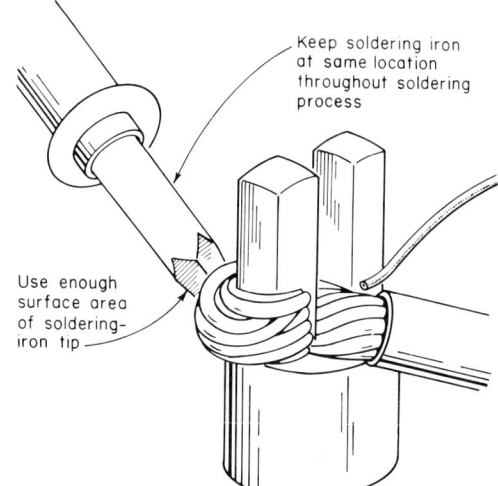

FIG. 6·6 Proper application of heat.

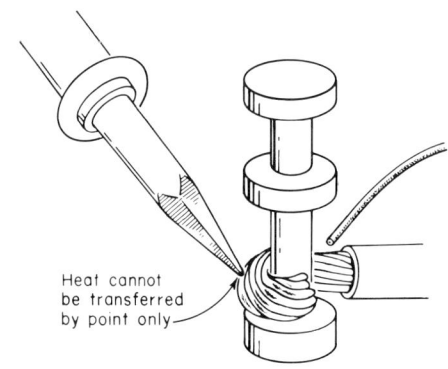

FIG. 6·7 Improper application of heat.

chanical connection is not sufficient when joining conductors. Soldering, with its molecular penetration, can accomplish this electrical union, and is therefore employed for this purpose.

6·4 THE SOLDERING PROCESS

HEAT APPLICATION

The soldering-iron tip is used to impart heat to the surfaces to be soldered. The proper transfer of heat is illustrated in Fig. 6·6. The soldering iron should rest on one side of the terminal or wire. It should be kept there until the total joint to be soldered is hot enough to melt the solder.

The transfer of heat requires a large conduction surface. Figure 6·7 illustrates an inefficient way to transfer heat from a soldering-iron tip.

SOLDER APPLICATION

Solder should be melted by the joint to be soldered, *not by the soldering-iron tip*. If the joint is heated properly, the solder will melt when the joint is touched. Solder should never be applied to the hot soldering-iron tip and allowed to run in a molten state on to the joint to be soldered. Solder should be applied on the side of the joint opposite to that of heat application (see Fig. 6·8).

When gaps in the wire wrap or in the terminal exist, they should not be filled with solder unless it is specified (see Fig. 6·9).

Proper soldering results are obtained only when the amount of solder used is sufficient to accomplish a good electrical connection. An example of good soldering results is shown in Fig. 6·10. Solder should not run into the wire insulation, nor should the insulation be allowed to touch the terminal.

CONTOUR SOLDERING

When the minimum amount of solder is properly applied, the contour of the wire, or other surface

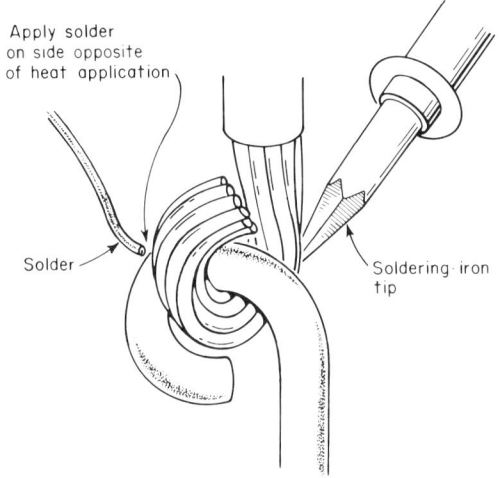

FIG. 6·8 Proper application of solder.

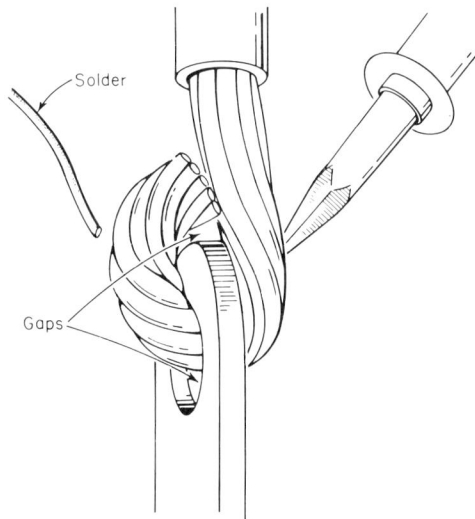

FIG. 6·9 Solder should not be used to fill gaps.

being soldered, will be clearly visible. This is called contour soldering, illustrated in Fig. 6·11.

IMPROPER SOLDERING RESULTS

Many factors contribute to unacceptable soldering results. In most cases improper soldering results are clearly visible, but in some cases critical inspection is required to reveal a faulty soldered joint. Compare the illustrations of Figs. 6·11 and 6·12. The following factors contribute to a poor solder job: (1) blowing to quickly cool a hot joint, resulting in too rapid solidifying of the molten solder; (2) improper transfer of heat or the lack of sufficient heat; (3) disturbing or moving a soldered joint before the solder solidifies; (4) a dirty or oxidized surface prior to the application of the solder; (5) insufficient flux; and (6) the use of an untinned or improperly kept soldering-iron tip. Figure 6·13 illustrates four examples of bad soldering.

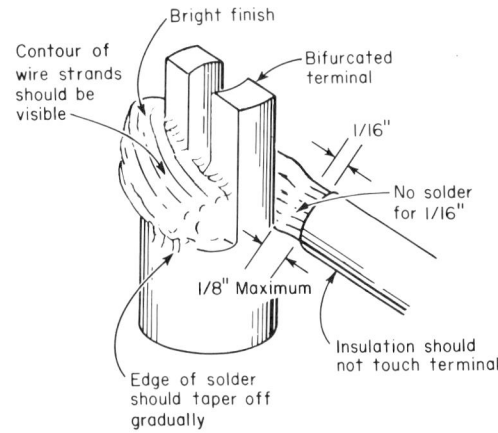

FIG. 6·10 Proper soldering results.

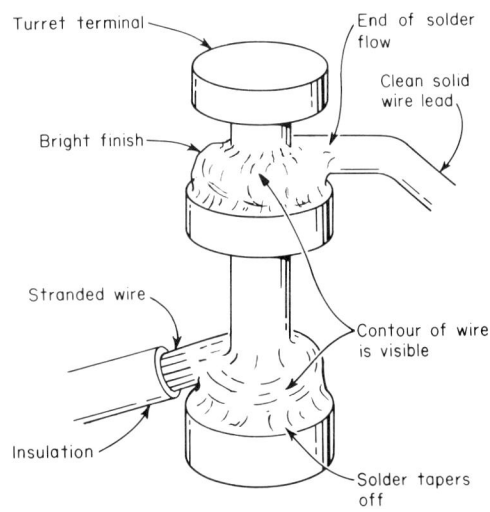

FIG. 6·11 Contour soldering. (Note: Wire wrap does not meet NASA standards due to the lead being wrapped around the terminal one full turn, even though soldering is excellent.)

93

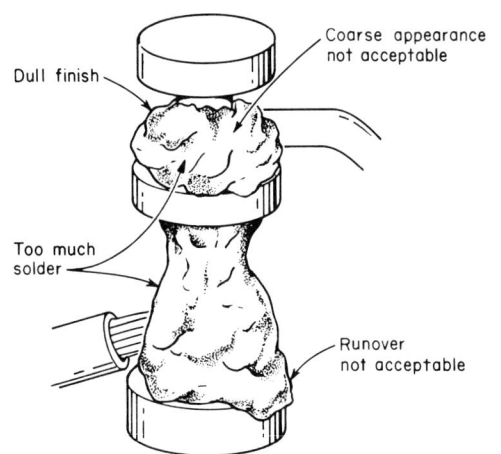

FIG. 6·12 Improper soldering results.

6·5 DAMAGE BY HEAT

The high temperatures developed by soldering irons can damage the insulation of wires or render delicate component parts useless. Proper heat application and timing can prevent the heat from running up the wire conductor and thus prevent damage to the insulation. However, protection of heat-sensitive components requires additional precautions.

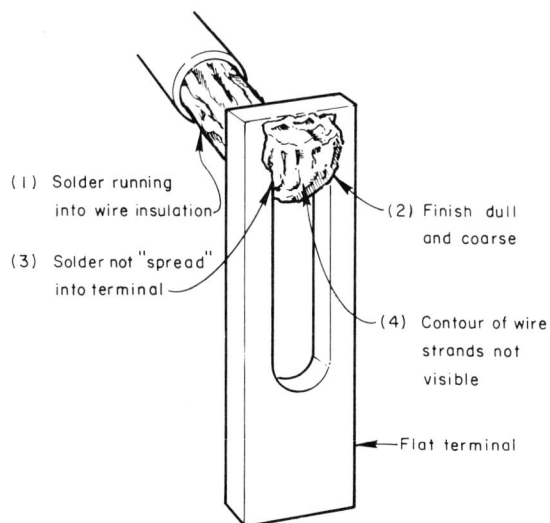

FIG. 6·13 Four examples of bad soldering. However, *wire-wrap* meets NASA standards.

THE HEAT SINK

A heat sink is a small metallic clamp designed to draw heat away from its point of contact with a hot wire lead (see Fig. 6·14). An alligator clip, shown in Fig. 9·13c, may be employed as a heat

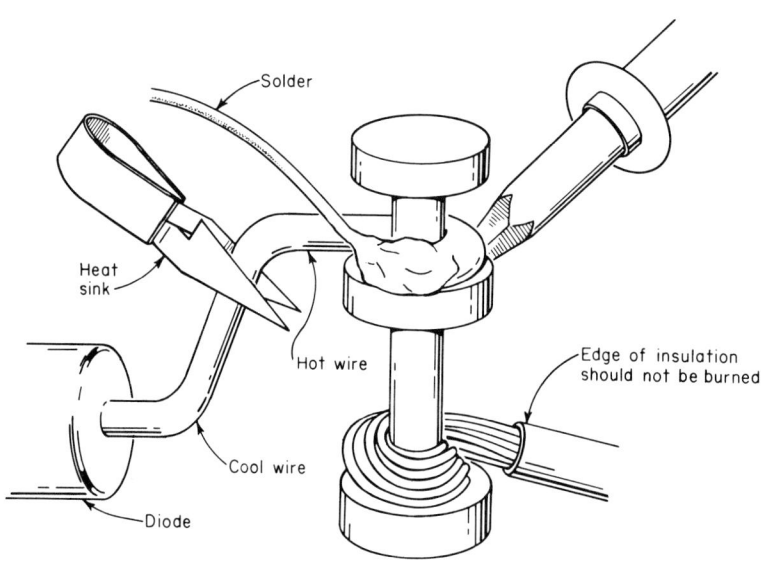

FIG. 6·14 Use of the heat sink.

sink also. A pair of long-nose pliers can also be employed as a heat sink, although it may prove to be an awkward device. A heat sink should not be detached immediately after the removal of the soldering iron, but should be allowed to remain for at least 30 seconds longer.

Heat-sensitive parts include semiconductors, small capacitors, small resistors, small inductors, and terminals that protrude through a glass seal.

6·6 SOLDERING ON CONNECTOR PINS

Some cable connectors employ *solder pots* (also known as *solder cups*), which are the rear end of the *contact pins*. Into these pots a wire is inserted and soldered. Before actual soldering, the pots must be prepared by partially filling them with a small amount of solder in a molten state. This molten solder may be introduced into the solder pot as shown in Fig. 6·15. Another way to place the solder in the solder pot is to cut a piece of solder of a length less than the pot height and, while it is in a cold state, drop it into the pot. The solder pot may then be heated, and the solder will melt inside. In no event should the amount of molten solder completely fill the pot without the wire being inserted.

A wire that has previously been cut to size and tinned may now be inserted into the solder pot by reheating and melting the solder in the pot. A properly soldered connector solder pot, with a wire inserted, is illustrated in Fig. 6·16. For proper wire length refer to Fig. 8·16.

In order to obtain a good solder connection without too much difficulty, especially on a connector solder pot, the use of an *anti-wicking tool* (Fig.

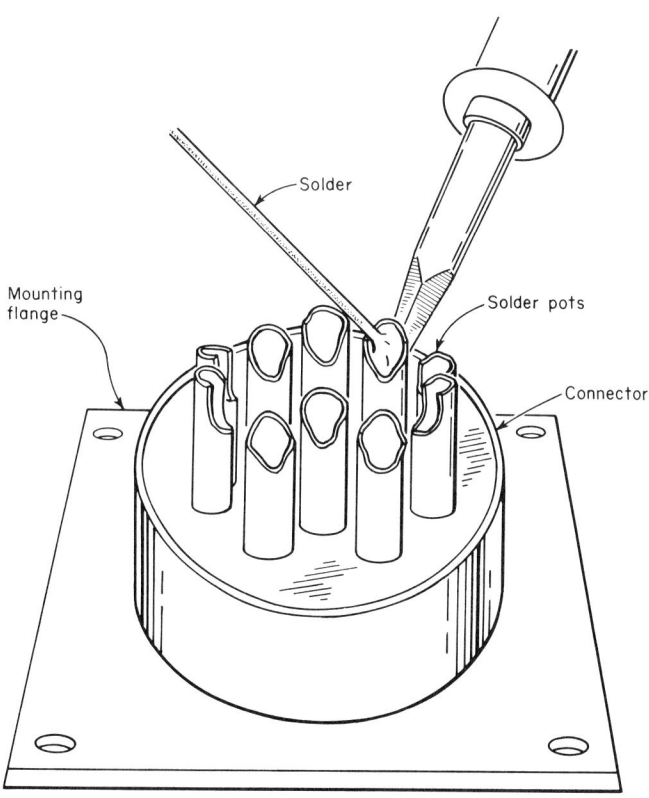

FIG. 6·15 Preparation for soldering on connector solder pots (cups).

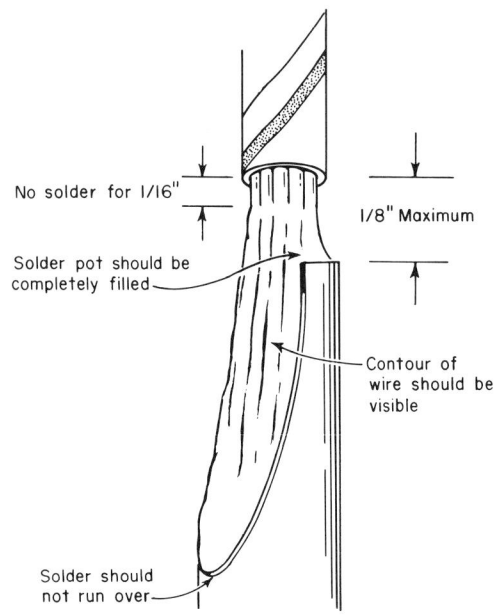

FIG. 6·16 A properly soldered connector solder pot (cup).

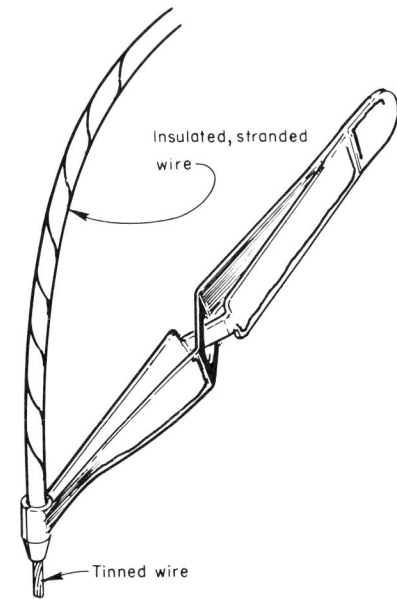

FIG. 6·17 Anti-wicking tool.

6·17) is recommended. This tool is designed to prevent solder from flowing beneath the insulation (*wicking*) of a stranded wire during tinning and soldering operations. The tool acts as a heat sink (*heat shunt*). When wicking occurs, the wire becomes stiff and will easily break during normal flexing.

6·7 SOLDERING ON PRINTED-CIRCUIT BOARDS

When soldering on printed-circuit boards, standard soldering techniques are employed. However, greater emphasis is placed on the dangers of excessive and prolonged heat application. Heat damage can be done either to the thin copper foil or to the miniature components or semiconductors normally associated with printed circuitry.

Prior to soldering, the lead protruding through the plated-through hole should be clinched, or bent, to provide the necessary mechanical support (see Figs. 3·5 and 6·18).

HEAT DAMAGE TO PRINTED CIRCUITS

When excessive or prolonged heat is applied to a printed circuit, either the *pad* or the copper-foil

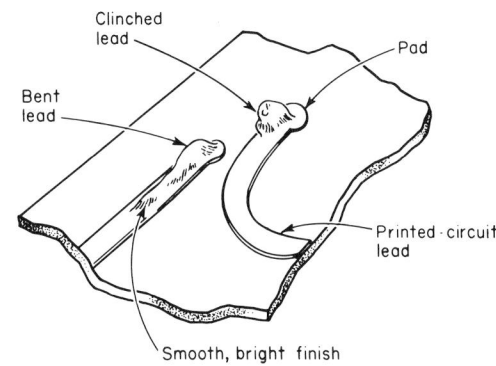

FIG. 6·18 Acceptable soldered leads on printed circuits.

lead may become separated from the laminated plastic board.

A small soldering iron with a rating of approximately 30 watts should be used. All surfaces should be cleaned, and a 60/40 solder alloy used, to ensure a fast solder job. An ink eraser may be used to clean the pad and copper-foil lead as necessary.

Contour-soldering techniques should be applied as shown in Fig. 6·18.

HEAT DAMAGE TO COMPONENTS

Components not easily damaged by heat should have their leads clinched and soldered in a way that completely fills the hole. Soldering should be administered from the clinched-lead side of the board. (See Fig. 3·4.)

Soldering of heat-sensitive components requires use of a heat sink between the plated-through hole and the component. To ensure a minimum of heat conduction to the component, the lead is bent and soldered at the bottom side without allowing solder into the plated-through hole. (See Fig. 3·5.)

6·8 DESOLDERING FROM PRINTED-CIRCUIT BOARDS

At times it is necessary to remove a component that has been soldered from a printed-circuit board. Assuming that it is a resistor that has to be removed, the best thing to do is to first cut the resistor in half with a pair of diagonal cutters. If this is not desirable, the next best procedure is to cut at least one of the leads, leaving enough wire soldered to the printed-circuit board so that after desoldering it can be grabbed with long-nose pliers. See Fig. 6·19.

To unsolder the wire lead from the pad, it is necessary to execute all *soldering* operations: use a clean, hot, tinned soldering iron; apply a drop of resin-core solder to the tip of the soldering iron; apply the tip of the soldering iron to the soldered junction, and with a pair of long-nose pliers *carefully* pull on the wire lead. (It may be necessary to carefully unclinch the component lead first.) Safety glasses should be worn during this operation because molten solder may accidentally splash towards the operator's eyes when the wire lead becomes unfused. To remove multiple lead components, special soldering iron accessories are available.

Solder which has been left on the pad of the printed-circuit board can be safely removed by *drawing* it up with the use of a thin length of braided shield as shown in Fig. 6·20, or by using a specially designed *desoldering* tool as shown in Fig. 6·21. The braid uses the *wicking* principle to remove the solder, and the desoldering tool illustrated uses the *suction* principle. Nineteen-strand wire is a good substitute for the braid. However, for best results the braid and the stranded wire should be immersed in liquid resin prior to their use as wicking agents. After desoldering has been completed, resoldering of a replacement component can be undertaken using the same procedures as *first-time* soldering.

6·9 CLEANING SOLDERED JOINTS

After soldering, flux (resin) residue is often visible. To remove this residue, use a small cotton-tipped

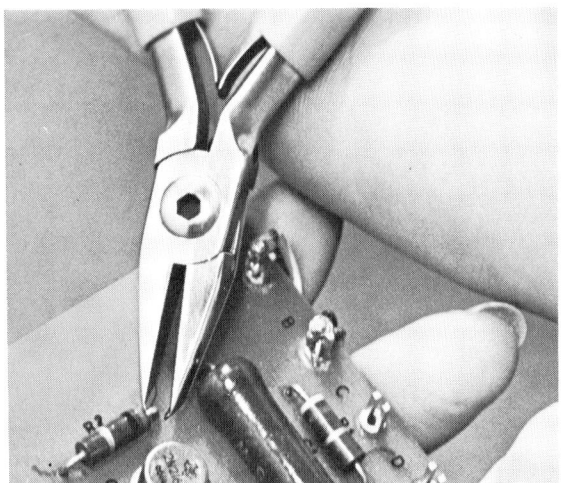

FIG. 6·19 Cutting component leads from printed-circuit boards.

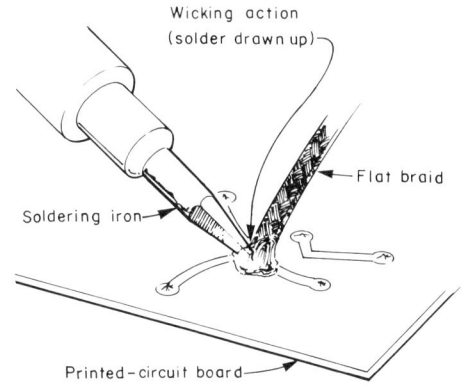

FIG. 6·20 Removing solder from a printed-circuit board soldered connection.

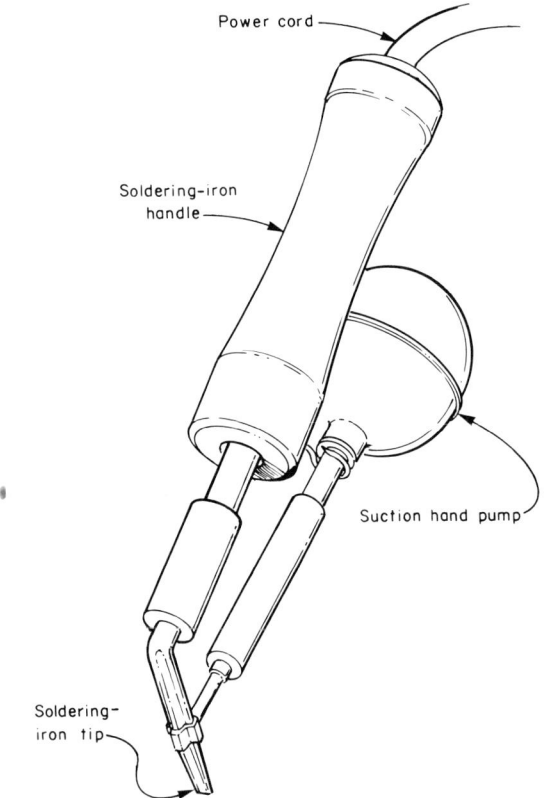

FIG. 6·21 Desoldering tool.

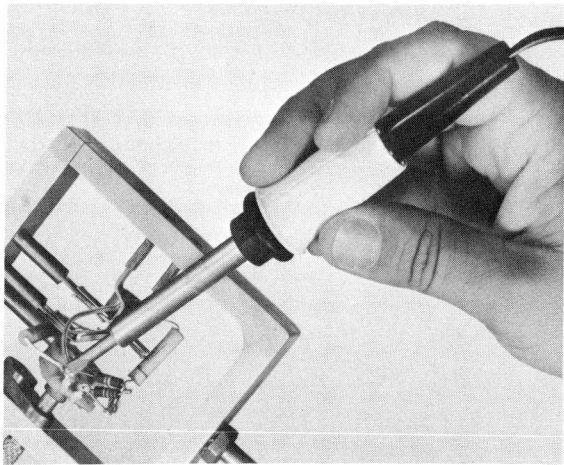

FIG. 6·22 Bit soldering.

probe (or a small piece of cloth) which has been soaked in alcohol or some other approved noncorrosive solvent. Removal of this residue allows better visual observation during inspection. It also prevents dust or other foreign matter from settling on the residue and later causing a high-resistance short-circuit.

If the product being fabricated is for use in NASA projects, a medium-stiff bristle brush dipped into alcohol or other approved solvent shall be used to clean soldered joints.

6·10 MASS-PRODUCTION SOLDERING METHODS

Soldering done with a hand-held soldering iron is often referred to as *bit soldering*. See Fig. 6·22. However, although bit soldering is done extensively, much of production soldering is accomplished on a mass basis with the aid of machines. This is especially true of printed circuits.

AUTOMATED WAVESOLDERING

Soldering by machine requires that all components be mounted on a laminated or ceramic printed-circuit board prior to soldering. Having mounted the parts on the boards, with their leads protruding through the plated-through holes in the boards, the leads are trimmed to an appropriate length for soldering by a component lead-cutting machine as shown in Fig. 6·23. The leads are then *all soldered at the same time* by a mass production soldering method known as *wavesoldering*.

FIG. 6·23 Component lead-cutting machine.

Automated wavesoldering combines into one production-line operation the individual fluxing, preheating, soldering, and cleaning functions necessary for proper solder processing. The printed-circuit boards are placed in specially designed carriers, which in turn are placed on a conveyor. Many carriers with their printed-circuit boards are placed on this conveyor. The conveyor takes the printed-circuit boards first across a *wave* of liquid flux where a thin, uniform coating of flux is applied to the underside of the board. The board then passes over a preheating station where the flux solvent is evaporated and the board thermally conditioned for soldering. The board continues at a steady speed onto the soldering station. Here, a unit pumps molten solder upward, forming a precisely controlled, constant-temperature *wave*. The crest of this wave barely touches the underside of the board as it passes, forming a perfect solder joint without damage to components. Following the soldering operation, the boards are washed and cleaned by the same machine to remove any flux residues. Moisture is evaporated at another drying station. The boards are then ready to be taken to a final assembly workstation.

Dip-soldering is very similar to the above operation, except that some of the operations may not be so automated.

ELECTRICAL RESISTANCE SOLDERING

In bit soldering, the hand-held soldering iron gets its heat due to an electric current flowing through a heating element within the iron. By contrast, a system whereby heat for soldering is developed *outside* of a soldering machine, is known as *electrical resistance soldering*. Actually, sometimes this type of soldering is done with portable electrodes that resemble a soldering iron. Regardless if a fixed machine is used, or if the portable electrode is used, the principle is the same. The abbreviated term, *resistance soldering*, is often used.

The principle of operation is this: when an electric current flows through a resistor, the resistor gets hot. This then is where heat for this type of electrical soldering comes from. However, it differs from the regular type of soldering iron in that *the part to be soldered becomes part of that resistor*; at least on the surface of the part. The metal part to be soldered is placed between two electrodes which are fed an electric current provided by a low-voltage transformer. In this case, carbon electrodes are used in a machine that *resembles* that shown in Fig. 6·25. When the metal part is gripped between the electrodes, an electric circuit is completed and the part is heated.

In the portable version, a carbon pencil is used as one of the electrodes while the metal part being soldered becomes the other electrode. See Fig. 6·24. When the carbon pencil is made to touch the part to be soldered, immediate intense heat is generated *at the point of contact*. Because resistance soldering generates heat directly in the metal area to be soldered, it affords a means of localizing or confining the heat to a particular area. This method is very applicable to the soldering of connectors. By the very principle of operation, and by easily confining the heat, the assembly of electrical connectors is speeded up tremendously.

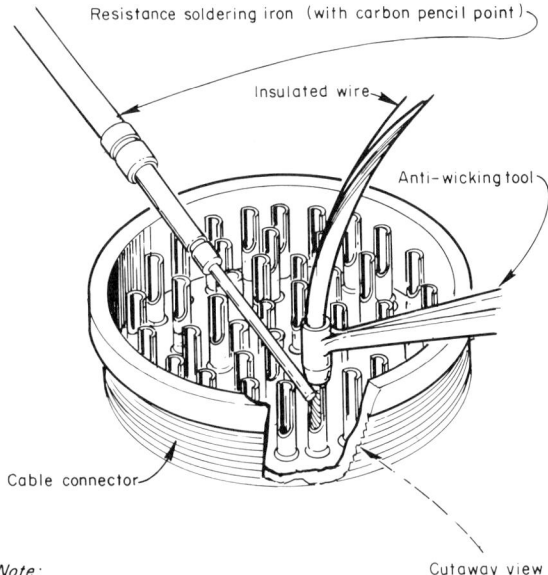

Note: Ground lead of soldering iron attached to pin being soldered. Attached on opposite side of cable connector.

FIG. 6·24 Electrical resistance soldering.

ELECTRONIC WELDING

When soldering two electrical wires together, a third metal (alloy) is needed, namely *solder* itself. This is all right when the basic conductors are made of copper. But electrical solder will not unite copper to a dissimilar metal, such as aluminum, and still maintain a reliable electrical connection. In fact, it is almost impossible to get electrical solder to unite anything except copper. Yet, with the electronics industry presently using a variety of dissimilar metals, it is often necessary to make good, reliable electrical connection between them. One answer to this problem is *electronic welding*. Although welding is not a form of soldering, nevertheless, this subject finds acceptance in this chapter since it substitutes for soldering in many phases of the industry.

Electronics welding finds popularity in miniaturized packages, such as the cordwood module illustrated in Figure 3·1. The technique uses the principle of *forging* the parts together. The parts can be lead to lead or lead to flat surface, without regard to the similarity of the metals involved. An illustration of an electronic welder can be seen in Fig. 6·25.

Such a welding machine generally operates in the following sequence. Electrical energy is accumulated and stored in the power supply. Then the operator brings the joint to be welded up between the two electrodes, causing the joint to touch both electrodes. Then, by actuating an electrical switch, usually a foot control, the electrical energy is rapidly discharged *through the joint being soldered*. The time required for this discharge is approximately 1/1000 of a second, referred to as one millisecond. Since all of the stored energy passes through the joint in such a short period of time, the joint gets hot enough to effect the forging. By using electronic welding, uniform electrical connections are made over and over again, without extra bonding material (solder) or flux as in soldering or other welding techniques.

PERCUSSION WELDING

Percussion welding is another substitute for soldering which is used in the electronics industry. This

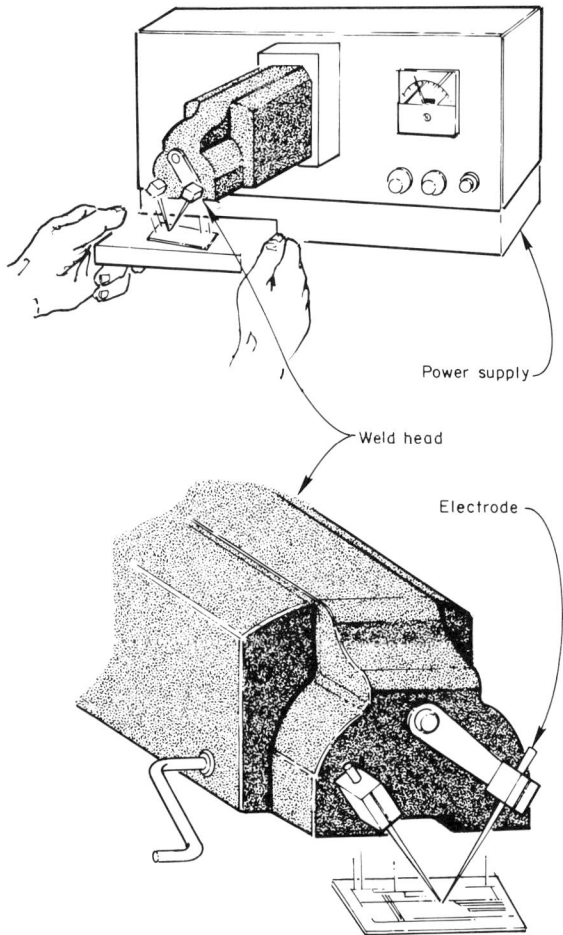

FIG. 6·25 Electronic welder.

type welding is quite similar to electronics welding, using equipment almost identical (at least in principle) to that shown in Fig. 6·25. The main difference being that one of the electrodes vibrates lengthwise and therefore acts like a hammer during the energy discharge period. By doing this, the process of joining metal to metal more closely resembles forge welding, which is the oldest method of welding used by man. All of the advantages and techniques discussed above under electronic welding apply to percussion welding. Materials such as aluminum, steel, nickel, and molybdenum can be welded to themselves or to other alloys; such as copper to aluminum, silver to steel, or any other combination. As with electronic welding, this technique has great advantage where heat is a problem since the heat is concentrated at the point of contact and only for approximately 1 millisecond. *The problem of heat* is always present in microminiature electronics.

6·11 SUMMARY: SOLDERING PRINCIPLES

In summary, soldering can be considered one of the most important phases of electronics fabrication. Hand soldering (bit soldering), the process used by individuals to unite two or more wire conductors to each other, or to unite wire conductors to flat surfaces, is the most popular method of conductor unification. However, machines are often used to accomplish certain soldering operations on devices that can be mass-produced. Electronic welding of one form or another is also used to unite conductors together, especially in the area of microelectronics.

This textbook gives emphasis to hand soldering since it is in this area that everyone connected with electronics assembly or repair will find need for the personal skills and techniques developed during training. Besides, this is one of the skills most demanded by prospective employers.

As to soldering itself, it requires that all surfaces involved be thoroughly cleaned and tinned; that a good hot soldering iron be used and that this heat be transferred properly; and finally, that a solder alloy having the right amount of tin-to-lead content, along with a resin core, be used. Soldering techniques employed on printed circuits are essentially the same as those for general soldering, but require a more exacting approach. Extreme care should be exercised in the application of heat for soldering purposes because printed-circuit boards and many of their mounted components are heat-sensitive.

The following jobs will provide a practical approach to mastering the art of soldering.

EXERCISES

JOB 6·1 How to tin the tip of a soldering iron

OBJECTIVES

1. To learn how to prepare a soldering iron prior to soldering
2. To become acquainted with materials used in soldering

MATERIALS REQUIRED

Power source
117-volt, 60-cycle alternating current

Equipment
Soldering iron with iron-clad tip
Soldering-iron stand
Sponge (do not use a synthetic sponge)

Supplies
Wire solder with resin core

PROCEDURE

1. Lightly dampen the sponge. DO NOT USE TOO MUCH WATER.
2. Place the soldering iron on the stand with the tip pointing toward you (see Fig. 6·3).
3. Plug the line cord of the soldering iron into an appropriate power source, usually 117-volt, 60-cycle alternating current.
4. When the tip of the soldering iron is hot, apply the wire solder to the point of the tip until it melts (see Fig. 6·4). Rotate the soldering iron until the tip is completely covered with a film of solder, all around at the point.
5. *Gently* wipe off the tip of the soldering iron with the sponge. Follow the method illustrated in Fig.

6•26, pulling back with the soldering iron so as to wipe toward the point.

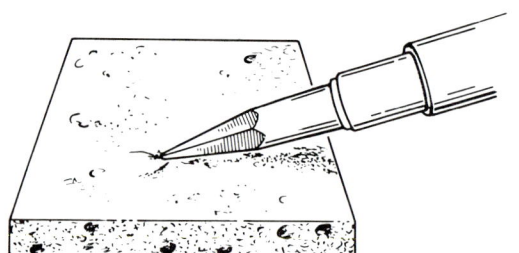

FIG. 6•26 Use of a sponge to wipe the tip of a soldering iron.

6. If necessary, repeat the application of solder.
7. Inspect to see if the film of molten solder is smooth and shiny, completely covering the surface of the area that has had solder applied. Remove any excess solder with the sponge as directed above.
8. If the area covered with solder passes inspection as described in step 7, then this area is said to be tinned.
9. An area that cannot be tinned properly may not have been clean enough, as explained in Sec. 6•1.
10. No filing of the tip is necessary when it is iron-clad. However, if a plain copper tip is to be tinned, it must be filed smooth before application of power to the soldering iron.

QUESTIONS

1. Were you able to see any resin? If so, where?
2. Why must resin be employed when tinning a surface?
3. Explain the importance of cleanliness in connection with the tinning process.
4. What type of solder alloy did you use?
5. Why is the tinning of a soldering-iron tip necessary?
6. How were you able to tell that the tip of the soldering iron had reached the proper temperature?

JOB 6•2 How to tin wire with a soldering iron

OBJECTIVES
1. To learn how to tin the surface of a copper wire
2. To become acquainted with solder identification
3. To become acquainted with the necessities of soldering
4. To learn to recognize a 100-watt soldering iron

MATERIALS REQUIRED

Power source
117-volt, 60-cycle alternating current

Equipment
Soldering iron, 100-watt
Soldering-iron stand
Damp sponge (do not use a synthetic sponge)
Wire stripper

Supplies
Solid copper wire, No. 12 gauge, 12 inches long
Wire solder, 50/50 alloy, 0.062-inch diameter, resin core
Fine sandpaper or emery cloth

PROCEDURE
1. Tin the tip of the soldering iron.
2. Remove the insulation from the wire, using the wire stripper if necessary.
3. If the wire is oxidized, use fine sandpaper or emery cloth to clean the surface.
4. Place the soldering iron on the stand in the manner illustrated in Fig. 6•3.
5. Wipe the tip of the soldering iron with a damp sponge, using the same procedure as outlined in Job 6•1.
6. Apply a small amount of solder to the soldering-iron tip. This will ensure that enough heat is transferred to the wire to be tinned by providing a soft cushion of molten metal into which the wire can be partly imbedded.

7. Lay the clean copper wire on top of the soldering-iron tip (see Fig. 6·3).
8. Apply solder to the hot wire directly. DO NOT LET THE SOLDER WIRE TOUCH THE TIP OF THE SOLDERING IRON. See Fig. 6·3.
9. Apply solder to all clean areas of the copper wire to be tinned. This whole procedure of tinning must be accomplished quickly to avoid burning the insulation not removed. Further, prolonged heat application will cause the molten resin to drain away, which in turn will prevent the solder from spreading smoothly when the copper wire is removed.
10. Pull the solder wire away.
11. Carefully lift the copper wire from the soldering iron.
12. Examine the tinned copper wire. No sharp points should be visible. The film of solder should be smooth and shiny.
13. If a repetition of the tinning procedure is necessary, start with step 5.

QUESTIONS
1. List the basic soldering requirements employed in tinning.
2. Define *tinning*.
3. Explain why a 100-watt soldering iron develops more heat than a soldering iron that is rated at 25 watts.

JOB 6·3 How to splice wires

OBJECTIVES
1. To learn how to join wires
2. To appreciate soldering
3. To learn to recognize a 25-watt soldering iron

MATERIALS REQUIRED

Equipment
Soldering iron, 25-watt
Assorted hand tools
Damp sponge (do not use a synthetic sponge)

Supplies
Solid copper wire, No. 20 gauge, three lengths (12 inches per length)
Wire solder, 60/40 alloy, No. 20 gauge, resin core
Fine sandpaper or emery cloth

PROCEDURE
1. Tin the soldering iron.
2. Remove all insulation from the two ends of the copper wire.
3. Tin the clean ends of the copper wire.
4. Form the copper wire into rings and mechanically join the ends with three different splices as illustrated in Fig. 6·5.
5. The wrap of each splice should be as tight as possible. However, do not damage the wire by applying too much pressure.
6. Each wrap of the hook splice should not be more than one complete turn.
7. Place the wire ring so that it is steady and self-supporting.
8. Wipe the soldering iron with the damp sponge.
9. Apply a small amount of solder to the tip of the soldering iron for proper heat transfer.
10. Apply the soldering iron to one side of each splice while applying the wire solder to the opposite side of the splice (refer to Fig. 6·8).
11. Do not allow the wire ring to move until the soldered joint has cooled enough for the solder to solidify.
12. Do not blow cold air on the soldered joint to speed up the cooling process because this will lower the quality of the soldering job.
13. Inspect the soldered joint to see if it is smooth and shiny. No sharp points should be visible.

QUESTIONS
1. Compare this job with Job 6·2.

2. Why use a small-diameter wire solder with a 25-watt iron?
3. Define soldering.
4. Explain why a mechanical connection not soldered may interfere with an electric current flowing through a splice.
5. How does soldering reduce the electrical resistance at a wire joint?

JOB 6•4 How to make a set of practice soldering boards

OBJECTIVES
1. To provide each student with practice soldering devices that contribute to the motivation for *practice*
2. To give the student an opportunity to use hand tools to create something useful

MATERIALS REQUIRED

Equipment
Woodworking hand tools
Bench vise
6-inch, diagonal-cutting pliers
Tape measure
Scissors, 8-inch size

Supplies
2 ¼-inch plywood, 6 × 10 inches
4 Soft pine-wood blocks, ¾ × 1 × 6 inches
4 Wire staples
8 Thumbtacks
2 #14 solid copper wire, bare, 10 inches long
1 7-mil sheet copper, 5 × 10 inches
   Box nails, ¾ inch long
   Wood glue

PROCEDURE
1. If the material supplied is not precut to exact dimensions, cut it now.
2. Assemble the blocks and plywood as illustrated in Figs. 6•27 and 6•28. The plywood is to serve as the baseboard. Use wood glue between the blocks and baseboard. Nail the assembly from the baseboard into the blocks.
3. Mount the #14 solid copper wire on the board, using wire staples as illustrated in Fig. 6•27a. Keep the wire tight and straight. (*Note:* If the wire supplied has plastic or similar insulation, remove the insulation prior to mounting the wire on the board.)
4. Mount the sheet copper on the second board, using thumbtacks as illustrated in Fig. 6•28a. Keep the sheet copper tight and without wrinkles.
5. These two board assemblies will serve as soldering practice boards in future jobs.

QUESTIONS
1. What does *7 mil* refer to as in the description of the sheet copper used?
2. List by *correct name* the woodworking tools that you used.
3. If the sheet copper becomes damaged with use, will it be necessary to make another complete practice board? Explain.

JOB 6•5 How to practice soldering wire

OBJECTIVES
1. To learn the fundamentals of soldering through practice
2. To learn the fundamentals of wrapping wire around a terminal

MATERIALS REQUIRED

Equipment
Hand tools for electrical assembly
Soldering iron, approximately 50 watt
Soldering practice board (see Fig. 6•27)

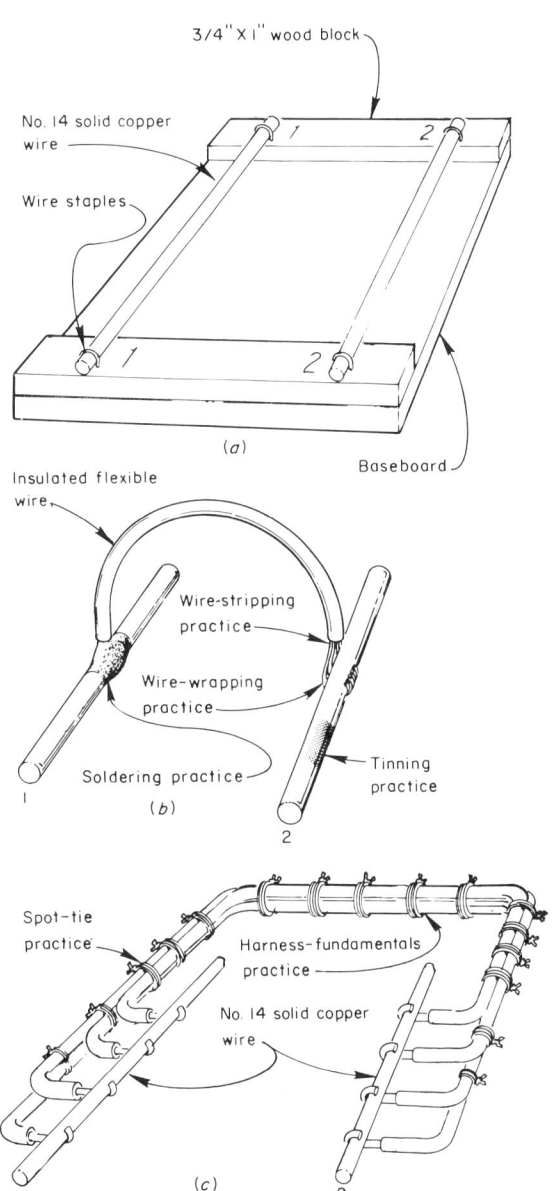

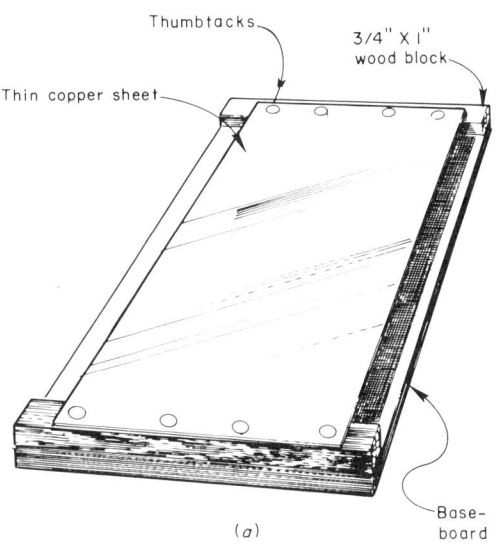

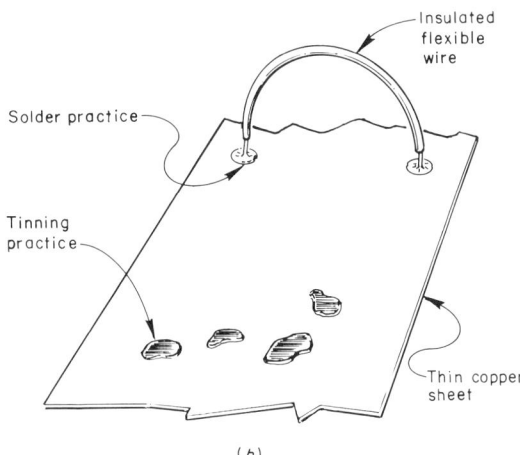

FIG. 6·27 Soldering practice board. (*a*) Basic board. (*b*) Use of board for solder practice. (*c*) Use of board for harness-making practice.

FIG. 6·28 Solder-control practice board. (*a*) Basic board. (*b*) Possible applications for printed-circuit soldering practice.

Supplies

Wire solder, 60/40 alloy, No. 20 gauge, resin core

Fine sandpaper

Assorted wire types, with assorted insulation and wire diameters (lengths to be at least 12 inches)

PROCEDURE

1. Tin the soldering iron.
2. Remove all insulation from the ends of the wires. Use the wire strippers and sandpaper as neces-

105

sary. Care should be used when removing enamel-type insulation with the sandpaper. Too much use of the sandpaper will reduce the diameter of the wire and render it useless for normal electrical use. (See Fig. 7•11.)
3. Tin the ends of the wires to be soldered.
4. Tin the total length of the #14 solid copper wires on the soldering practice board. Be sure to tin the wire completely around.
5. Mount the wires supplied with this job between the two #14 solid wires of the practice board. See Fig. 6•27b. First wrap the tinned wire ends as shown in Fig. 6•27b, and then solder these ends to the #14 wire.
6. Your soldering results should compare favorably with that illustrated in Fig. 6•11. Soldering such as illustrated in Fig. 6•12 is not acceptable.
7. If you are having trouble, review the soldering procedures listed in Job 6•3. If you continue to experience difficulty ask your instructor for help.
8. After having mounted the wires supplied to you on the soldering practice board as directed above, remove them by unsoldering them.
9. Cut off the ends of the wires and do this whole job over again, starting with step 1. *A good technique for soldering can only be acquired through practice.*
10. Continue repeating all steps listed above until your instructor directs you to continue with the next job.

QUESTIONS
1. What soldering operations gave you the most trouble?
2. In your opinion, what is the most essential part of soldering?
3. Summarize the factors necessary for a good soldering job.

JOB 6•6 How to practice soldering to sheet copper

OBJECTIVES
1. To learn the fundamentals for soldering on printed-circuit boards
2. To learn to control the spread of solder

MATERIALS REQUIRED

Equipment
Hand tools for electrical assembly
Soldering iron, approximately 50-watt
Soldering practice board (see Fig. 6•28)

Supplies
Wire solder, 60/40 alloy, No. 20 gauge, resin core
Assorted wire types, with assorted insulation and wire diameters (lengths to be at least 12 inches)

PROCEDURE
1. Tin the soldering iron.
2. Remove all insulation from the ends of the wires. Use the wire strippers and sandpaper as necessary. (Careful with too much use of the sandpaper.)
3. Tin the ends of the wires to be soldered.
4. Tin various spots on the surface of the sheet copper mounted on the practice board. See Fig. 6•28b. The tinned spots should be smooth and shiny. The overall thickness of the tinned spots should appear as though the spots were painted on with thick paint. The spots should not be thick and bulky. *The thinner the tinned spots on the sheet copper, the better.* There should be no points sticking up after you remove the soldering iron. Do not blow on the hot tinned spot to cool it off. Let it cool by itself.
5. Solder the wires prepared in steps 2 and 3 to the sheet copper as illustrated in Fig. 6•28b. Solder these wires on the spots tinned during step 4. Again, the finished soldering job will be smooth and shiny (bright). Contour soldering as shown in Fig. 6•11 and Fig. 6•18 shall prevail.
6. If you are having trouble, review the soldering procedures listed in Job 6•3. If you continue to experience difficulty ask your instructor for help.
7. After having mounted the wires supplied to you onto the sheet copper of the practice board as directed above, remove them by unsoldering them.
8. Cut off the ends of the wires and do this whole job over again, starting with step 1. *Remember: "practice makes perfect."*

9. Continue repeating all steps listed above until your instructor directs you to continue with the next job.

QUESTIONS
1. What soldering operations gave you the most trouble?
2. What relation does this job have to soldering on printed-circuit boards?

JOB 6·7 How to remove solder from a soldered joint

OBJECTIVES
1. To learn the techniques of removing solder from a soldered joint
2. To appreciate the presence of resin in solder
3. To recognize a bad soldered joint

MATERIALS REQUIRED

Equipment
Same as in Job 6·3

Supplies
Wire ring made in Job 6·3

PROCEDURE
1. Purposely add an excessive amount of solder to the hook splice soldered in Job 6·3. Allow the solder to melt all around the joint so that when it cools it resembles the soldering of Fig. 6·12.
2. Wipe the soldering iron with the damp sponge as in previous jobs.
3. Place the soldering iron on the stand.
4. Hold the wire ring so that the soldered joint rests on top of the hot soldering-iron tip.
5. Solder should now *drain* onto the hot tip.
6. Wipe the soldering-iron tip clean with the sponge.
7. Notice that a small sharp point of solder is present at the hook splice where the solder drained to the soldering-iron tip. This is caused by a lack of resin. Resin helps solder to spread smoothly.
8. Repeat steps 4 to 6 until most of the solder has been drained away.
9. In order to remove the final sharp point of solder, apply a little of the resin-core solder to the soldering-iron tip and to the joint from which you are removing the solder. The object is to replace some of the resin that drained away, even at the expense of adding solder. Some of this additional solder will drain onto the hot tip anyway. The resin should help to reduce the sharp solder point, allowing the solder to spread smooth. It may be necessary to repeat this procedure several times.

QUESTIONS
1. Why is it important to know how to remove solder from a soldered joint?
2. Explain any problems encountered in the first step of this job.
3. Were you able to remove the sharp solder point completely?
4. State the advantages of using resin-core solder for soldering.
5. Why must the tip of the soldering iron be placed under the soldered joint when removing solder?

JOB 6·8 How to inspect the quality of a soldered joint

OBJECTIVES
1. To be able to evaluate one's own progress in soldering
2. To appreciate a good soldered connection

MATERIALS REQUIRED

Equipment
Hand-held magnifying glass
Lamp or suitable lighting

Supplies
Wire ring made in Job 6·3

PROCEDURE
1. Carefully compare the illustrations of Figs. 6·11 and 6·12.
2. With the aid of the magnifying glass, inspect all splices soldered to make the wire ring of Job 6·3.
3. See if the following factors prevail: soldering is smooth, bright, and shiny; the solder feathers smoothly onto the untinned portions of the wire; there are no sharp points anyplace; you can clearly see the contour of all wires soldered; a thin layer of solder covers all areas of the soldered joint.

4. If any of the soldered joints (splices) does not meet all of the factors specified in step 3, the soldered joint is not satisfactory.
5. The above inspection procedure should be followed throughout your training period whenever you are required to solder. After that, experience will normally point out unacceptable soldered joints at a glance. Further, while employed as an electronics assembler, trained inspectors will constantly check your work. If they reject too many of your soldered joints, you probably will lose your job.

QUESTIONS
1. After making the above inspection of the soldered joints on your ring, do you think it best to resolder these splices?
2. How many of the factors specified in step 3 did your soldering job on the ring fail? Which factors?
3. If you inspected a soldered joint made by a close friend of yours and it did not meet all of the factors pointed out in step 3, would you ask him to do it over again? If not, would you do it for him? Why?

JOB 6·9 How to install components on printed-circuit boards

OBJECTIVES
1. To appreciate a printed-circuit board
2. To learn to solder on printed-circuit boards
3. To appreciate specifications

MATERIALS REQUIRED

Equipment
30-watt soldering iron
Small hand tools
Ink eraser or lead cleaning tool
Razor blade, single edge

Supplies
Solder, 60/40 alloy, 0.036-inch diameter, resin core
Printed-circuit board for practice
Assorted resistors for solder practice
Assorted semiconductor diodes for solder practice

PROCEDURE
1. Compare the printed-circuit board supplied with the printed-circuit board illustrated in Fig. 3·2.
2. Use the razor blade to pry up a section of the copper-foil conductor. Inspect the thickness of the conductor.
3. With the ink eraser, clean a few of the pads around the plated-through hole.
4. Practice tinning these clean pads. This will help you to acquire techniques in applying solder to a printed circuit.
5. Clean the leads of a resistor. Do not rub too hard, just enough to remove any foreign material from the tinned surface of the lead (see Figs. 7·9 and 7·10).
6. Bend the leads of the resistors so that they may be inserted into the plated-through holes of the printed-circuit board. Follow the principles illustrated in Figs. 8·20 and 8·21.
7. Clean a pair of pads that have not been tinned and insert the resistor leads into the plated-through hole. Clinch and cut the leads as illustrated in Fig. 8·19.
8. Solder the resistor leads to the pads from the clinched-lead side of the printed-circuit board. Allow the solder to fill the hole completely as illustrated in Fig. 8·19.
9. The completed solder job should appear as illustrated in Fig. 6·18.
10. Practice mounting resistors on printed circuits until you become proficient at this operation.
11. Using the techniques developed above, proceed to install the semiconductor diodes on the printed-circuit board. Refer to Fig. 8·22 for the specifications relating to mounting heat-sensitive components. Be sure to use a heat sink as illustrated in Fig. 6·14.

QUESTIONS
1. Describe a printed-circuit board.
2. List the specifications to be observed when mounting a resistor on a printed-circuit board.
3. List the specifications to be observed when mounting a semiconductor diode on a printed-circuit board.

4. List the advantages and disadvantages of a printed-circuit board.
5. List the problems and their solutions experienced in performing this job.

JOB 6·10 How to replace components on printed-circuit boards

OBJECTIVES
1. To learn to replace components on printed-circuit boards
2. To appreciate a soldering aid

MATERIALS REQUIRED

Equipment
30-watt soldering iron
Small hand tools, including soldering aid

Supplies
Solder, 60/40 alloy, 0.036-inch diameter, resin core
Printed-circuit board used in Job 6·9 with resistors and diodes mounted
10-inch length of small-diameter copper braid

PROCEDURE
1. Heat and tin the soldering iron in the normal manner.
2. Apply a small amount of solder to the tip of the soldering iron for heat transfer.
3. Apply the tip of the soldering iron to the soldered, clinched lead of a resistor, using the soldering approach. This will melt the solder in the plated-through hole.
4. Use the soldering aid to pry the resistor lead out of the hole. A soldering aid is shown in Fig. 5·3. The long-nose pliers can also be used to pull the lead out.
5. To simplify the operation, cut the resistor in half, or cut one of its leads with a pair of diagonal cutters and unsolder one lead at a time. See Fig. 6·19.
6. Wipe the tip of the soldering iron and proceed to remove as much solder as possible from the plated-through hole in preparation for mounting a new resistor. In this solder-removal operation, the copper-braided shield supplied is helpful in drawing off the molten solder. Apply the braid in a manner similar to applying solder when soldering. The molten solder will travel up the braid. See Fig. 6·20.
7. Use the same procedures for removing the semiconductor diodes. If the diode is defective, no heat sink will be necessary. However, if the diode is to be tested and used again, use of the heat sink is mandatory.
8. When the solder is removed from the plated-through hole, the printed-circuit board is ready for installation of another component.

QUESTIONS
1. Why must the soldering iron be in a tinned condition when unsoldering a joint?
2. Why is solder used on the tip of a soldering iron in order to transfer heat properly to a joint already soldered?
3. List the advantages and disadvantages of a soldering aid as shown in Fig. 5·3.
4. Explain the procedure involved in removing solder with a copper braid.
5. Why is it acceptable to cut a resistor in half when replacing it?
6. Compare this job with Job 6·7.

7

WIRE PREPARATION AND HARNESS ASSEMBLY

Soft copper wire is used to interconnect resistors, capacitors, inductors, etc. These wires normally are covered with insulation such as enamel, plastic, or rubber.

A wire provides for the movement of electrons in an electrical circuit. Insulation covering the wire serves to confine the moving electrons (current) within the wire. In working with electricity one must know how to handle wire, prepare it for use, and its limitations and applications.

7·1 TYPES OF WIRE

Wire is manufactured as a single thread of solid copper. A solid wire is shown in Fig. 7·1. Wire is also manufactured as groups of copper threads twisted together, as illustrated in Fig. 7·12a.

Solid wire is used for the leads of components such as resistors and capacitors. It is also used in the coils of transformers. Solid wire is easy to handle, but cannot withstand much flexing.

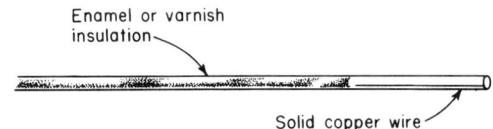

FIG. 7·1 Solid copper wire insulated with enamel or varnish materials. A popular insulation within this class is known as *Formvar*.

The group of wire threads seen in Fig. 7·12a is known as a *flexible wire*. The individual wire threads are referred to as *strands*. Flexible wires are used for the leads of transformers and for other applications where flexibility is important.

7·2 TYPES OF WIRE INSULATION

Most wires are protected by an insulating covering to prevent short circuits. Solid wires, when insulated, are protected by enamel paint or varnish (see Fig. 7·1) or by materials such as plastic or rubber. Flexible stranded wire is never insulated with enamel or varnish. Stranded wire has a more elaborate covering, of rubber, plastic, woven fiber glass, or similar material (see Fig. 2·1).

Some electrical conditions require that several layers of insulation cover the wire (see Figs. 7·2, 7·3, and 7·6).

7·3 WIRE BRAIDS AND SHIELDS

A *braid* is composed of woven segments of stranded wire. Braids are used as conductors of heavy current. They are also used to *shield* a conductor from electrical interference. The braid shown in Fig. 7·4 is flat in form. The flat braid is used as a conductor of heavy current and is highly flexible.

The application of a braid for shielding a wire is seen in Fig. 7·5. The braid can also be used as a second conductor, as in a *coaxial cable*, illustrated in Fig. 7·6.

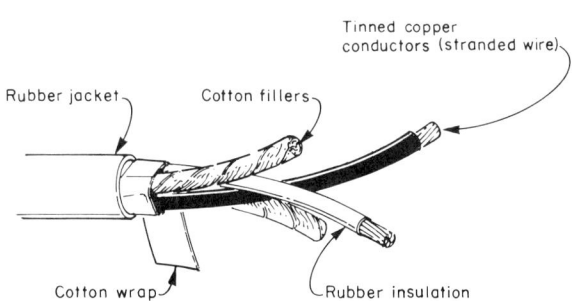

FIG. 7·2 Multiple-insulated wire conductors. (Electrical power and control cable.)

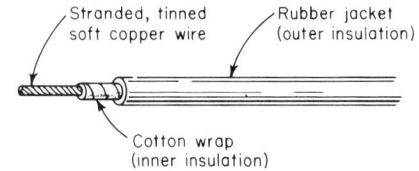

FIG. 7·3 Test lead wire. (*Note:* Vinyl or Teflon is also used often for insulation.)

FIG. 7·4 Flat braid.

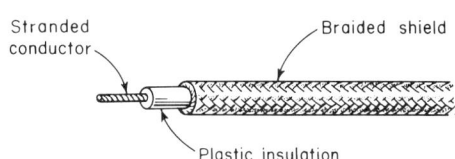

FIG. 7·5 Shielded plastic-covered wire.

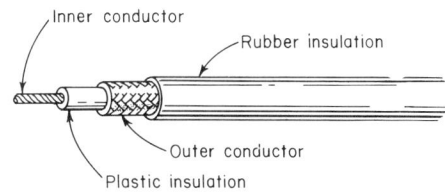

FIG. 7·6 Coaxial cable.

7·4 MULTIPLE-WIRE CABLES

When two or more wires are bundled together they form a cable. Cables are often developed by electrical or electronics assemblers and are also prefabricated by the manufacturers of wire products. This chapter will illustrate only the prefabricated cable. Typical multiconductor cables are shown in Figs. 7·7 and 7·8.

7·5 PREPARING WIRES FOR SOLDERING

Wires may be united by soldering. The art of soldering involves the unification of the copper conductors, and not the insulation. Therefore, all in-

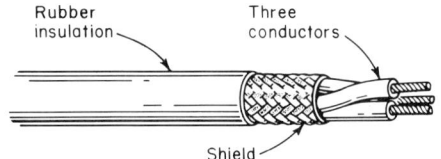

FIG. 7·7 Shielded three-wire cable.

FIG. 7·8 Special-purpose interconnecting cable.

sulation in the immediate area to be soldered should be removed from a wire before soldering. All oil, dirt, and oxidation must also be removed.

CLEANING SOLID WIRES

Solid-wire leads used on resistors can be cleaned with an ink eraser as illustrated in Fig. 7·9, or with a special lead-cleaning tool that uses flat copper braid as the abrasive surface. See Fig. 7·10.

To remove the insulation from an enamel-covered wire, a more abrasive cleaner must be employed. A piece of fine sandpaper works nicely, as illustrated in Figure 7·11. However, care must be taken not to remove too much copper when sanding the enamel off, or the diameter of the wire will be reduced, weakening the wire and also restricting its current-carrying capabilities. Paint-type insulation can also be removed (*stripped*) by using special chemicals which are available for this purpose. However, the use of chemicals should always be undertaken with care. A special problem that may arise by the use of chemicals for wire stripping is that the corrosive action involved may continue long after the stripping operation, eventually working its way through the conductor as well, and thereby causing the wire to break.

WIRE STRIPPING

As indicated above, removal of insulation from a wire is called *wire stripping*. Wire stripping refers to the removal of any type insulation which normally protects the copper conductor from becoming

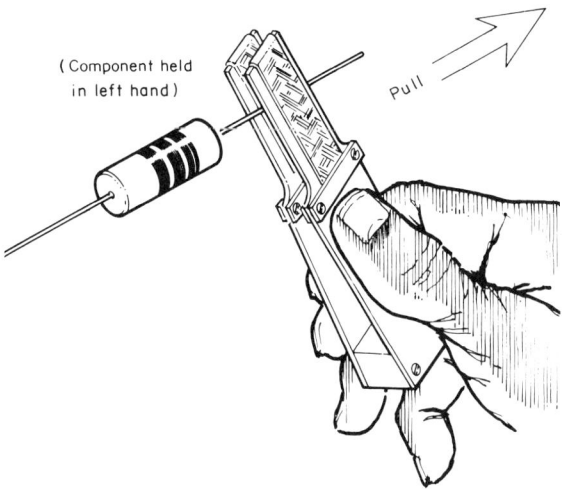

FIG. 7·10 Use of the lead-cleaning tool.

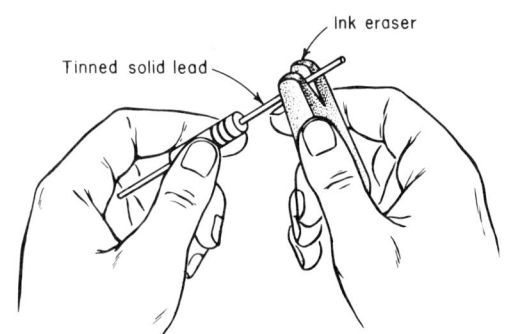

FIG. 7·9 Cleaning component tinned leads.

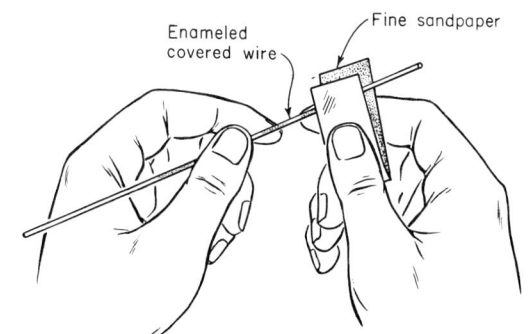

FIG. 7·11 Removing enamel insulation.

a *short circuit*. Figure 7·12a illustrates a *flexible* wire that has been properly stripped of its insulation. By contrast, an unacceptably stripped flexible wire is illustrated in Fig. 7·12b. Reasons why the wire stripping job is unacceptable are indicated in the illustration.

Tools and techniques used in wire stripping operations are illustrated in Figs. 5·7, 7·13, and 7·14. Each of these representative types has its merits. However, NASA specifies that wire strippers to be used on products designed for use in the aerospace industry must be of the precision or thermal types.

7·6 SOLDERLESS CONNECTORS

Fabrication of electrical devices does not require that all wires be soldered to terminals or to other wires. Solderless terminals are also used, as well as crimped and screw-type lugs. Figure 7·15 illustrates a workstation where *solderless wire wrapping* is done on special terminals designed for this type of electrical connection. This type of operation basically requires that a stripped wire be tightly

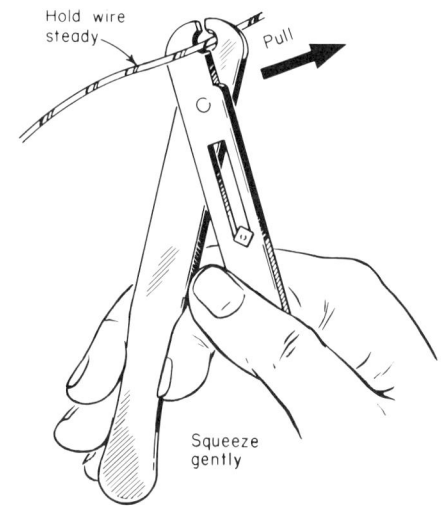

FIG. 7·13 The mechanical wire stripper.

wrapped around a square terminal. A specially designed Wire-Wrap gun is used in order to insure a uniform tight wrap. (*Note:* "Wire-Wrap" is a registered trademark of the Gardner-Denver Company.) The gun may be powered by an electric motor, by an air pressure (*pneumatic*) motor, or manually. Since use of the motor greatly expedites making an electrical connection, the wire leads to be wrapped are often precut and prestripped. For fast access to these wires, they are stored in convenient wire bins.

Solderless terminal lugs require the use of a *crimping tool*. See Fig. 7·16. In this case, the conductor is inserted into the lug and the lug squeezed firmly to make the electrical connection. The lug uses a screw-type terminal strip to make further electrical connections, as shown in Fig. 7·17. Crimping is also used to attach wires to pins used in some cable connectors. Crimping is further used to splice two wires to each other by using union fittings designed for this purpose. Crimping tools are available in many sizes and formats. Some are even powered by air pressure. A cable-connector pin-crimping tool is illustrated in Fig. 7·18. Crimping tools should be considered hazardous to the hands since great pressure is involved.

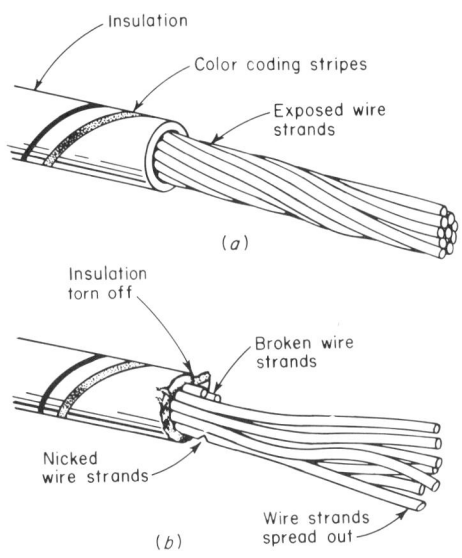

FIG. 7·12 Flexible stranded wire. (a) Acceptable stripped flexible wire. (b) Unacceptable stripped flexible wire.

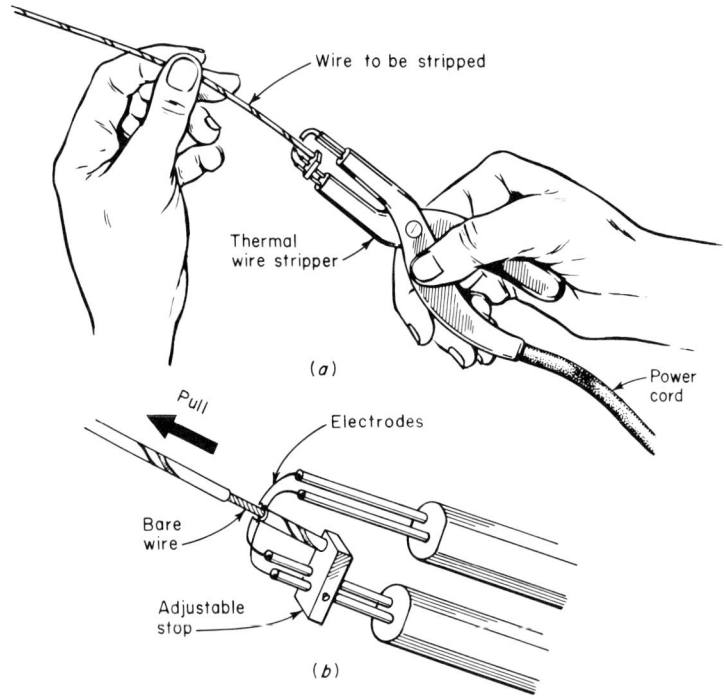

FIG. 7·14 The thermal wire stripper. (a) How to handle a thermal wire stripper. (b) Stripping wire of its insulation.

7·7 TINNING WITH A SOLDER POT

Prior to soldering, a wire must be *tinned*. To tin a wire is to apply a fine film of solder alloy on its surface. (More information on tinning is given in Chap. 6.) One method used in tinning employs a *solder pot* and falls within the category of wire preparation. For this reason tinning with a solder pot is introduced here.

A solder pot is a metal container heated electrically to melt a solder alloy. Such a solder pot is shown in Fig. 7·19. Also shown is a bar of solder alloy before melting. The soldering flux shown in Fig. 7·19 is a liquid resin, used as a cleaning agent to prepare a wire for tinning.

A wire to be tinned is first stripped of its insulation. It is then immersed in the liquid resin and finally in the molten solder. The process of tinning a wire with a solder pot is illustrated in Fig. 7·20.

7·8 CONNECTING LEADS TO BRAIDED SHIELDS

The braid of a shielded cable is required to be connected to the electrical ground of most circuits. Because of this, a shield must terminate in a lead that can be soldered into a circuit. A flexible stranded wire can be attached to the braid, or the braid itself can be processed to serve as its own lead.

A flexible stranded wire can be attached to the braid by one of two methods. The stranded lead can either be soldered to the braid or attached by pressure. When the flexible lead is soldered to the braided shield, care must be taken not to damage the insulation between the inner conductor and the shield (see Fig. 7·21a).

The pressure method of attaching the shield lead is preferred since no heat is necessary for application. However, a special crimping tool and special supplies are necessary. A shield lead attached by crimping is illustrated in Fig. 7·21b.

When the braided shield is itself used as its own lead, the individual wire strands can be combed out with a pointed instrument, and these strands then twisted to form the ground lead (see Fig.

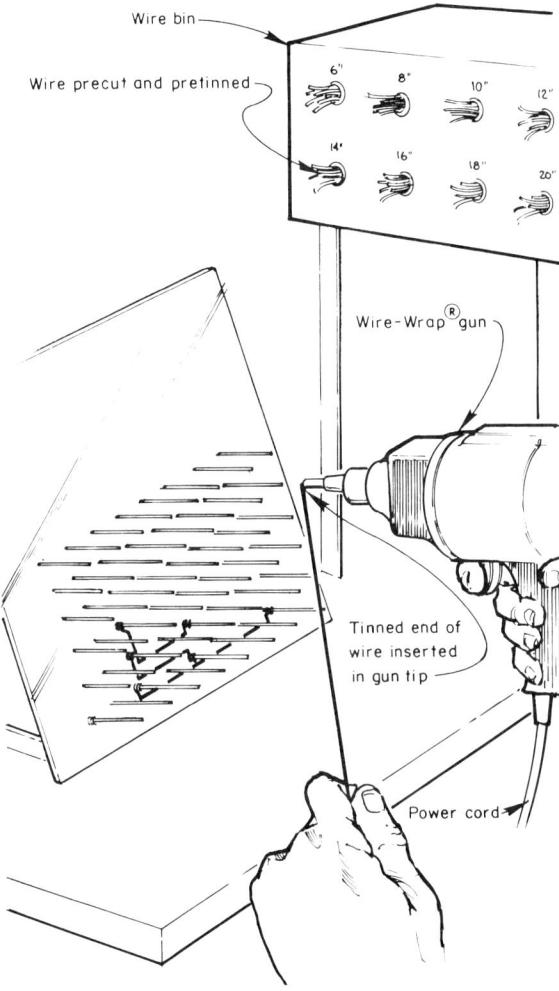

FIG. 7·15 Solderless wire connections.

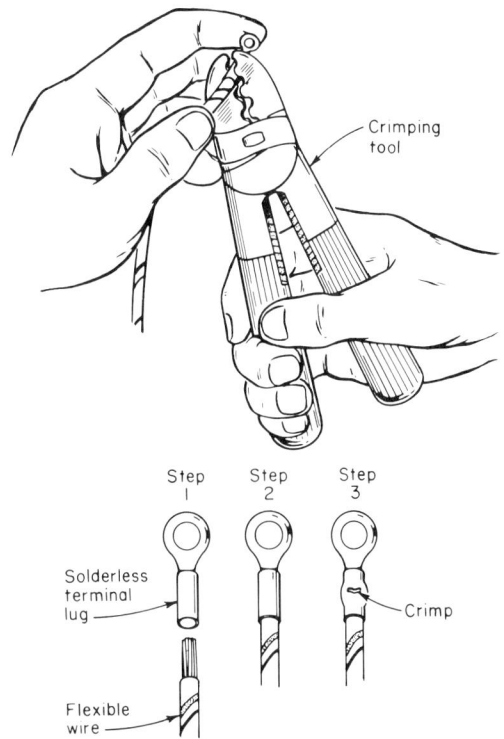

FIG. 7·16 The use of a crimping tool.

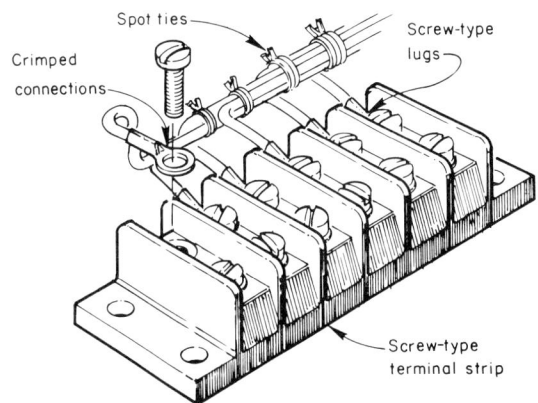

FIG. 7·17 Complete solderless connections, using crimping and screw-type lugs.

7·22). Another way to separate the inner conductor from the braided outer conductor is illustrated in Fig. 7·23. This is known as the *poke-through* method for producing a ground lead in a shielded cable. Whenever work has to be done on braided shield coverings, care must be taken not to break any of the braid strands.

7·9 WIRE SIZES

The current-carrying capability of a solid-wire conductor is determined by the diameter of the wire. A wire with a large diameter is able to handle more current than a wire with a small diameter. That is why solid wire should not be nicked or sanded

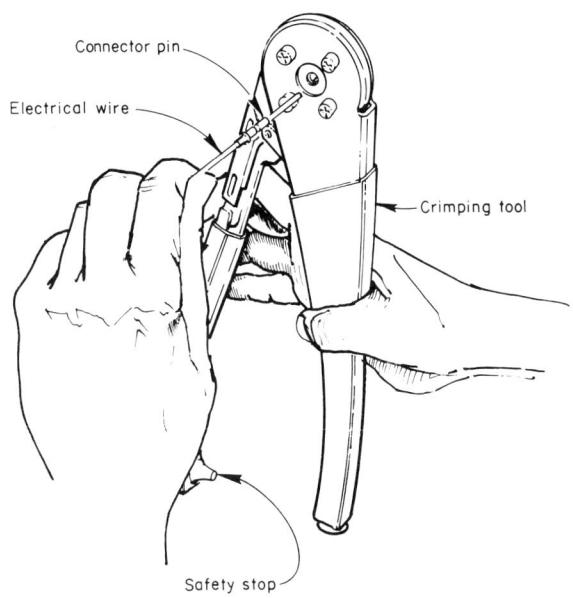

FIG. 7·18 Cable-connector pin-crimping tool.

down, and why wire strands from flexible or braided conductors should not be broken. Besides, damaging wire or breaking of strands weakens the conductor; and this may cause the conductor to break during flexing.

SOLID-WIRE DIAMETERS

The diameter of solid wire is designated by code numbers. For example, the solid wire often used in home wiring has a diameter of 0.08081 inch, but is simply identified as a No. 12 wire. Fine wire, often as fine as human hair, has a diameter of 0.00157 inch. It is designated as a No. 46 wire. Capacitors and resistors often have No. 18 or No. 20 solid wire for their leads. The diameter of a No. 18 wire is 0.04030 inch, and the No. 20 solid wire has a diameter of 0.03196 inch.

STRANDED-WIRE DIAMETERS

Stranded wire is flexible because it is made up of many individual strands of fine solid wire. The more fine strands it contains, the greater the flexibility of the stranded conductor. The overall diameter of the stranded wire is determined by the number of strands times the cross-sectional area of each strand.

A No. 22 size stranded-flexible-wire conductor is composed of 10 strands of No. 32 wire. The No. 22 size can also be composed of other combinations of strands, for example, 16 strands of No. 34 wire or 26 strands of No. 36 wire. All of these combinations *approximately* equal the cross-sectional area of number 22 solid wire.

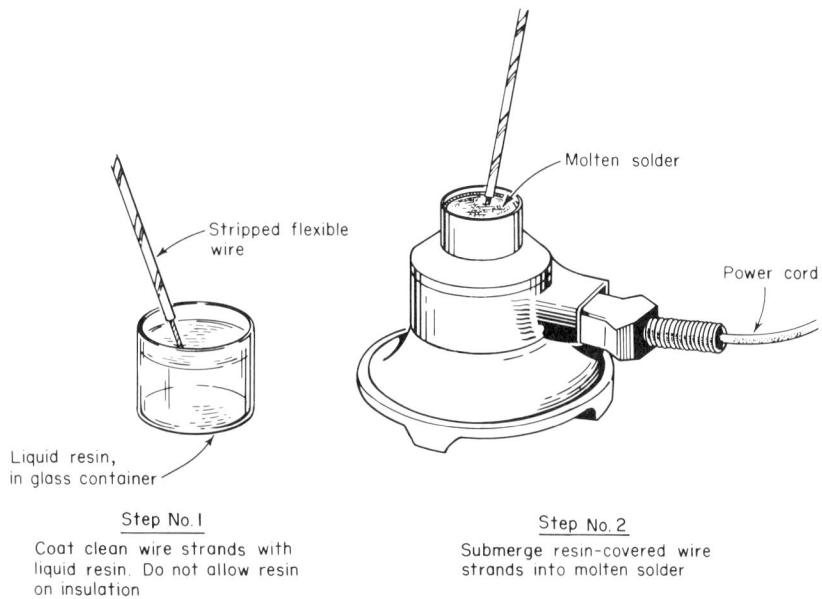

FIG. 7·20 Tinning wire with a solder pot.

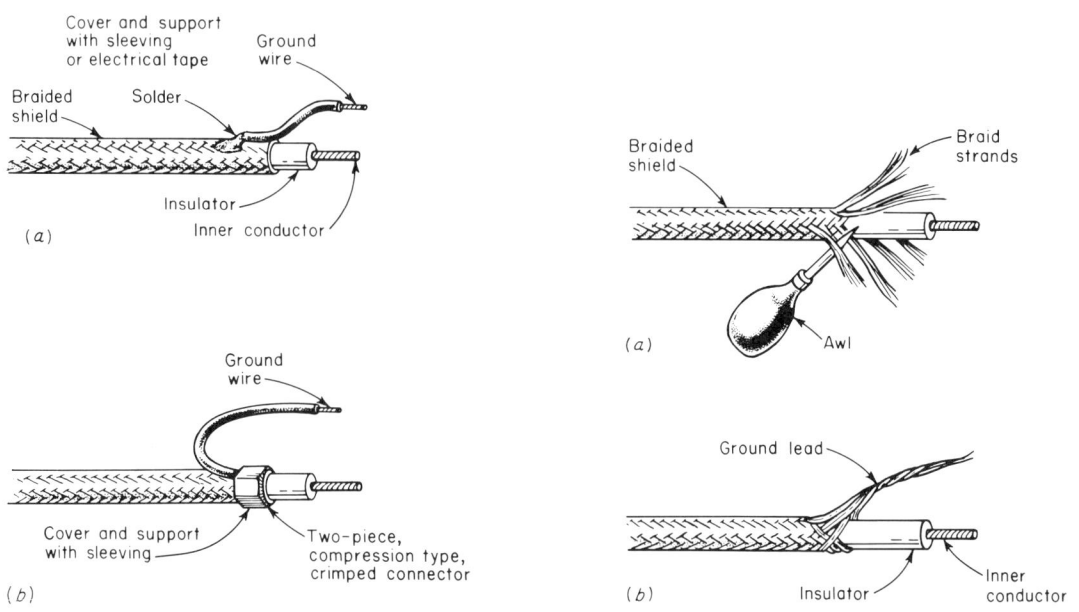

FIG. 7·21 Attaching the ground wire. (a) Soldering the ground wire. (b) Crimping the ground wire.

FIG. 7·22 (a) Combing the braid strands. (b) Use of the braid strands for the ground lead.

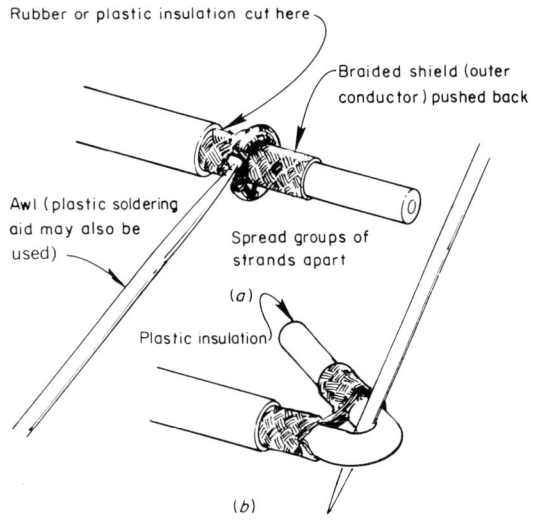

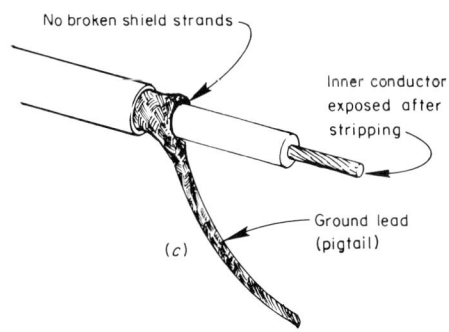

FIG. 7·23 The *poke-through* method for producing a ground lead in a shielded cable.

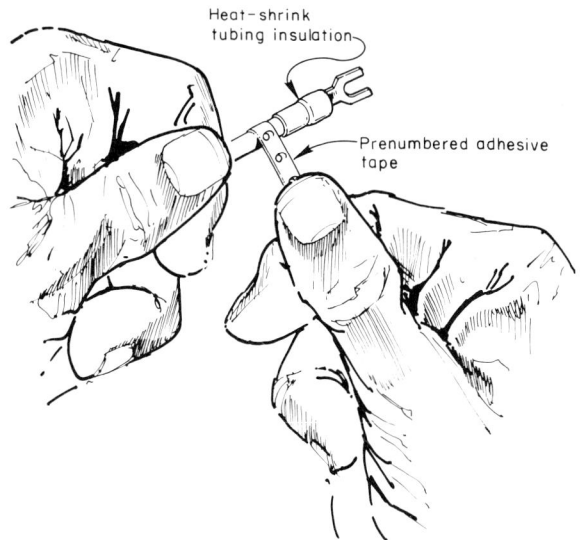

FIG. 7·24 The numbering of leads.

7·10 WIRE COLOR CODING AND IDENTIFICATION

Insulated wire is often identified by a number stamped on its insulation, or by the use of prenumbered sections of adhesive tape. This numbering may appear periodically throughout the length of the wire, or it may appear only at the ends. Figure 7·24 illustrates the numbering of wire leads.

Another method used to identify wires is by color coding. In this method, the code may involve the impregnating of the insulation with solid color dyes, or it can be coded in the form of color stripes extending the length of the wire. See Fig. 7·12.

7·11 HARNESSING WIRES INTO BUNDLES

Wires that interconnect various electrical and electronic components are often routed in a group. This group of wires, tied into a neat bundle, is known as a *wire harness*. A typical wire harness is illustrated in Fig. 7·25. A wire harness may be developed in a chassis, or it may be prefabricated on a jig board. (See Fig. 7·40.) A wire harness is similar to a multiple-wire cable. However, a wire harness is not usually a linear cable, but provides for the extrusion of individual wires throughout its length as required by the circuits associated with this harness. Normally, a wire harness terminates at a terminal board or at a cable connector.

7·12 PRODUCTION TECHNIQUES OF WIRE HARNESSES

Several devices are employed for binding wires into a harness. Among these are spot ties, cable lacings, nylon belts, and wrapping tape.

SPOT TIES

Spot ties are individual ties produced with linen lacing cord or with nylon lacing tape. Spot ties

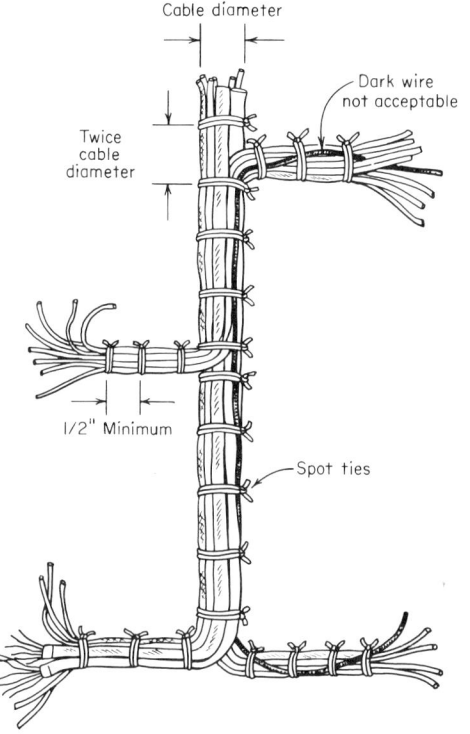

FIG. 7·25 A typical wire harness.

are used on the harness of Fig. 7·25. Linen lacing cord and nylon lacing tape are shown in Fig. 7·26.

Spot ties are illustrated in detail in Fig. 7·27. The principles of a square knot are followed to secure the tie. The steps necessary for the production of a spot tie are shown in Fig. 7·28. Special attention is called to the square knot. Notice how, in Fig. 7·28b, A and B lie alongside each other. The same is true of cords C and D. Compare the square knot of Fig. 7·28b and the knot used to finish the spot tie of Fig. 7·27.

CABLE LACING

Cable lacing is the term applied to the production of many ties with a single length of lacing cord or tape. Cable lacing is illustrated in Fig. 7·29. When lacing cable, care should be exercised that each tie is snug and self-locking; otherwise the lacing will be inefficient. To guard against their slippage, lacing cord and tape are impregnated with a small amount of wax.

NYLON CABLE TIES

Nylon cable ties are small belts or straps used to bind a wire harness individually. Figure 7·26 shows

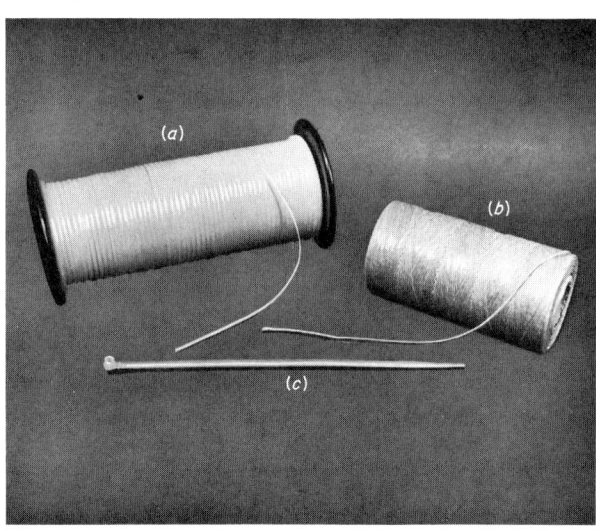

FIG. 7·26 Materials used to lace or spot-tie. (a) Nylon lacing tape. (b) Linen lacing cord. (c) Nylon cable tie.

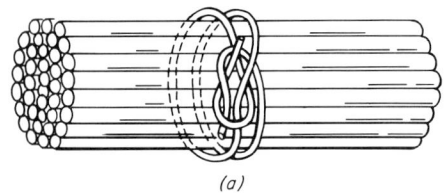

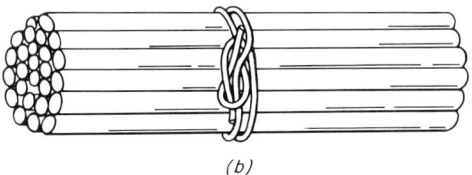

FIG. 7·27 The spot tie. (a) Expanded view. (b) Tighter view.

a typical nylon cable tie. The use of these cable ties is illustrated in Fig. 7·30. Since no special skill is required to install these nylon cable ties, no further discussion of them will follow. Figure 7·31 illustrates the use of special tools during cable tie application.

CABLE WRAPPING TAPE

The simplest method of binding a wire harness is to use a wide, flat tape. The use of flat tape for cable wrapping is illustrated in Fig. 7·32. The tape employed may be adhesive, such as ordinary electrical tape, or it may be nonadhesive. Nonadhesive tape may be plain cotton, fiber glass, varnished cambric, or any similar material.

SPACING BETWEEN TIES

The minimum spacing between adjacent ties of a wire harness should be no less than ½ inch. The maximum spacing is determined by the diameter of the harness, but twice the diameter of a harness is the proper spacing between the ties of larger harnesses (see Fig. 7·25).

HARNESS NEATNESS

A harness should be neat and orderly. All wires should run parallel to each other. Wires should not weave in and out as illustrated by the dark wire of Fig. 7·25. All spot tying and lacing should also be uniform.

HARNESS PROTECTION

A harness that leans against a hard edge, such as a mounting bracket, should be protected against rubbing. For this the cushioned clamp shown in Fig. 9·9b is available. A flexible plastic tubing known as *sleeving* is also available. This plastic sleeving is placed around the harness and held in place with spot ties. Cable protection with plastic sleeving is illustrated in Fig. 7·33.

7·13 HARNESS TERMINATION

Most wire harnesses will either originate or terminate at a terminal board, as in Fig. 7·30, or at a cable connector, as in Fig. 7·34. The wires of the harness may be soldered or crimped at its termination. Reference to Chap. 8 on terminals and terminal boards will be useful here.

CABLE CONNECTORS

A great variety of cable connectors are encountered in the electronics field. Typical cable connectors are illustrated in Fig. 7·34. Some connectors use round pins to make contact, while other connectors use a flat type of contacting device.

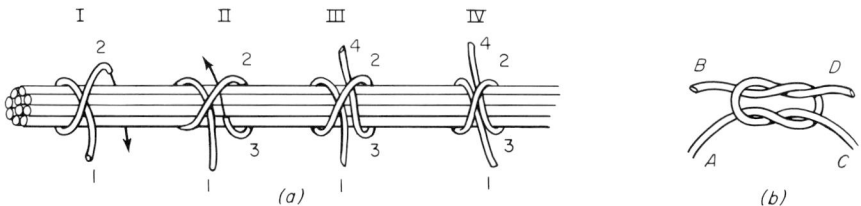

FIG. 7·28 Producing the spot tie. (a) Primary steps for making a spot tie. (b) Square knot.

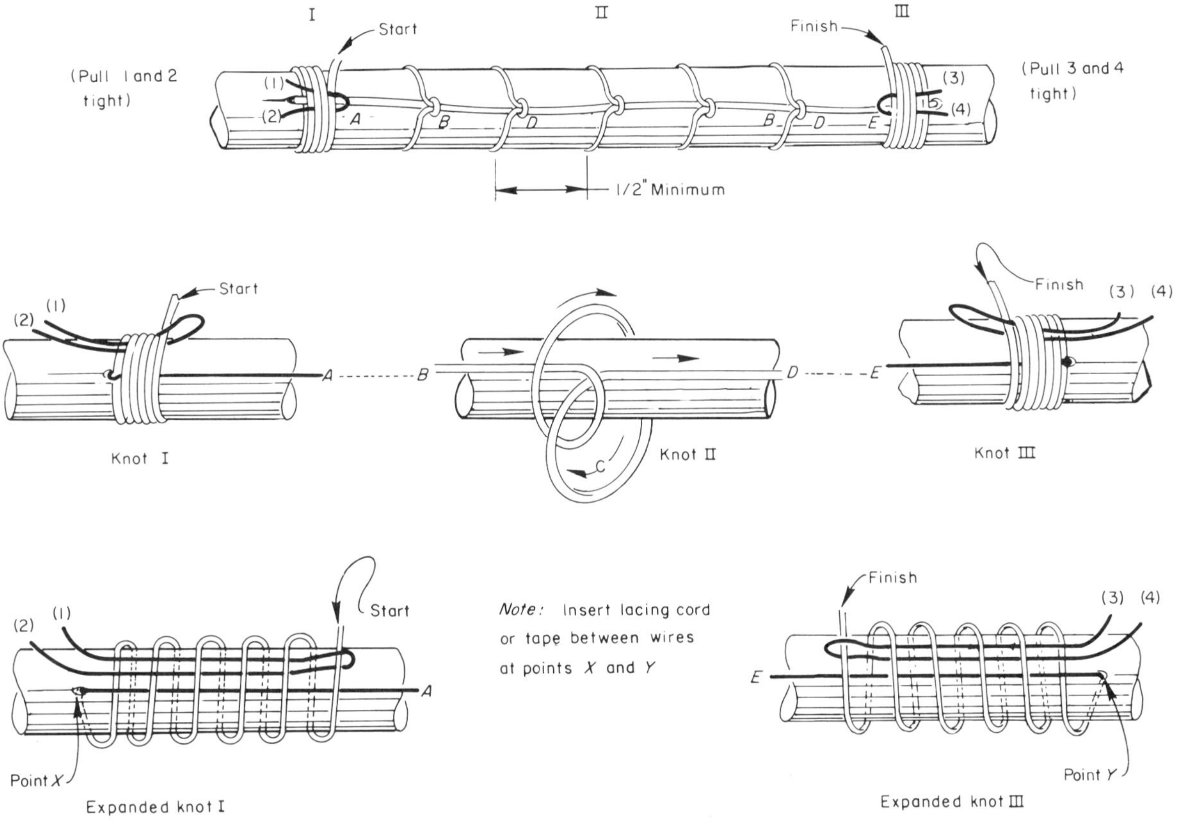

FIG. 7·29 Cable lacing procedures.

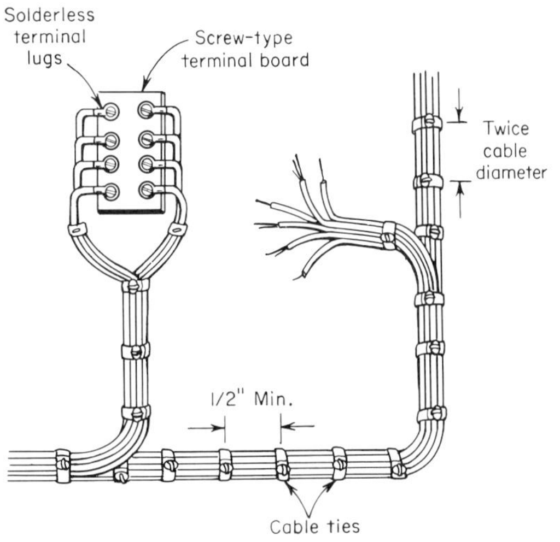

FIG. 7·30 Use of the nylon cable tie.

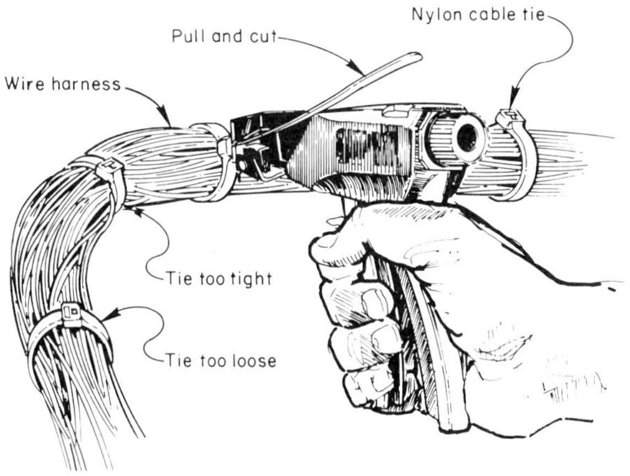

FIG. 7·31 The cable-tie applicator tool.

121

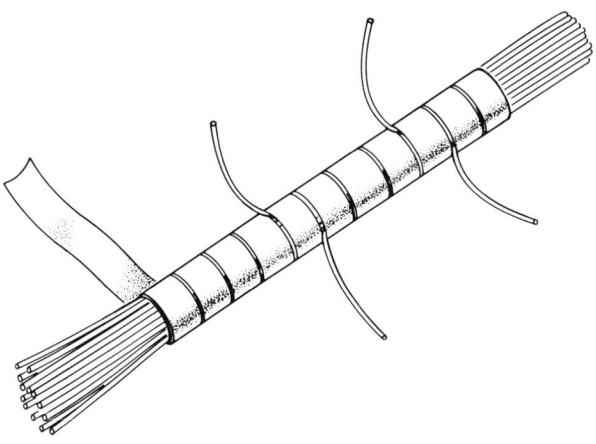

FIG. 7·32 Cable-wrapping tape.

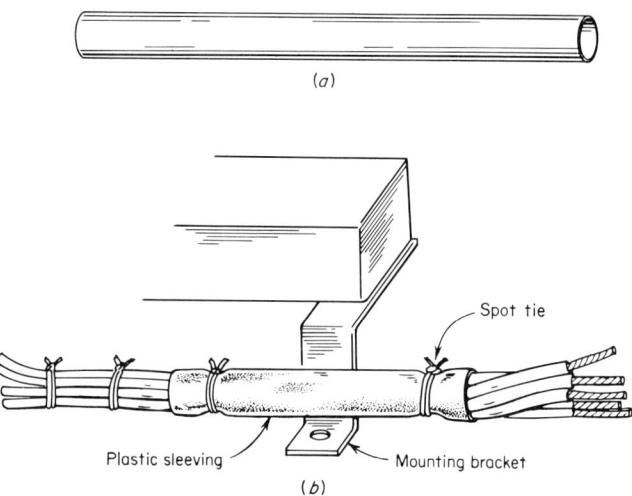

FIG. 7·33 Cable protection. (a) Flexible plastic sleeving. (b) A method for securing sleeving to the cable.

All cable connectors require a mate. The male connector will mate with a female half. To ensure proper connections, cable connectors are polarized by the use of *keys*, which enable the plug and the receptacle, or socket, to mate properly.

If a connector is to be attached to a chassis, a mounting flange is provided.

PRINTED-CIRCUIT CONNECTORS
Printed-circuit modules slip into a printed-circuit connector as shown in Fig. 7·34c. These connectors are the termination of a harness.

RACK-AND-PANEL CONNECTORS
In many areas of electronics it is necessary to disconnect a chassis quickly from its mounting rack for rapid test or replacement. When this is the case, a quick-disconnect type of connector is employed. A typical rack-and-panel connector is shown in Fig. 7·35.

CARE IN WIRING CONNECTORS
Most cable connectors are quite compact. For this reason extra care should be exercised when wrapping the wire leads on the connector terminals.

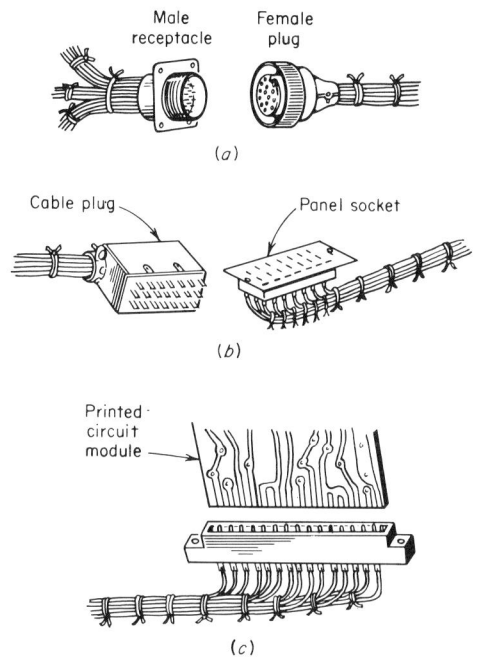

FIG. 7·34 Typical cable connectors. (a) Pin-contact type of connector. (b) Flat-contact type of connector. (c) Printed-circuit connector.

Solder should also be kept to a minimum. As additional protection, a short length of sleeving is slipped over the soldered connector terminal (see Figs. 7·24 and 7·36). Tubing insulation that shrinks with heat is excellent for this purpose.

7·14 FLAT PRINTED-CIRCUIT CABLE

Cable is also made in the printed-circuit variety as illustrated in Fig. 7·37. The fact that it is thin and flexible allows this type of cable to find many applications. The cable in Fig. 7·37a is a nine-conductor cable. The conductors are thin flat lengths of copper imbedded in a plastic film. Flat printed-circuit cable needs no spot tying or lacing.

The connectors shown in Fig. 7·37 are engineered solely for flat printed-circuit cables.

7·15 SYMBOLS FOR CABLES AND CONNECTORS

Wiring diagrams refer to a harness as a cable. Typical of the symbols used in wiring diagrams is the five-conductor cable shown in Fig. 7·38a. Also shown are the symbols used for the cable connectors. A simple application of these symbols is shown

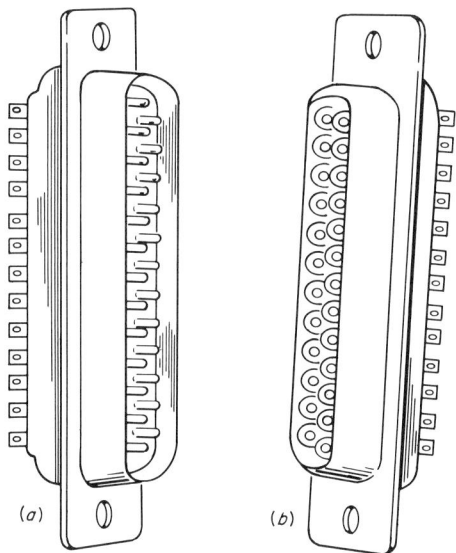

FIG. 7·35 Rack-and-panel connectors. (a) Male. (b) Female.

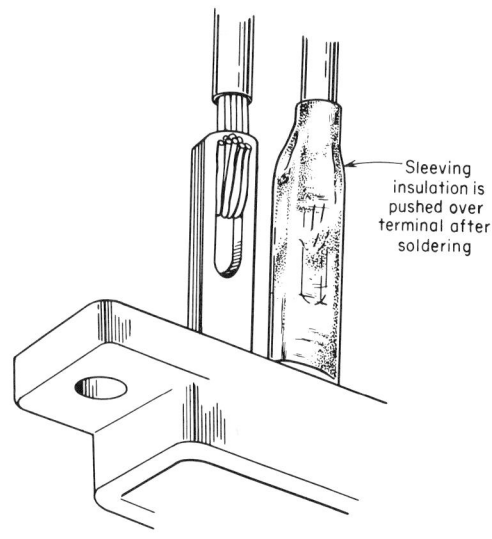

FIG. 7·36 Proper methods for wrapping wire leads on flat connector terminals, and for insulating the soldered joint.

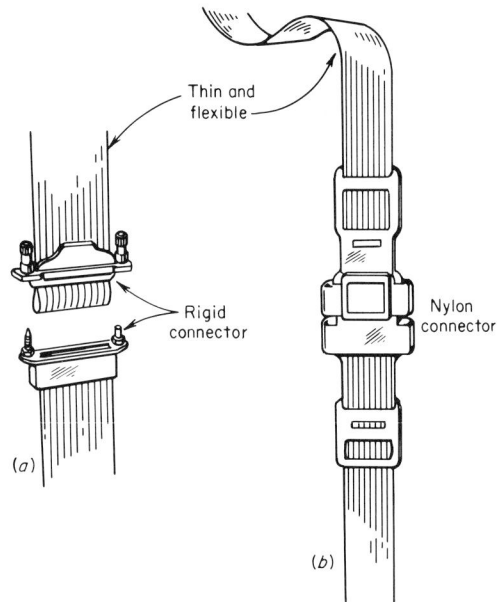

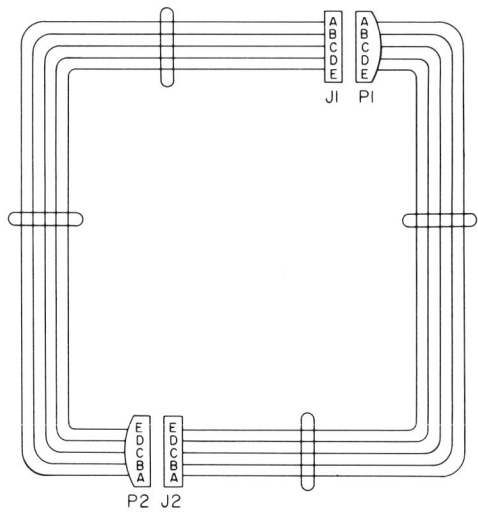

FIG. 7·39 Cable-wiring diagram.

7·16 SUMMARY: WIRE PREPARATION AND HARNESS ASSEMBLY

In summary, copper wires are used as conductors of electricity. These conductors are either solid or stranded. All insulation must be removed before soldering.

The greater the diameter of a wire conductor, the greater will be its current-carrying capacity. Wires are often identified by color code or by a number stamped on the insulation.

This chapter can also be summarized by stating that wires are bound into harnesses. These harnesses terminate at cable connectors or terminal boards. Spot tying and cable lacing are methods used to bundle the wire groups in a harness.

The following jobs are designed to better acquaint the student with wire conductors and their preparation for use.

FIG. 7·37 Flat printed-circuit cable. (a) Nine-conductor cable. (b) Eight-conductor cable.

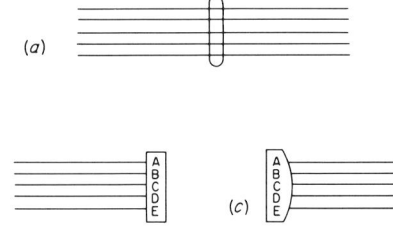

FIG. 7·38 Symbols for use with cables and connectors. (a) Five-conductor cable. (b) Five-contact jack (connector receptacle. (c) Five-contact connector plug.

in the cable-wiring diagram of Fig. 7·39. This wiring diagram depicts two identical five-conductor cables mated to produce five closed circuits.

EXERCISES

JOB 7·1 How to identify wire sizes

OBJECTIVES
1. To learn how to identify the size of a wire
2. To appreciate the need for many different sizes of wires
3. To distinguish between solid and stranded wire

124

MATERIALS REQUIRED

Equipment
A 1-inch micrometer
A wire-size table

Supplies
A spool of No. 20 solid wire
A spool of stranded wire

PROCEDURE

1. Have your instructor check you out on the use of a micrometer.
2. Measure the diameter of the solid wire supplied.
3. The diameter should be 0.03196 inch.
4. On the wire-size table this diameter should appear opposite the No. 20, indicating that a wire of this diameter will be identified as a No. 20 wire.
5. The wire size will also appear on the side of the wire spool. Verify this by checking the sides of the wire spool in your possession. The information is placed there by the wire manufacturer for convenience.
6. Notice, on the wire table, that wires with smaller diameters are identified by larger wire size numbers. Therefore, a wire with a large wire size number will not carry as much electric current as a wire with a smaller wire size number, since the diameter is not as great.
7. Compare the solid wire with the stranded wire supplied.
8. Count the number of strands.
9. Notice that each strand of wire is not insulated from the others. This in effect makes a parallel circuit of all wire strands throughout the length of the total conductor. As such, all wire strands combine to produce what is known as stranded wire.
10. The code size of the total stranded wire is printed on the side of the wire spool. Although stranded-wire tables are available, the average electronics worker seldom has to measure and count each wire strand to determine the size of the total stranded wire.
11. Included with the wire-size information is the number of strands contained by the total conductor. Verify this by inspecting the information on the side of the spool.
12. From this it can be seen that it is unwise to obliterate the information contained on the side of the wire spool. Do anything practical to preserve this information.

QUESTIONS

1. How many *mils* does a No. 20 wire contain?
2. Besides the use of a micrometer, in what other ways can the diameter of a solid wire be determined?
3. What is the advantage of using stranded wire?
4. What are the specifications of the stranded wire supplied? (Obtain this information from the wire spool.)

JOB 7·2 How to strip wire

OBJECTIVES

1. To learn how to remove different types of insulation from a wire
2. To learn to identify wire strippers
3. To learn how to use a wire stripper

MATERIALS REQUIRED

Equipment
An assortment of wire strippers
Scissors

Supplies
Solid wire, enamel-covered
Stranded wire that has been insulated with plastic or rubber
Fine sandpaper

PROCEDURE

1. With the aid of scissors, cut a small piece of fine sandpaper approximately ½ inch wide and 2 inches long.
2. Fold the sandpaper into a V as illustrated in Fig. 7·11.
3. From a length of the enamel-covered solid wire supplied remove the enamel insulation as illustrated in Fig. 7·11.
4. Use a sliding motion back and forth along the end of the wire to be cleaned. After a little practice the proper pressure can be determined.
5. Do not sand the wire to the point of reducing the diameter of the wire. Sand just enough to remove completely the enamel from the area to be cleaned.
6. Identify each of the wire strippers supplied.
7. Use each type of wire stripper to perform the following operations.
8. Use a wire stripper to remove the insulation from a length of stranded wire. Study Figs. 5·7, 7·13, and 7·14.
9. An illustration of a properly stripped wire appears in Fig. 7·12a.
10. Wire strands should not be broken, and they should not be nicked (see Fig. 7·12b).
11. The insulation should be cut clean, and not torn off (see Fig. 7·12).
12. When using the mechanical wire stripper, hold the wire steady and gently squeeze the wire stripper. Release the stripper and revolve the wire one-quarter turn and squeeze the wire stripper again.
13. When you are satisfied that most of the insulation has been cut, and while holding the wire stripper in a squeezed position, pull the insulation off gently, as illustrated in Fig. 7·13.
14. The procedures outlined above are applicable to all wire strippers. However, small modifications may be in order according to need. (The precision wire stripper of Fig. 5·7 is rather automatic.)

QUESTIONS

1. What will happen if a coarse sandpaper is used when removing enamel insulation?
2. What happens to the electric current when the diameter of a wire conductor is reduced?
3. List by name the types of wire strippers supplied.
4. What will happen to the electric current if a strand of wire in a conductor is broken?
5. Why is nicked wire objectionable?
6. Why is torn insulation objectionable?

JOB 7·3 How to tin wire with a solder pot

OBJECTIVES

1. To learn about coating a metal surface with solder
2. To learn to use the solder pot

MATERIALS REQUIRED

Power source
117-volt, 60-cycle alternating current

Equipment
Soldering pot

Supplies
Liquid resin
Bar solder alloy
Stranded wire

PROCEDURE

1. Obtain the supplies illustrated in Fig. 7·19.
2. Apply electrical power to the solder pot.
3. If the solder pot is empty or low in molten solder, fill by cutting a piece of bar solder and placing it in the pot to melt. BE CAREFUL NOT TO SPLASH THE MOLTEN SOLDER.
4. Strip one end of the stranded wire.
5. Submerge the clean wire strands into the liquid resin as illustrated in Fig. 7·20.
6. Submerge the resin-covered wire strands into the molten solder for an instant as illustrated in Fig. 7·20. AVOID LETTING ANY WATER COME IN CONTACT WITH THE MOLTEN SOLDER. Water will cause the molten solder to splatter.
7. Resin and solder should not come in contact with the insulation.
8. Inspect the stranded wire after it is removed from the solder pot.
9. Notice the smooth, shiny appearance.

10. Notice that all strands have been united. This is the basic principle of soldering. However, the specific act just performed is called tinning. It prepares a stranded wire for future soldering.
11. If a solid wire had been used, its surface would also have been tinned, although no unification of strands would have been involved.

QUESTIONS
1. Define tinning.
2. Describe a solder pot.
3. What are the dangers involved when using a solder pot?

JOB 7·4 How to use the wire color code

OBJECTIVES
1. To appreciate the color coding of wire
2. To recognize methods of wire color coding

MATERIALS REQUIRED

Supplies
White stranded wire, ten 6-foot lengths
Stranded wire, different-colored and color-coded, ten 6-foot lengths
Various lengths of string

PROCEDURE
1. Tie the *white* lengths of stranded wire into a bundle.
2. After the bundle is made, attempt to identify the corresponding ends of each wire.
3. Make a separate wire bundle, using the colored stranded wire.
4. After the bundle is made, identify the corresponding ends of each wire.
5. Notice that a color coding can be placed on a white wire by using colored stripes, called *tracers* (see Fig. 7·12a).
6. When numerical figures are assigned to color-coded wire, the color used corresponds to the numerical figure used. Refer to the color-code table used in Job 2·2.

QUESTIONS
1. Were you able to identify positively the corresponding ends of the white wires in step 2?
2. Were you able to identify positively the corresponding ends of the color-coded wires in step 4?
3. What number would be assigned to a green wire?

JOB 7·5 How to attach solderless terminal lugs

OBJECTIVES
1. To become acquainted with solderless terminal lugs
2. To become acquainted with crimping tools
3. To learn how to attach a solderless terminal lug to stranded wire

MATERIALS REQUIRED

Equipment
Crimping tool
Wire stripper
Pair of pliers

Supplies
Stranded wire, one 10-inch length
Solderless terminal lugs, of a size to match the wire

PROCEDURE
1. Study the illustrations of Fig. 7·16.
2. Strip one end of the stranded wire so that, when it is inserted into the solderless lug, it does not protrude to interfere with the terminal mounting washer.
3. Insert the stripped wire into the barrel until the insulation is at least 1/32 inch beyond the crimp barrel for inspection purposes.
4. Place a crimp on the barrel of the solderless terminal lug, using the method illustrated in Fig. 7·16.
5. Test the connection by gripping the solderless terminal lug with a pair of pliers and pulling on the wire.

QUESTIONS
1. List the advantages of using solderless terminal lugs.

127

2. List the disadvantages of a solderless terminal lug.
3. Investigate the various types of crimping tools available and make a report on them.

JOB 7·6   How to make a spot tie

OBJECTIVES
1. To learn to recognize a spot tie
2. To learn to make a spot tie

MATERIALS REQUIRED

Equipment
Diagonal cutters (dykes)
Wooden shaft, round, 1-inch diameter, 2 feet long
6-inch ruler

Supplies
Wrapping twine, smooth surface, approximately 3/32-inch diameter

PROCEDURE
1. Inspect the spot-tie illustrations in Figs. 7·25 and 7·33.
2. Normally, the material used to make spot ties is the nylon lacing tape and linen lacing cord shown in Fig. 7·26. This job uses the wrapping twine and the wooden shaft for purposes of training.
3. Inspect the spot-tie illustrations in Fig. 7·27.
4. Practice making a spot tie with the wrapping twine on the wooden shaft supplied. Use Fig. 7·28 as a guide. Be sure to finish the tie with a square knot.
5. Make a series of spot ties on the wooden shaft. Follow the specifications given in Fig. 7·25.

QUESTIONS
1. What is a spot tie used for?
2. Why is a square knot necessary to terminate a spot tie?
3. During the procedure of step 5, how much distance did you leave between spot ties? Why?
4. Were all spot ties made in step 5 in a straight line as illustrated?

JOB 7·7   How to lace cable

OBJECTIVES
1. To learn how to lace cable
2. To learn to distinguish between spot ties and cable lacing

MATERIALS REQUIRED

Equipment
Same as in Job 7·6

Supplies
Same as in Job 7·6

PROCEDURE
1. Inspect the cable-lacing illustrations of Fig. 7·29.
2. Practice making one of the ties used in cable lacing by following the illustration (knot II) in Fig. 7·29. Use the wrapping twine and the wooden shaft supplied. In actual applications, the lacing tape and cord shown in Fig. 7·26 would be used.
3. Make a series of running ties as illustrated in Fig. 7·29. Pull tight on all ties to ensure a snug wrap. The specifications relating to distance between the ties are the same for cable lacing as they are for spot ties.
4. A special knot is used before and after the regular ties to prevent the cord from slipping.

QUESTIONS
1. What is the difference between spot tying and cable lacing?
2. List the advantages and the disadvantages of each.

JOB 7·8   How to make a wire harness

OBJECTIVES
1. To learn to make a wire harness
2. To appreciate the use of spot ties and cable lacing
3. To recognize materials used in harness making

MATERIALS REQUIRED

Equipment
Diagonal cutters (dykes)
Soldering practice board (see Fig. 6·27)
Soldering equipment (same as Job 6·5)

Supplies

Nylon lacing tape
Linen lacing cord
20 lengths of assorted-colored wire, 2 feet long, various diameters
Wire solder, 60/40 alloy, No. 20 gauge, resin core

PROCEDURE

1. Make a wire harness, similar to the one illustrated in Fig. 6·27c. Use spot ties made of nylon lacing tape. Refer to Job 7·6.
2. When completed, have your instructor check your wire harness.
3. Cut all the spot ties, unsolder the wires and straighten all wire for use again.
4. Make a wire harness (again on the practice board) similar to the one illustrated in Fig. 7·25, but *use the cable-lacing method.* Use the linen lacing cord for this procedure. Refer to Job 7·7.
5. When completed, have your instructor check the wire harness.
6. Cut all ties, unsolder the wires, and straighten all wires for future use.

QUESTIONS

1. List the problems encountered while performing step 1.
2. List the problems encountered while performing step 4.
3. Which of the two wire harnesses you made satisfied you the most? Why?

JOB 7·9 How to protect a harness

OBJECTIVES

1. To learn to protect a wire harness
2. To appreciate the importance of wire insulation

MATERIALS REQUIRED

Equipment
Scissors

Supplies
A ready-made wire harness
Plastic sleeving, flexible, 3 inches long, diameter 1¼ times that of the wire harness used

PROCEDURE

1. Inspect the illustrations in Fig. 7·33.
2. Cut the plastic sleeving supplied in a straight line throughout the length of the sleeving.
3. The length of the sleeving depends upon the amount of wire harness that needs protection.
4. Open the sleeving and embrace the wire harness in the manner illustrated in Fig. 7·33. The sleeving should overlap since its diameter is greater than the diameter of the wire harness.
5. Secure the plastic sleeving in place with two spot ties as illustrated in Fig. 7·33. If the wire harness makes a bend at the place of protection, or if it must be protected for a long span, several spot ties may have to be used.

QUESTIONS

1. Why is protection of the wire harness in Fig. 7·33 necessary?
2. List other ways of protecting the cable in Fig. 7·33.
3. What happens when the insulation of a wire wears through?

JOB 7·10 How to check the polarity of a cable connector

OBJECTIVES

1. To become acquainted with cable connectors
2. To appreciate a cable connector
3. To learn how to identify the pins and jacks in a cable connector

MATERIALS REQUIRED

Equipment
Ohmmeter

Supplies
One set of pin-contact-type cable connector

PROCEDURE

1. Compare the cable connector supplied with the connector illustrated in Fig. 7·34a.
2. The set should contain a *male* section and a *female* section.
3. Note that the two halves of the connector can engage only when the key of one inserts into the slot of the other. Refer to Job 2·24 and Fig. 2·45.

4. Identify the key and slot of the cable-connector set. The polarity of the connector is determined from these key points.
5. Remove enough of the connector shell to expose the solder terminals of both the male and female sections.
6. Notice that each pin and jack is identified with a small alphabetical letter molded in the insulator assembly.
7. Notice that, if it were not for the key and slot, it would be possible to mate the wrong opposing terminals of the connector.
8. Engage the two halves of the connector completely.
9. With the ohmmeter, check for continuity between the two A terminals.
10. Repeat step 9 for all sets of terminals.
11. On completion of step 10, reassemble the shells of both halves of the connector.
12. Compare the cable connector used with the schematic symbols in Fig. 7·38.

QUESTIONS

1. When is a cable connector used?
2. What does the word polarity mean as used in this job?
3. Explain the meaning of male and female as applied to cable connectors.
4. What did the continuity test prove?
5. Where is the key for the proper engagement of the connector in Fig. 7·34b?
6. Explain the schematic symbols of Fig. 7·38.

JOB 7·11 How to make a wire-harness jig board

OBJECTIVES

1. To acquaint the student with jig boards used in mass production
2. To assure the student that classroom training is relevant to industrial needs and practices

MATERIALS REQUIRED

Equipment
Woodworking hand tools
Drafting tools

Supplies
1   Heavy paper or poster board, 10 × 14 inches
1   Plywood board, ¾ × 10 × 14 inches
8-doz   Finish nails, 1½ inches long, 6p
5   Steel springs, ¼ × ½ inches
9   Steel springs, ¼ × 1 inches
28   Wood screws, ½ inch long, #6
28   Flat washers, size to fit screws
        Rubber cement

PROCEDURE

1. Closely view the wire-harness jig board of Fig. 7·40.
2. Since this type of jig board is used extensively in industry to fabricate wire harnesses, it is important that the electronics assembler trainee have such a jig board to practice making wire harnesses. Therefore, you will provide yourself with such a board by making one with the supplies provided.
3. Start by drawing on the heavy paper (or poster board) a pattern of the wire harness you wish to build. See Fig. 7·40. Assign each wire that you plan to include in the harness a *number*. Since every wire will have two ends, show with the number assigned the wire, where the ends of the wire will protrude out of the harness. You may want to include on the pattern a designation of the arteries that carry a group of wires out of the harness with alphabetical *letters*.
4. After the pattern drawing is complete, cement it to the plywood baseboard.
5. With the wood screws and washers, install the steel springs at the ends of the wire arteries, as shown in Fig. 7·40. These springs will be used to hold the ends of the wires in place while the cable or harness is being assembled.
6. Drive the finish nails into the plywood baseboard, through the pattern drawing, in a way that the nails straddle the proposed layout of the wire harness. See Fig. 7·40.
7. The jig board that you have just completed will ensure that all wire harnesses built on it are uniform and identical.

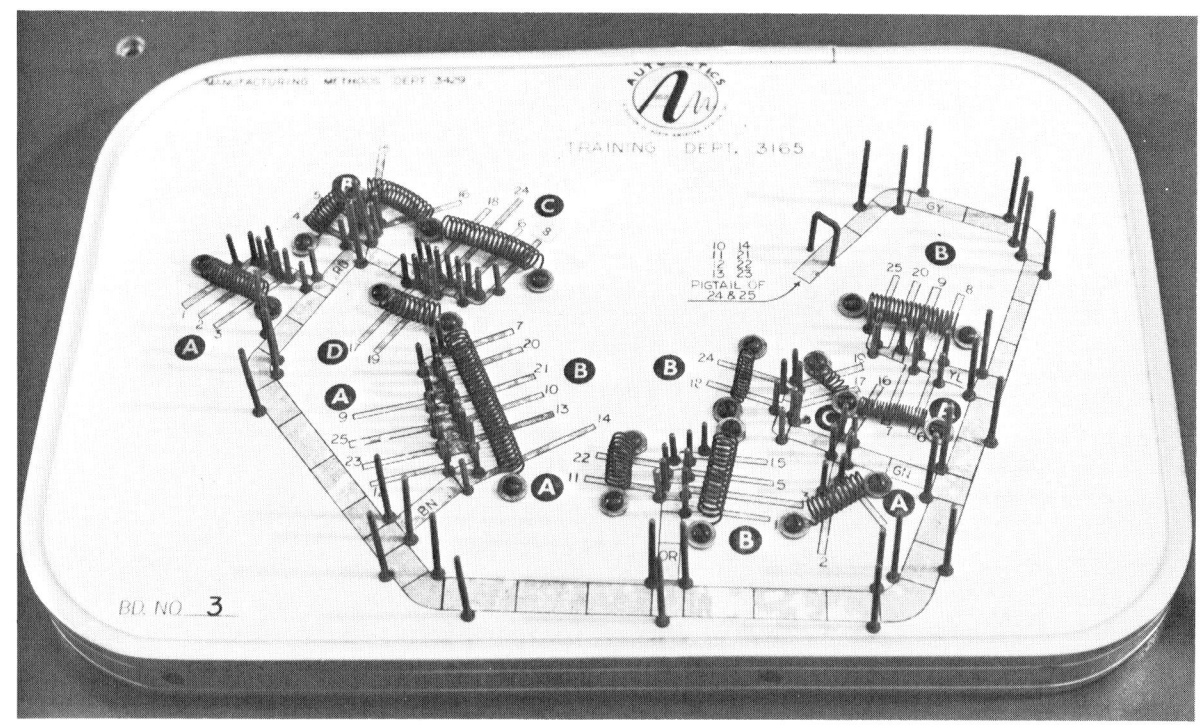

FIG. 7·40 Wire-harness jig board.

QUESTIONS
1. Define a jig board.
2. What are the advantages in the use of jig boards?
3. If you had to build one wire harness, would you build a jig board first?
4. Give examples of where jig boards are used, including the type of industry and the product manufactured.

JOB 7·12 How to use a wire-harness jig board

OBJECTIVES
1. To learn to use a wire-harness jig board
2. To appreciate uniformity in mass production

MATERIALS REQUIRED

Equipment
Hand tools for wire-harness production, including solder pot
Wire-harness jig board (see Fig. 7·40)

Supplies
Various spools of different color-coded flexible wire
500-yard spool of lacing tape, Alpha LC-134 or equal
Various rolls of prenumbered adhesive tape (see Fig. 7·24)
Liquid resin
Bar solder alloy

PROCEDURE
1. Review the wire harness of Fig. 7·25.
2. Review the wire-harness jig board of Figure 7·40.
3. You will make a wire harness on the jig board provided. The finished product will look similar to that illustrated in Fig. 7·25. However, the overall appearance of the harness you make has been predetermined by the form of the jig board you are to use.

4. Cut one wire at a time of a length that will go from one output artery on the jig board to another. The wires should extend beyond the holding springs on the jig board at least 1 inch. Select a different color-coded wire every time. Keep a record of where you install them.
5. When you run out of different color-coding for the wires, identify the remaining wires you use with prenumbered adhesive tape as shown in Fig. 7•24.
6. Normally you would follow a wiring table that would direct you as to which color codes or numbers to use, including at which output artery the wires should protrude from the harness. Further, this same table would give you the length of the wire to use, and the wires would often be precut and pretinned.
7. After having cut all the wires to size, remove them from the jig board and strip the insulation from the ends.
8. Tin the exposed leads by using the solder pot. Refer to Job 7•3.
9. Put the wires back in the harness jig board according to the record you made in step 4. The springs will hold the ends.
10. Proceed to spot tie or lace the harness, depending on the instructions given by your instructor. Refer to Jobs 7•6 or 7•7 as necessary.
11. When the wire harness is completely tied in a bundle, gently lift the harness off the jig board.
12. With the aid of the record you made during step 4, you are now able to make identical additional wire-harnesses quite rapidly, especially if you included in the record the length of every wire you cut.

QUESTIONS

1. Is a wire-harness jig board a good investment for a company that manufactures electrical or electronic products? Why?
2. Organize the records you kept in connection with this job into a table that could assist another person in duplicating the wire harness you made.

8

TERMINAL CONNECTIONS

Terminals serve as tie points for wire junctions. These terminals are mounted on insulating material, such as Bakelite or porcelain. An assortment of terminals may be seen in Fig. 8·1. Various terminal supports are illustrated in Fig. 8·2.

8·1 TURRET TERMINALS

Turret terminals find application in the mounting of component parts, such as resistors, capacitors, and semiconductor diodes. Terminal boards employing turret terminals are shown in Fig. 8·2c and 8·3.

MOUNTING COMPONENTS ON TERMINAL BOARDS
Components, such as resistors, are held in place on a terminal board by their leads. To ensure that the weight of the component does not break the lead or cause the lead to separate from the terminal under severe vibration and shock conditions, the following three precautions must be observed: (1) a wire lead must be protected from damage during installation; (2) the lead must be *wrapped* sufficiently around a terminal; (3) the lead must allow a small degree of flexibility.

Leads may be damaged by sharp bends. Sharp bends are made by bending the lead against the end of the component or by using sharp tools when making necessary bends. Figure 8·3 indicates where and how the component leads are to be bent.

The slight curve on the leads between the resistor and the terminals in Fig. 8·3 serves to absorb some of the shock experienced with vibrations. This fea-

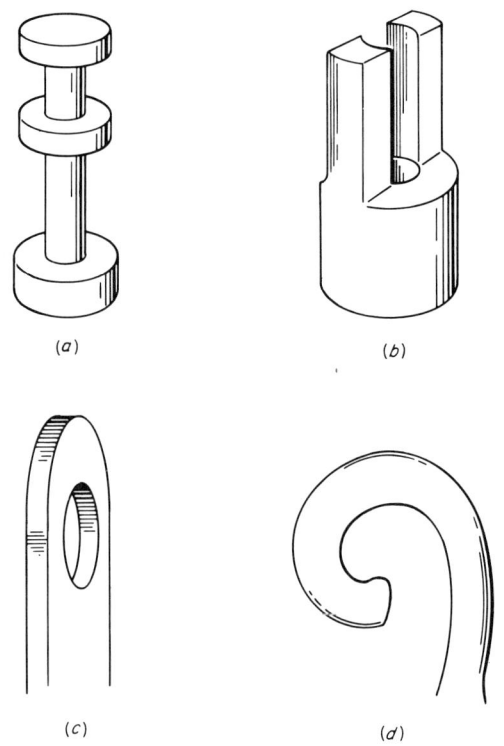

FIG. 8·1 An assortment of terminals. (a) Turret terminal. (b) Bifurcated terminal. (c) Flat terminal. (d) Hook terminal.

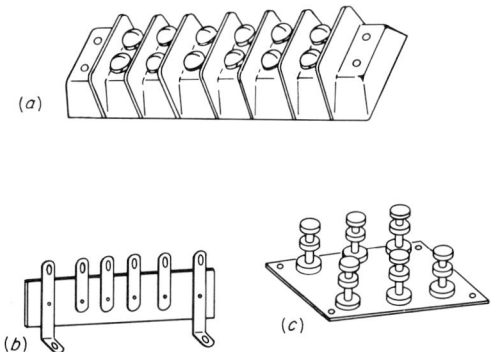

FIG. 8·2 Types of terminal boards and terminal strips. (a) Barrier, screw-type terminal strip. (b) Solder-type terminal strip. (c) Terminal board.

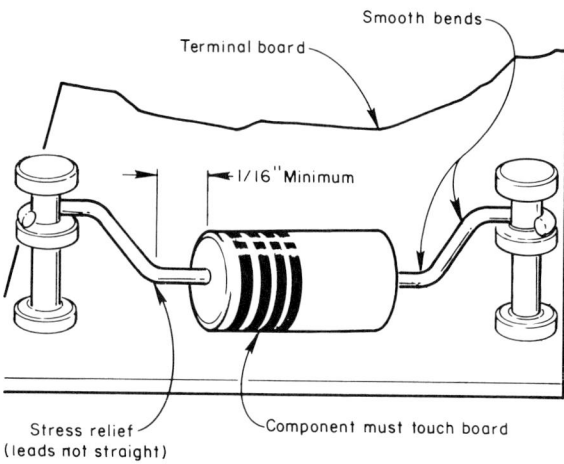

FIG. 8·3 NASA requirements for mounting tubular components on terminal boards.

Another standard practice is to use the upper section of a turret terminal for the leads of components while reserving the lower section for the interconnecting wires.

When insulated wire is used, a section of insulation should be removed before wrapping the conductor around the terminal. The amount of insulation to be removed is determined by the finished wrap. See Figs. 8·4 and 8·5.

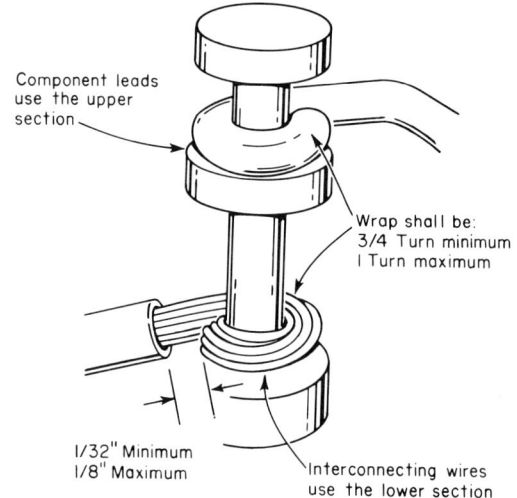

FIG. 8·4 General specifications for wrapping wire around turret terminals.

ture is extremely important in aircraft and space vehicles. Leads installed in this manner are called a *stress relief*.

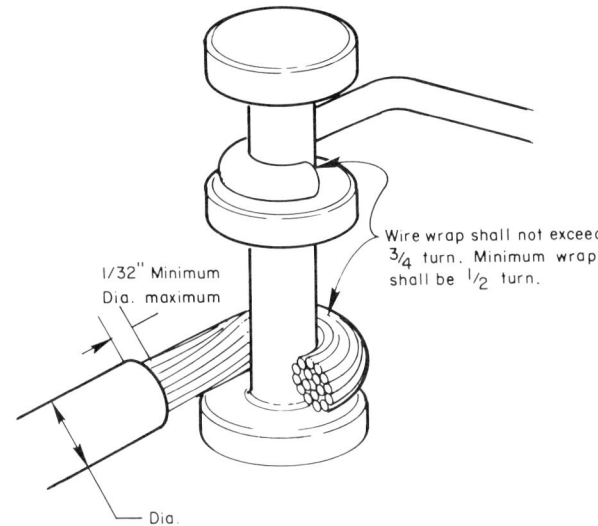

FIG. 8·5 NASA specifications for wrapping wire around turret terminals.

8·2 BIFURCATED TERMINALS

The use of bifurcated terminals parallels the application of the turret terminal. However, according to construction and design, different methods of wrapping are employed. Various methods employed in wrapping bifurcated terminals are illustrated in Figs. 8·6 to 8·10.

Bifurcated terminals are often used on terminal boards of encased transformers and other inductors.

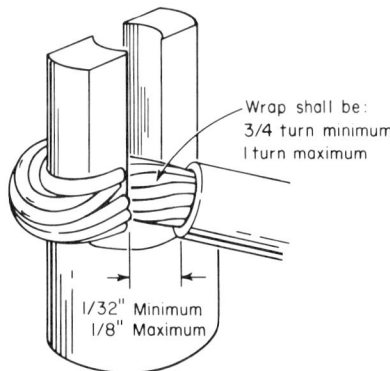

FIG. 8·7 General specifications for wrapping wire around one prong of a bifurcated terminal.

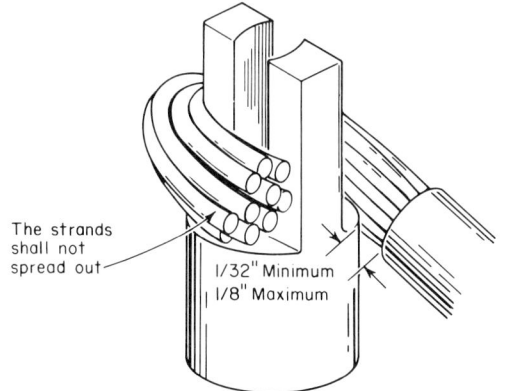

FIG. 8·6 General specifications for wrapping wire around both prongs of a bifurcated terminal.

8·3 FLAT TERMINALS

Flat terminals are found on vacuum tube sockets, on terminal strips, on some cable connectors, and on many other devices.

The wires or leads may approach the flat terminals from any angle. Three methods of wire approach and wrap are illustrated in Figs. 8·11 to 8·13.

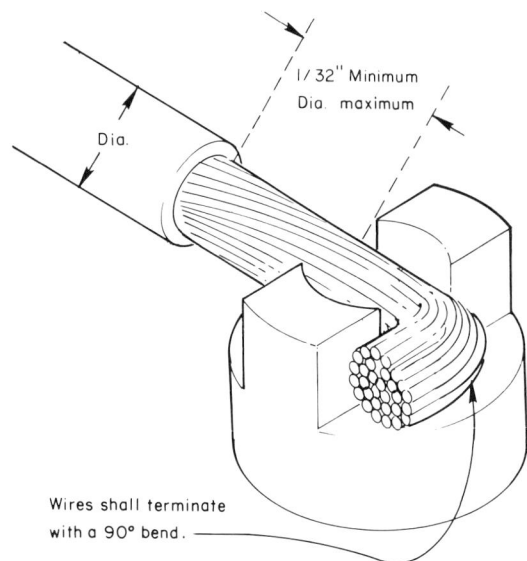

FIG. 8·8 NASA specifications for wrapping wire around bifurcated terminals.

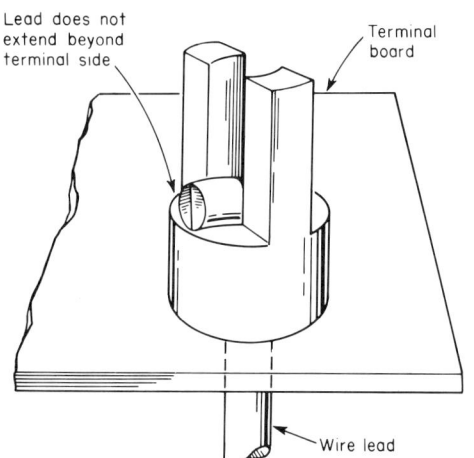

FIG. 8·9 Terminating a lead through a bifurcated terminal (NASA specification).

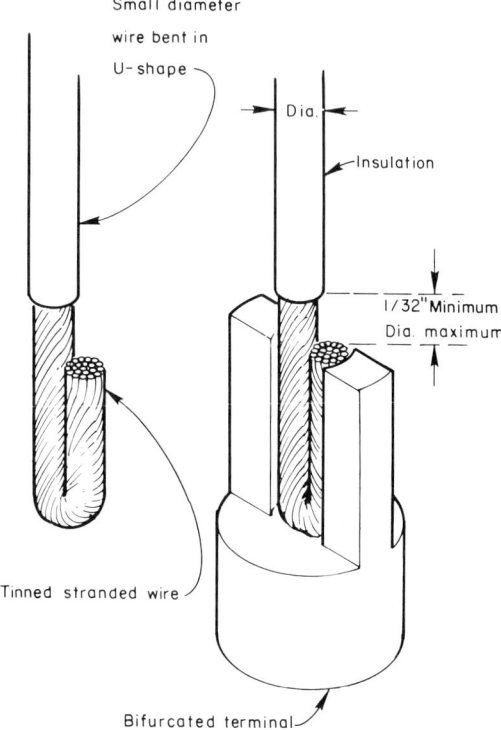

FIG. 8·10 Top-route connection on bifurcated terminal (NASA specification).

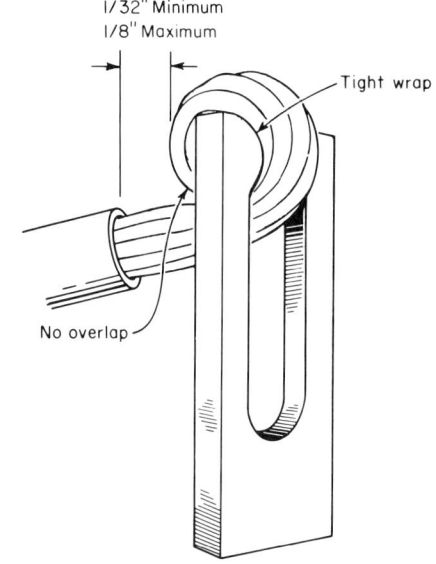

FIG. 8·11 General specifications for wrapping wire on a flat terminal when approached from a side.

8·4 HOOK TERMINALS

The hook terminal finds wide application in terminal boards of encased relays. The same wrapping requirements described above for other terminals apply for the hook terminal (see Figs. 8·14 and 8·15).

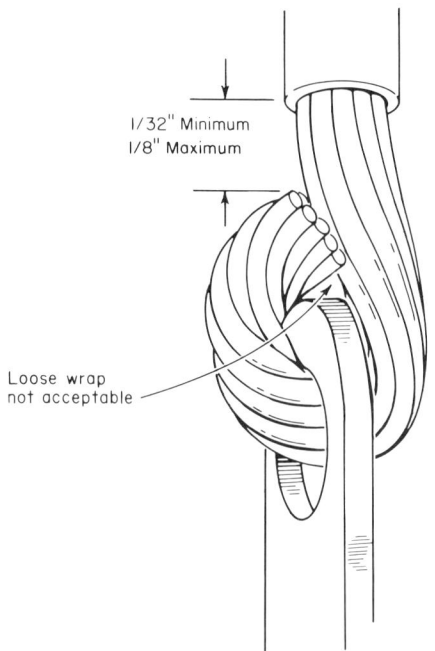

FIG. 8·12 General specifications for wrapping wire on a flat terminal when approached from the end.

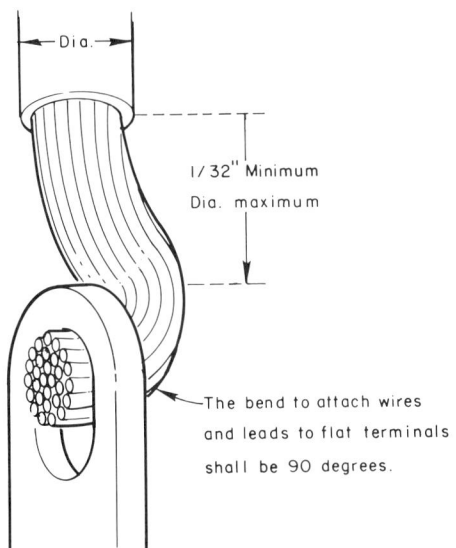

FIG. 8·13 NASA specifications for wrapping wire around a flat terminal when approached from the end.

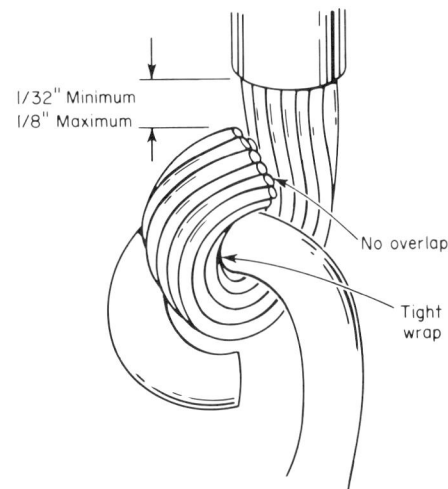

FIG. 8·14 General requirements for wrapping wire around hook terminals.

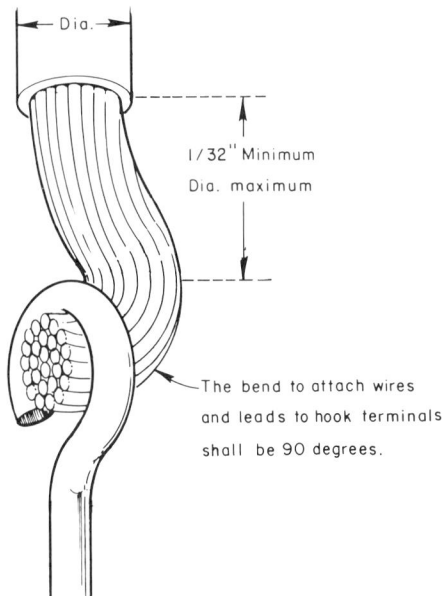

FIG. 8·15 NASA specifications for wrapping wire around hook terminals.

8·5 CABLE CONNECTORS

Wires of cables often terminate in connectors as illustrated in Fig. 7·34. If the connector employs solder pots, the wire is not wrapped, but inserted

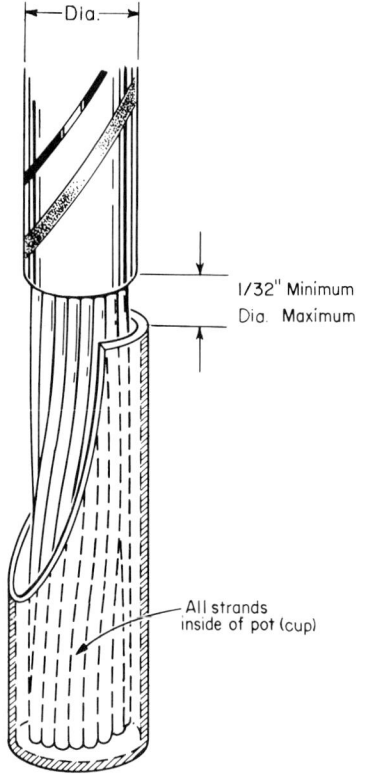

FIG. 8·16 How a stranded wire should appear within a connector solder pot (cup) (NASA specification).

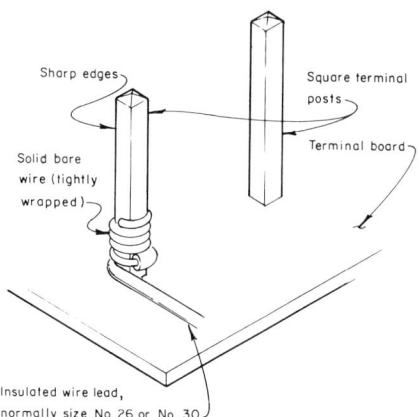

FIG. 8·17 Terminating a solderless wire connection.

into this type terminal. The placement of a wire in such a pot is illustrated in Fig. 8·16. Multistrand wire is always used with this type of terminal since flexibility is quite important.

8·6 SOLDERLESS TERMINALS

The *screw-type* terminal of Fig. 7·17 is about the simplest type of solderless terminal. However, it is necessary for the screw of the terminal to be tight to ensure a good electrical connection. Another type of solderless terminal in wide use in commercial products not destined for air or spacecraft is the *Wire-Wrap*® terminal. A Wire-Wrap® terminal is illustrated in Fig. 8·17.

8·7 MOUNTING COMPONENTS ON PRINTED-CIRCUIT BOARDS

When mounting a component on a terminal board, whether it be a printed-circuit board or any other type, attention to the soldering procedure alone is never enough. The continued utility of the component under all conditions must be a strong consideration during fabrication. Care should be exercised not to damage the leads of components since this weakens the lead, and under extreme vibration or acceleration, may cause a break. Special care should also be exercised when installing heat-sensitive semiconductors and other miniature components, especially during the soldering processes. An assortment of these small components is illustrated in Fig. 8·18.

An additional consideration is that of component replacement during service and repair functions. If it is anticipated that certain components may have to be removed for repair or adjustment during the normal life of the product, a *service loop* should be provided. A service loop is nothing more than a longer-than-necessary lead used during installation. The service loop allows for cutting off of an end of the lead that may have been damaged during removal of the component from the chassis. A service loop is seldom longer than an inch. However, it may be shorter or longer as may be applicable to the circumstances involved.

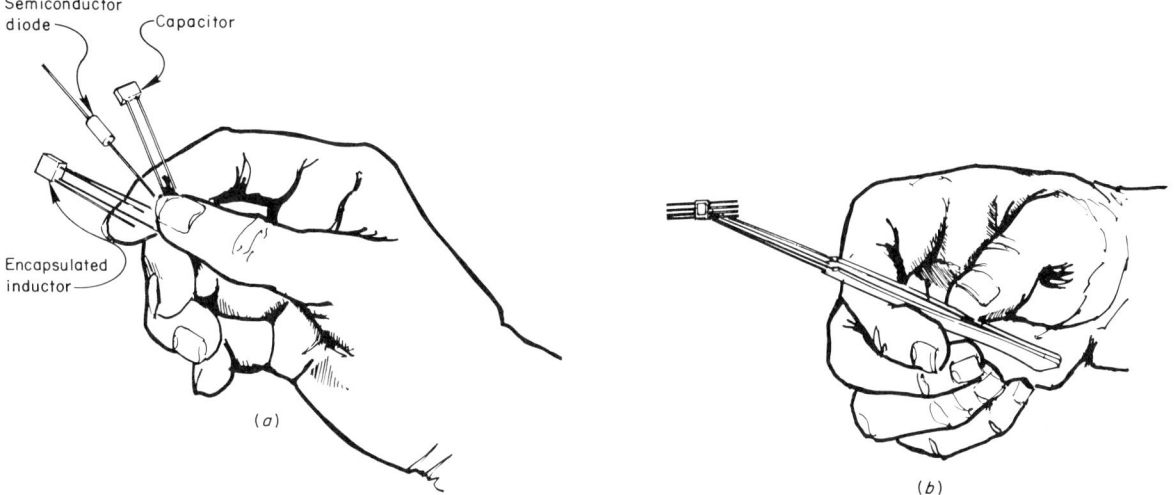

FIG. 8·18 Small components demand special handling and mounting procedures. (a) Miniature discrete components. (b) Integrated circuit.

SUPPORT

Most small components are held in place by their connecting leads. It is therefore important that these leads be sound. They should not be pulled tight. A tight lead maintains a stress at all times. As shown in Fig. 8·19, *round bends* of the wire are required. The basis for bending component leads before mounting is shown in Fig. 8·20. A component lead should never be bent at the point of egress, but should proceed straight out for at least $1/16$ of an inch. Components which have a welded lead, such as liquid electrolytic tantalum capacitors, should have the start of the bend begin at least $1/16$ of an inch *from the weld*.

Special tools are available to simplify bending of component leads. The component lead bending gauge shown in Fig. 8·21 is designed for fast, uniform bending of leads. The round-nose pliers illustrated in Fig. 5·4 are also excellent for this purpose.

Since small components become relatively heavy during acceleration or during high-speed maneuvers, components should touch the board at least at one point for constant support. A maximum clearance of $1/64$ of an inch is acceptable for any other high point of the component. By such an installation, the leads of the component are relieved of sudden stresses. See Fig. 8·19.

Fragile components, such as glass-bodied semiconductor diodes, should *not* be allowed to touch the board for they may rattle against the board during vibration and break. However, a clearance of $1/32$ of an inch is maximum. See Fig. 8·22. Actually, it is best to secure *all* components to the printed-circuit board by using an appropriate clamp or potting compound whenever possible. Glass-encased components should be enclosed in

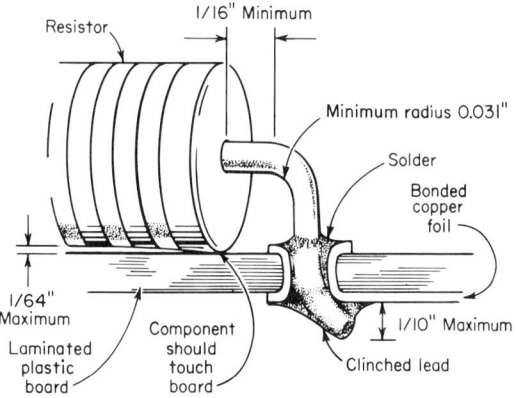

FIG. 8·19 Specifications for mounting components on printed-circuit boards (NASA specifications).

139

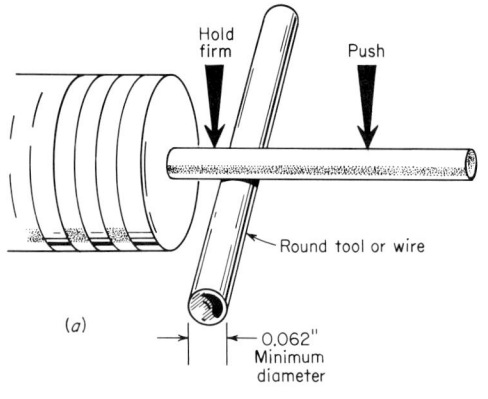

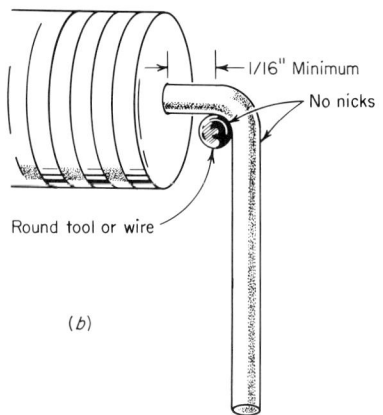

FIG. 8·20 Basics for bending component leads.

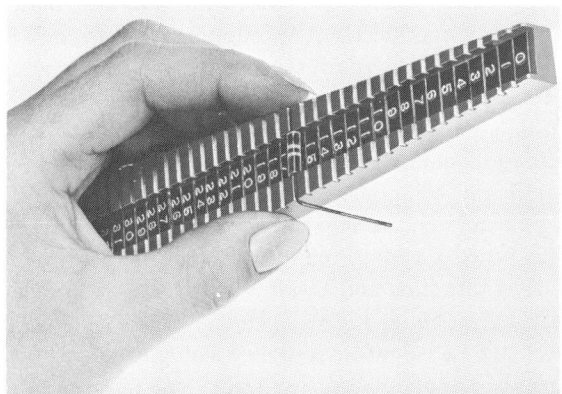

FIG. 8·21 Component lead bending gauge. (*Hunter Tools, Santa Fe Springs, California.*)

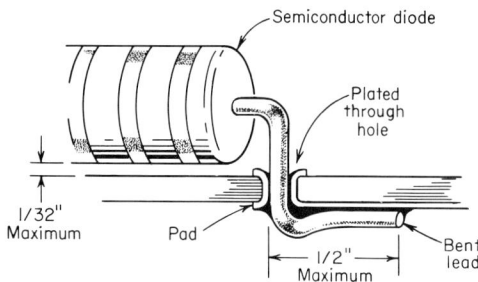

FIG. 8·22 Specifications for mounting heat-sensitive components on printed-circuit boards (NASA specifications).

flexible tubing prior to mounting and encapsulation for protection against thermal expansion.

MOUNTING ACROSS CONDUCTIVE LINES
Although components mounted on printed-circuit boards are normally mounted on the side opposite the printed circuitry (and in contact with the printed-circuit board unless otherwise specified), there are times when it is necessary to mount components across conductive lines on the printed-circuit side. When this is the case, the component needs to be insulated by enclosing it within a transparent flexible tubing, allowing only the leads to be exposed. The tubing used should be able to withstand voltage and temperature environments applicable to the area.

MECHANICAL CONNECTIONS ON
PRINTED-CIRCUIT BOARDS
As illustrated in Figs. 8·19 and 8·22, the leads of the components are inserted through the plated-through holes of the printed-circuit board, *clinched* or *bent*, and soldered on the opposite side. Actually, it is the clinching or bending of the lead that gives the primary mechanical connection or support. When these leads are clinched or bent, a nonmetallic tool should be used so as not to damage the leads in the process. Such a tool is shown in Fig. 8·23. The clinching or bending is made in the direction of the circuit pattern. Bending of the wire instead of clinching it, allows for a longer component lead. This is is important during the soldering operation so that the minimum amount

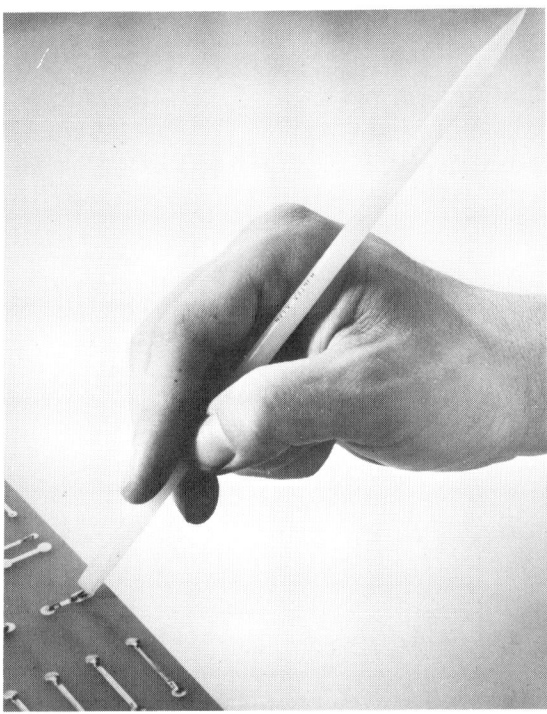

FIG. 8·23 Lead clinching tool. (*Hunter Tools, Santa Fe Springs, California.*)

of heat reaches the heat-sensitive components. Another technique used to minimize the amount of heat reaching a heat-sensitive component is to use a sponge saturated with alcohol on the area to be kept cool during the soldering operation.

USE OF SOLDERING GUNS

Since the application of heat during soldering is an exacting procedure, especially when mounting components on printed-circuit boards, use of a soldering gun is to be discouraged. It is almost impossible to control the application of heat to the joint being soldered with a soldering gun since the soldering tip of the gun is normally cold until the spring-loaded switch that applies power is depressed. (See Figure 8·24.) Any momentary relaxation of the "trigger finger" causes the tip to begin to cool off. Soldering guns are an excellent tool for general field work, but should not be used on printed-circuit boards or where exacting soldering

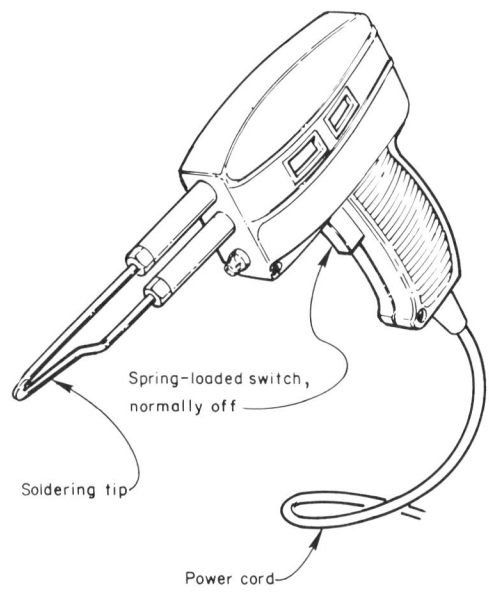

FIG. 8·24 The soldering gun.

requirements are necessary for reliable soldered connections. Soldering guns do not meet the requirements of NASA. It is not so much the gun as it is the operator. Besides, soldering guns are designed for intermittent operation and not for continuous use.

8·8 USE OF SOCKETS AS TERMINALS

With the advent of integrated circuits and miniature electronic components, it has become increasingly difficult to use small component leads for interconnections. To meet this problem, the use of sockets is finding wide acceptance; especially standardized sockets that are centered around the dual-in-line package. Sockets such as the one illustrated in Fig. 8·25 are used as separate items, or are used in banks mounted on laminated printed-circuit boards. Into these sockets will be inserted complete integrated circuits (Fig. 2·35c), or discrete components assembled into network packages such as shown in Fig. 8·26. These *networks of components* are flexible in that by cutting off some of the leads, only the desirable values of resistors are made available to the circuit. Another technique is to

141

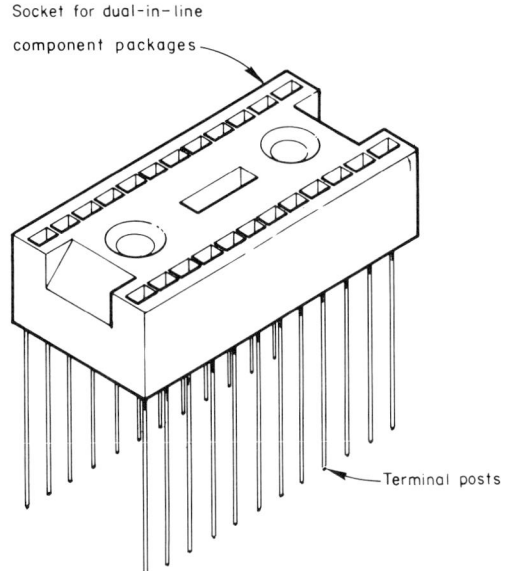

FIG. 8·25 Sockets as terminals.

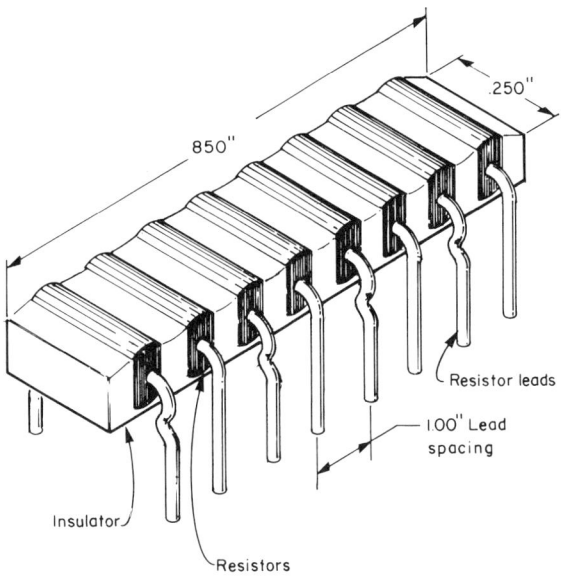

FIG. 8·26 Resistor network in dual-in-line package.

connect the leads of the resistors in these networks into series, parallel, or series-parallel arrangements; and then, once again leave only the appropriate leads available for insertion into the socket.

Precision variable resistors are also packaged in the dual-in-line format so as to make use of the standard socket illustrated in Fig. 8·25. Although the reader will notice that the variable resistor illustrated in Fig. 8·27 only has four leads, nevertheless, the spacing between the leads are such that the variable resistor slips nicely into the socket.

Another example of miniature discrete components mounted in dual-in-line packages is the relay illustrated in Fig. 8·28. As can be seen, this package will also fit the socket of Fig. 8·25.

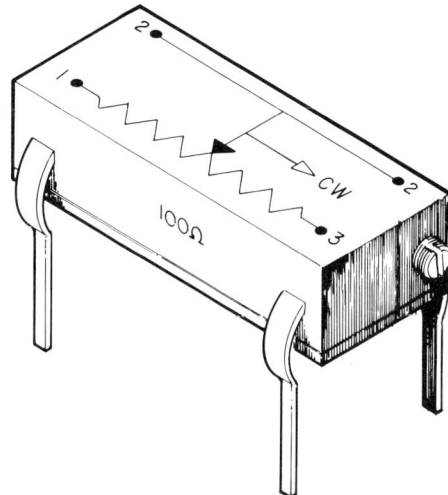

FIG. 8·27 Variable resistor mounted in a dual-in-line package.

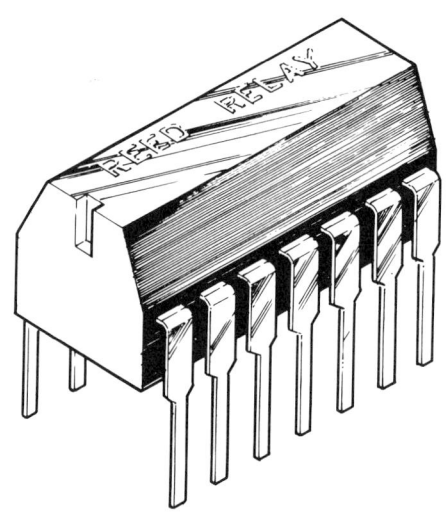

FIG. 8·28 Relay mounted in a dual-in-line package.

8·9 SUMMARY: TERMINAL CONNECTIONS

To summarize this chapter, it should be stated that terminals are used primarily to tie two or more wires electrically. Terminals are also used to support small component parts, such as resistors, capacitors, and semiconductor diodes. Although terminals are found right on larger components, such as transformers, usually they are found on *terminal boards* mounted in a chassis at strategic areas. Terminals come in many forms and sizes. However, the printed-circuit type terminal demands the most attention from an electronics assembler during fabrication processes. Finally, with the advent of electronics miniaturization, standardized sockets (for use with integrated circuits and very small discrete components) are finding popularity.

The following jobs will help the student become better acquainted with the use of terminals.

EXERCISES

JOB 8·1 How to appreciate the use of terminals

OBJECTIVES
1. To recognize terminals
2. To appreciate the usefulness of terminals

MATERIALS REQUIRED

Equipment
New vacuum-tube-type radio receiver (as used in home)
Used military electronics equipment
Vacuum-tube manual

PROCEDURE
1. Remove the cabinet from the radio receiver. A new radio receiver is employed to ensure proper wiring practices.
2. Do not apply power to any equipment while performing this job.
3. Identify each terminal used in the radio receiver.
4. Terminals on vacuum tube sockets are often used as tie points for two or more wire leads, provided the related vacuum tube does not need the particular terminal. With the aid of the tube manual determine if this condition exists in the radio receiver supplied.
5. Remove the cabinet from the military equipment. A used piece of equipment is satisfactory here since military restrictions require all electronics equipment to be replaced in the original condition whenever it is repaired.
6. Compare the wiring arrangement in military equipment with that found in the home-type radio receiver.
7. Identify each type of terminal used in the military equipment by its proper name.
8. Locate any junction of two or more wires that do not make use of a terminal.
9. Replace the cabinets on the receiver and military equipment.

QUESTIONS
1. How many different types of terminals were found on the radio receiver?
2. How many vacuum-tube socket terminals were used solely as tie points?
3. Was there any difference in the wiring arrangement found in the military equipment as compared with the home-type radio receiver?
4. How many of the terminals illustrated in Fig. 8·1 were you able to locate on the military equipment?
5. What is your opinion of the usefulness of terminals?
6. Define a terminal.

JOB 8·2 How to identify terminals

OBJECTIVES
1. To practice calling terminals by their proper names
2. To become acquainted with the variety of terminals available and where each type is commonly used

MATERIALS REQUIRED

Supplies
Notebook paper
Pencil
Rubber cement

PROCEDURE
1. Start keeping a record of every type terminal that you encounter during the rest of this school semester-period and submit the record to your instructor for credit before taking the final examination.
2. Include in the record the proper name of the terminal; a sketch of the terminal or a pictorial clipping cemented to the notebook paper; what device it was attached to; and how it is normally used (soldered, solderless, etc.).

QUESTIONS (Submit answers to your instructor now.)
1. What types of terminals do you expect to encounter in common use? Where?
2. Will you be looking for terminals mounted on component parts? Why?

JOB 8·3 How to make a terminal practice board

OBJECTIVES
1. To provide a project board for the practice of wire wrapping and soldering on terminals
2. To encourage the student to practice working on terminals without the fear of damage to a component

MATERIALS REQUIRED

Equipment
Electrical hand tools

Supplies
Plywood baseboard, ¾ × 6 × 10 inches
Wire solder
Assorted wood screws, ½ inch long
Access to junk pile of electronic chassis and components

PROCEDURE
1. Unsolder and remove used tube sockets from the electronic chassis made available for this purpose. Select two in good condition and mount them approximately two inches apart on the plywood baseboard provided. Mount them with the terminals upward, about one inch from the edge of the board with wood screws. You now have a makeshift *flat terminal practice board*. If you damage the socket during practice, replace it.
2. Look in the junk pile for terminal boards such as shown in Fig. 8·2. These may be found attached to a chassis, or they may be found mounted on cases of components such as transformers. When you locate some, unsolder them, clean them, and mount them on the practice board as in step 1 above.
3. Using the same procedure as in step 2, locate some cable connectors such as seen in Fig. 6·15. Clean them and mount them on your practice board.
4. If some terminal strips, boards, or connectors do not have readily available mounting holes, drill mounting holes in them. Use your ingenuity.
5. Try to get as wide a variety of terminals to practice on as possible.

QUESTIONS
1. List the different types of terminals represented on your practice board.
2. Look at a parts catalog and determine the cost of each terminal board or connector mounted on your practice board.

JOB 8·4 How to wrap wires on terminals

OBJECTIVES
1. To become acquainted with proper procedures for wrapping wire around a terminal prior to soldering
2. To practice preparing wire junctions for soldering

MATERIALS REQUIRED

Equipment
Hand tools for electrical assembly
Multiple terminal board made for practice

Supplies
Assorted lengths of solid wire, various diameters

PROCEDURE
1. This job is sort of a *dry run,* single-concept type of exercise. Solid wires are used throughout to avoid the problems of wire stripping, broken strands, etc., which are often associated with stranded wire.
2. Study all of the illustrations in Chap. 6 and this chapter as they pertain to wire wrapping, and, on the multiple terminal practice board, attempt to duplicate the wire wraps. Check with your instructor to see if you have the right idea. Do not attempt to solder the wire wraps at this time. Pay special attention to the specifications called out.

QUESTIONS
1. Differentiate between a mechanical connection and an electrical connection.
2. Is a wire wrap primarily a mechanical connection or an electrical connection? Explain.
3. Give a summary of the various wire-wrap specifications you encountered by working this job. Why the difference?

JOB 8·5 How to interpret tolerances in the metric system

OBJECTIVES
1. To be able to interpret tolerances given in the English system of measurement in the metric system, and vice versa
2. To practice English-to-metric measuring systems conversions

MATERIALS REQUIRED

Supplies
Notebook paper and pencil

PROCEDURE
1. Convert *every* tolerance or dimension given in this book from the English system of measurement to the metric system. Write in the textbook the metric figure alongside the English stated quantity.
2. In making these conversions you are reminded that 1 inch equals 2.54 centimeters (2.54 cm). One inch is also equal to 25.4 millimeters (25.4 mm).
3. In the *questions* section of this job you will be asked to give samples of your computations, and during classroom examination sessions you will be asked to demonstrate your ability to make such conversions. Therefore, be sure that you understand what you are doing.

QUESTIONS
1. Why is it important that you be able to interpret tolerances and dimensions in both the English and metric systems of measurement?
2. Show the complete computation for converting the specification of Fig. 8·13 from the English to the metric system of measurement.
3. Show the complete computations for converting the specifications of Fig. 8·19 from the English to the metric system of measurement.
4. Show the complete computations for converting the dimensions of Fig. 8·26 from the English to the metric system of measurement.

JOB 8·6 How to bend leads of components

OBJECTIVES
1. To become familiar with safe practices in bending component leads so as not to damage or break the lead
2. To become familiar with NASA specifications for bending component leads

MATERIALS REQUIRED

Equipment
Plastic rod, 0.062-inch diameter
Round-nose pliers
Lead bending tool (see Fig. 8·21)

Supplies
Assortment of resistors with straight leads

PROCEDURE
1. With the plastic rod supplied, bend the leads of a resistor as illustrated in Fig. 8·20. Be sure to

hold the lead firm as indicated prior to making the actual bend. Observe all specifications as required by the National Aeronautics and Space Administration (NASA) for products to be used in space vehicles.
2. With the round-nose pliers supplied, bend the leads of a resistor. Observe all of the specifications called out in Fig. 8•20.
3. With the special lead bending tool designed for this purpose, bend the leads of a resistor as illustrated in Fig. 8•21. This tool is designed for the proper radius of the bend. However, you must select the section of the tool to use, depending on the size of the resistor and where the bend is to take place. However, you must continue to remember the 1/16-inch minimum distance between the component wall and the start of the bend.
4. Although resistors were selected for this exercise, the same principles apply when bending leads of capacitors and diodes.

QUESTIONS
1. Should you use regular long-nose pliers to bend the leads of components? Why?
2. Why should you not bend the component lead at the immediate point of egress?
3. Is it a good idea to bend a component lead without the aid of tools? Why?
4. What is the importance of a minimum radius at the bend?

JOB 8•7 How to install components on printed-circuit boards

OBJECTIVES
1. To learn the basics for printed-circuit board assembly
2. To coordinate previously learned assembly activities in the fabrication of a printed-circuit board assembly

MATERIALS REQUIRED

Equipment
Hand tools for electronics assembly, including 25-watt soldering iron
Printed-circuit holding vise (see Fig. 5•19)

Supplies
Assorted resistors
Assorted small-signal diodes (semiconductor)
Wire solder, 60/40 alloy, 0.032-inch diameter

PROCEDURE
1. Support the printed-circuit board in the special holding vise. If a vise such as shown in Fig. 5•19 is not available, a drill-press vise (Fig. 5•18b) or a bench vise (Fig. 5•17) will do. However, some sort of protective material over the jaws of the vise used will be necessary to prevent damage to the printed-circuit board. In addition, extreme care must be exercised so as not to tighten the jaws too much.
2. Carefully clean the solder pads to be used on the printed-circuit board with a soft pencil or ink eraser. Do not rub too hard.
3. Clean the leads of the resistors or diodes to be mounted as illustrated in Figs. 7•9 and 7•10.
4. Bend the leads of the components as illustrated in Figs. 8•20 or 8•21.
5. Install the resistors in the printed-circuit board and solder, as illustrated in Figs. 6•18 and 8•19. Prior to soldering, clinch the leads on the foil side of the board with a nonmetallic tool such as illustrated in Fig. 8•23. Observe all specifications indicated in Fig. 8•19.
6. When installing the diodes, follow the same procedures, but observe the specifications indicated in Fig. 8•22.

QUESTIONS
1. Why is there a difference in mounting specifications between Fig. 8•19 and Fig. 8•22?
2. What advantages does the holding vise of Fig. 5•19 have over the vises illustrated in Figs. 5•18b and 5•17?
3. If the solder pads of the printed-circuit board, and the leads of the components look clean, is it still necessary to clean them with an eraser? Why?
4. Why the need for a *nonmetallic* lead clinching tool?

JOB 8•8 How to solder on terminals

OBJECTIVES
1. To review about wire-wrapping requirements
2. To review about solder requirements for terminals

MATERIALS REQUIRED

Equipment
Soldering iron, 25-watt
Hand tools for electrical assembly
Multiple terminal board made for practice

Supplies
Short lengths of solid wire
Short lengths of stranded wire
Resin-core solder, 60/40 alloy, 0.032-inch diameter

PROCEDURE

1. Prepare for soldering each wire to be used.
2. Wrap the wire around the terminals as illustrated in this chapter.
3. Pay close attention to all specifications listed on the drawings.
4. Neatly solder the properly wrapped wires onto the respective terminals as illustrated in the appropriate illustrations of Chap. 6, with special attention to Fig. 6•11.
5. When soldering, remember to apply the solder from the side opposite the side of heat application. When the terminal is hot enough the solder will melt.
6. Have your instructor inspect your work.
7. When approved, remove all wire and solder from the terminals.
8. Have your instructor inspect the newly cleaned terminals.

QUESTIONS

1. What is wire wrapping?
2. What is contour soldering?
3. In simple statements explain the purpose of each specification called for when wrapping wire.
4. In simple statements explain faulty techniques to be avoided when soldering on a terminal.

9

HARDWARE AND MECHANICAL ASSEMBLY

Hardware associated with electronic equipment may provide either *physical* or *electrical support*. Items like screws, bolts, nuts, and clamps provide physical support, while electrical parts such as switches, lamps, fuses, sockets, and terminals support the circuits electrically.

9·1 PHYSICAL-SUPPORT HARDWARE

THE CHASSIS

The basic unit of all hardware is the chassis. A typical chassis is shown in Fig. 9·1. On such a chassis will be mounted the transformers, sockets, terminal strips, variable resistors, switches, and similar parts. Although the chassis physically supports these component parts, very often it also serves as a common electrical conductor. When the chassis serves in the latter capacity, it is referred to as *chassis ground* and is assigned the electrical symbol .

MACHINE SCREWS

The machine screw and hexagon nut illustrated in Fig. 9·2 are used extensively as a holding device in the physical support of electrical parts. The roundhead with the slot is popular. A blade type

FIG. 9·1 A typical electrical chassis.

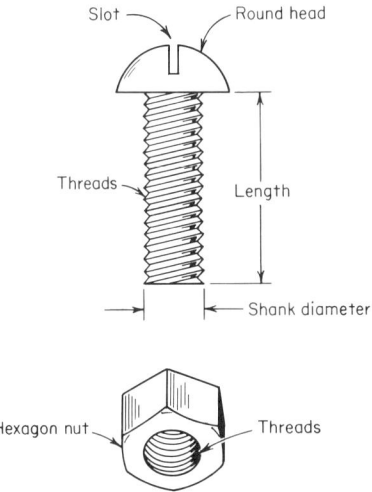

FIG. 9·2 Roundhead machine screw and nut.

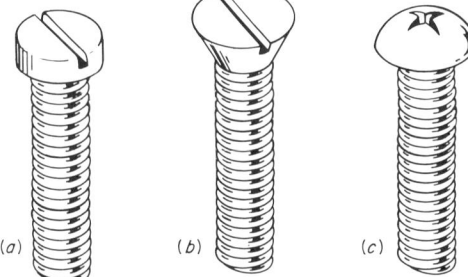

FIG. 9·3 Assorted machine screws: (a) Fillister head. (b) Flat head. (c) Phillips head.

of screwdriver is used to turn this roundhead screw. Other machine screws used are shown in Fig. 9·3.

Machine screws are classified according to the type of head employed. The four types of heads shown in Figs. 9·2 and 9·3 are representative. Other types of heads are also available.

Further identification of a screw is made by specifying the length and diameter of its shank and the quantity of threads each inch of shank contains. For example, the roundhead machine screw of Fig. 9·2 might have a code number 6 shank diameter and a ½-inch shank length. Within this length would be found 16 threads. A screw like this would be identified as a ½-inch 6-32 roundhead machine screw. The corresponding nut illustrated would be identified as a 6-32 hexagon nut.

Screw diameters are identified by code numbers, such as the number 6 above. The larger the diameter of the screw shank, the greater will be the code number. In electronics work screws with shank diameters having code numbers of 10 or less are encountered regularly.

NUTS

In addition to the plain hexagon nut seen in Fig. 9·2, a great variety of special-purpose nuts are available. Each nut has features that dictate its application. A small assortment of special-purpose nuts are shown in Fig. 9·4. The self-locking hexagon nut contains a few fiber threads to ensure continued support during vibration. The knurled thumb nut allows for quick finger tightening. The

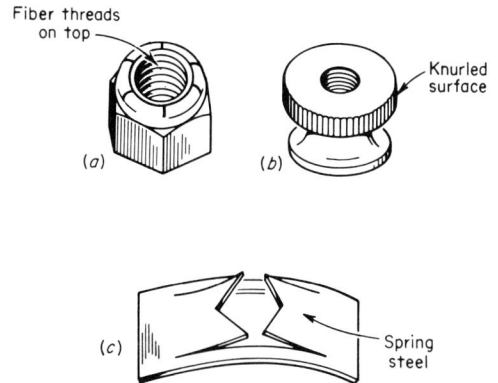

FIG. 9·4 Special-purpose nuts: (a) Self-locking hexagon nut. (b) Knurled thumb nut. (c) Speed nut.

speed nut is not threaded. It is used with self-tapping sheet-metal screws, which are described below. Other types of nuts are also available.

WASHERS

Some washers are used to protect a surface from damage, while other washers are used to ensure continued engagement between a screw and its respective nut. Figure 9·5 illustrates an assortment of washers. The flat washer is used primarily for the protection of a surface; the other three washers are for *locking* purposes.

BOLTS

When two large heavy units have to be united and held together, a bolt is preferred to a machine screw. A bolt is illustrated in Fig. 9·6. The shank diameter is expressed in fractions of an inch instead of by the use of a code. The threads are quite coarse, compared with machine screws, and most bolts are not threaded right up to the head. The head allows for the use of a wrench instead of a screwdriver.

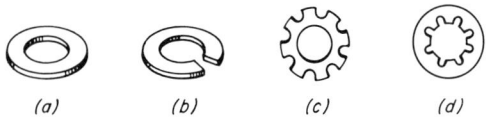

FIG. 9·5 Assorted washers: (a) Flat washer. (b) Spring-lock washer. (c) External-teeth, binding-lock washer. (d) Internal-teeth, binding-lock washer.

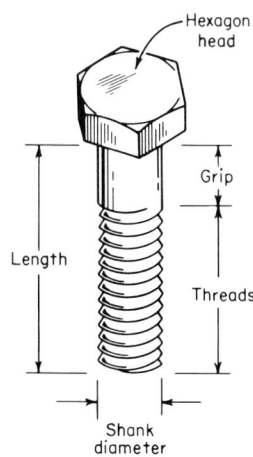

FIG. 9·6 Identification of a bolt.

COTTER PIN

Bolts and nuts are often secured to each other with the aid of cotter pins. To provide for the use of a cotter pin the bolt has a small hole drilled through the threaded shank. After the nut has been tightened in place, a cotter pin is inserted through the hole, and the ends of the pin spread apart. Such an arrangement prevents the nut from working itself loose during vibration. A cotter pin is illustrated in Fig. 9·7a.

RIVETS

Modern fabrication techniques continue to make use of the rivet for the permanent binding of two pieces of metal. Since the rivet is not threaded, its application can be quite simple and fast. This of course aids in the mass production of electronic devices. A rivet is illustrated in Fig. 9·7b.

SELF-TAPPING SHEET-METAL SCREWS

Another aid to mass production is the self-tapping screw used on sheet-metal products. Such a screw is illustrated in Fig. 9·8. Because of the self-tapping

FIG. 9·7 Miscellaneous locking devices: (a) Cotter pin. (b) Rivet.

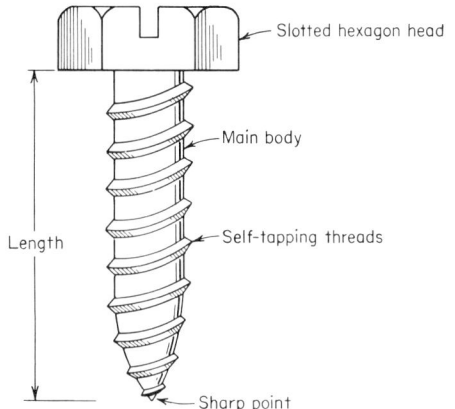

FIG. 9·8 Self-tapping sheet-metal screw.

feature, no threaded nuts or threaded holes are necessary. A hole of sufficient diameter to give clearance to the main body of the shank is all that is necessary. The threads of the screw are designed to cut their own corresponding threads in the sheet metal. When the use of a nut is necessary, the speed nut illustrated in Fig. 9·4c is used.

CLAMPS AND BRACKETS

When it becomes necessary to hold a cable or pipe in place, a clamp is used. Such clamps are illustrated in Fig. 9·9a and b. To protect the surface of the cable or pipe, the clamp may be covered with a section of rubber as shown in Fig. 9·9b. A protective clamp is known as a *cushioned clamp*.

A bracket is used as a mounting support for boxes, terminal strips, and transformers. Brackets are found in a variety of sizes and shapes. A simple right-angle mounting bracket is illustrated in Fig. 9·9c.

9·2 ELECTRICAL-SUPPORT HARDWARE

The following items contribute to the safety and convenience of operation and fabrication of electronic devices. They are not as essential as resistors and capacitors, yet without them electronics could not have progressed to its present state.

SOCKETS

Sockets are popular in vacuum-tube applications (see Fig. 2·37), but they find many other applications as well. A socket used with pilot lamps is illustrated in Fig. 9·10a. A pilot light supports an electrical device to indicate the presence of power at a particular circuit.

SWITCHES

Switches are manufactured in many forms. They are used to interconnect one or several circuits in various arrangements. Two switches frequently employed in electrical and electronic work are the *toggle switch* and the *rotary switch*. These switches are illustrated in Fig. 9·10b and c.

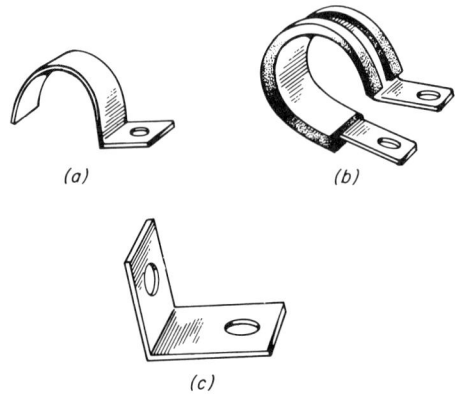

FIG. 9·9 Clamps and brackets: (a) Cable or pipe clamp. (b) Cushioned clamp. (c) Mounting bracket

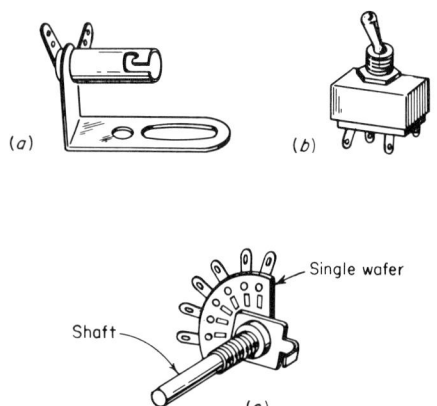

FIG. 9·10 Examples of electrical support hardware. (a) Pilot-light socket. (b) Toggle switch. (c) Rotary switch.

FUSES

The fuse is a safety device designed and utilized to protect good electrical and electronics components from damage when a fault appears in a circuit. A fuse is illustrated in Fig. 9·11a. The fuse element is designed to melt when an overload exists. In melting, this fuse element automatically opens the circuit and thus prevents continued application of electrical power to the circuit. Fuses are rated by the amount of current they can withstand.

Various types of fuse holders are available. The fuse block shown in Fig. 9·11b holds the fuse by means of spring pressure.

CIRCUIT BREAKER

The circuit breaker has the same function as a fuse. However, in a circuit breaker there is no element to melt. The circuit is broken by the separation of two or more contacts when an overload causes a heat-sensitive metal in the circuit breaker to bend. This action takes place within the *thermal type* of circuit breaker. Circuit breakers allow for *resetting* of the open contacts to a closed condition when the fault in the circuit has been eliminated. A thermal-type circuit breaker is shown in Fig. 9·11c. Not all circuit breakers use the thermal principle, but their objective is the same.

INSULATING AIDS

Often it is necessary to mount a component (such as a variable capacitor) without allowing the part to make contact electrically with the chassis. In order to do this insulators of different types and forms are available. Three such insulating aids are illustrated in Fig. 9·12. The porcelain standoff mounting support is threaded at both ends. The grommet is made of rubber. It is also used to protect wires or cables that pass through a metal hole. The fiber shoulder washer is a single unit that allows for centering a screw through a hole in a metal chassis, thus avoiding contact between the screw and the metal chassis.

WIRE-CONNECTION AIDS

Wire leads are normally soldered into a circuit. For this, various types of terminals are employed, as covered in Chap. 8. A single terminal is illustrated in Fig. 9·13a and is referred to as a *solder lug*. It is used as a tie point between a wire and a screw terminal.

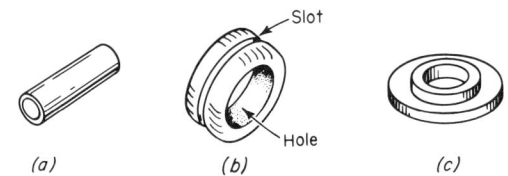

FIG. 9·12 Insulating aids: (a) Porcelain standoff mounting support. (b) Rubber grommet. (c) Fiber shoulder washer.

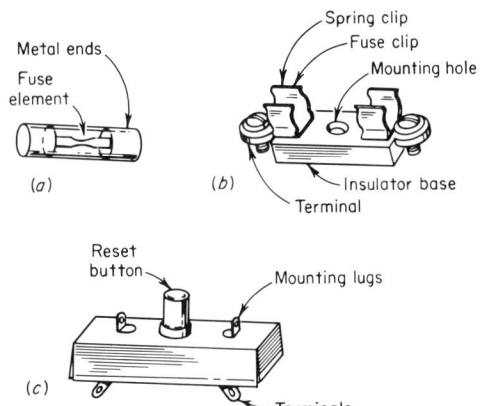

FIG. 9·11 Fuse and circuit breakers: (a) Fuse. (b) Fuse block. (c) Circuit breaker.

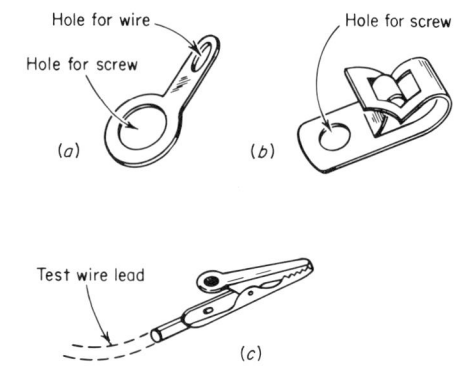

FIG. 9·13 Wire-connection aids: (a) Terminal solder lug. (b) Fahnestock clip. (c) Alligator clip.

The *Fahnestock clip* shown in Fig. 9·13b is used when a temporary tie point for wires is needed. Wires are held in place by spring pressure.

Test wire leads often terminate with an *alligator clip*, shown in Fig. 9·13c. This provides for a quick method to connect a wire into a circuit for test purposes.

9·3 MECHANICAL ASSEMBLY PRACTICES

A person working in electronics fabrication, or the electronics technician working in a repair shop, will quite often be required to install an extra switch, control, or transformer that was not anticipated. For example, let us assume that an added device has to be installed on a chassis. The task must be *planned*, and any holes needed must be *drilled* prior to installing the added part. Further, wrenches, screws, nuts, etc. must all be used. This will require that the person who is to make such installation know how to use the necessary equipment, *properly and safely*. Equipment and procedures used for this purpose are those normally associated with *mechanical assembly* functions.

PLANNING AND LAYOUT

For purposes of continuity, let us assume that a potentiometer (Fig. 2·13) is to be installed on a chassis. To start the task, it is necessary to *plan* where to drill the hole necessary for installing such a potentiometer. This planning must take into account the problem of connecting lead lengths. It is also important to figure out if there will be enough clearance between the potentiometer body, its terminals, and adjacent components or chassis walls. Further, the size and type of knob must be considered.

Having decided *approximately* where to install this potentiometer, it next becomes necessary to verify and finalize this decision by using a linear scale (Fig. 4·14) to *measure* the exact location of where the hole is to be drilled. Having made the necessary measurements, and having marked the exact point with a lead pencil or *scribe* (Fig. 5·15), the chassis has to be supported on a block in preparation for *center punching*. See Fig. 9·14. A center

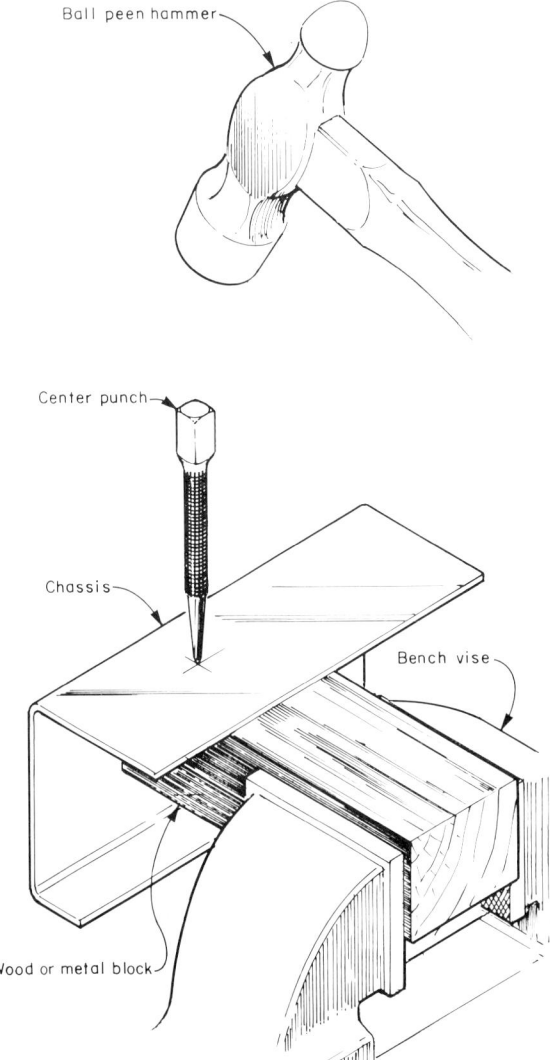

FIG. 9·14 Use of the center punch.

punch mark is necessary in order to ensure that the twist drill (Fig. 5·20) will make the hole at that location, and not run off to a side.

Measuring the diameter of the *threaded* portion of the potentiometer, with either a vernier caliper or a micrometer (Fig. 5·14), we find it necessary to drill a 0.468-inch ($^{15}/_{32}$-inch) diameter hole. This selection can be verified by the use of a drill-size gauge similar to that shown in Fig. 9·15. The con-

153

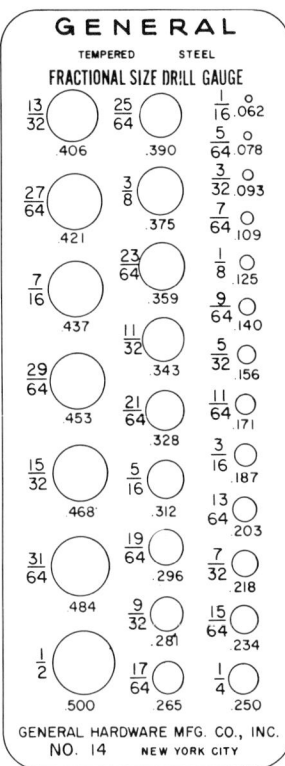

FIG. 9·15 Drill-size gauge. (Note that all hole sizes are listed in both common and decimal fractions of an inch.)

trol is inserted in the gauge, making sure that the hole selected will allow a good, snug fit.

DRILLING HOLES FOR MOUNTING CONTROLS
Having decided that a $15/32$-inch hole is necessary, it is best to drill a $3/16$-inch *pilot hole* first. However, before drilling any hole, the chassis should be secured in a bench vise (Fig. 9·14) or clamped to a tabletop with "C" clamps (Fig. 5·18a).

If a *portable drill* (Fig. 5·20) is to be used, disconnect the power cord of the drill and insert the $3/16$-inch twist drill in the *chuck* securely. Use the *chuck wrench* (Fig. 5·20) for this purpose. Next, put on *safety glasses* and proceed to drill the hole. If the chassis is unstable when you press, it may be necessary to further clamp-down the chassis.

Once the $3/16$-inch hole is drilled, and assuming the portable drill chuck can only accept up to a $3/8$-inch twist drill, proceed to enlarge the hole by using a $3/8$-inch twist drill in the same manner as when drilling the $3/16$-inch hole.

The hole must now be enlarged further to $15/32$-inch by using a round file (Fig. 5·22). A flat file (Fig. 5·22) may also be necessary to remove the burrs. A portable electric grinder (Fig. 5·23) will also be useful for enlarging holes. *Safety glasses should always be worn when working with metal, especially when power tools are being used.*

When the hole is big enough, and the burrs have been removed, mount the potentiometer with the necessary nuts on the chassis. Measure and *scribe* a mark where the shaft is to be cut so that the knob, when installed, is close to the chassis. Remove the potentiometer and clamp it *by the shaft* in a bench vise as illustrated in Fig. 9·16.

Using a hacksaw (Figure 9·16), proceed to cut the shaft at the scribed mark. Use both hands on the saw when cutting. However, you will have to hold the potentiometer with one hand while completing the cut with the other hand just prior to cutting through the shaft. This is necessary in order to keep the potentiometer from falling to the floor and being damaged. Now, using the flat file, *carefully* file the burr off the shaft. Do not use too much pressure when filing or damage to the potentiometer may result.

Install the potentiometer on the chassis, using a flat washer, internal-teeth binding lock washer, and hexagon nuts as necessary. However, before tightening the nuts, be sure to position the terminals of the potentiometer in their final orientation. Having installed the potentiometer, you are *now ready* to start the electrical wiring.

MAKING LARGE HOLES IN CHASSIS
Whenever there is need to make a large hole in a chassis, such as the holes necessary to mount sockets, the use of a *drawpunch* is advisable. The use of a drawpunch is illustrated in Fig. 9·17. Notice that four steps are necessary: (1) Scribe and center punch the location of the hole; (2) drill a pilot hole; (3) drill a hole large enough for the draw bolt; (4) use the drawpunch to *press* a cutout. Drawpunches are available that can cut out round, square, or

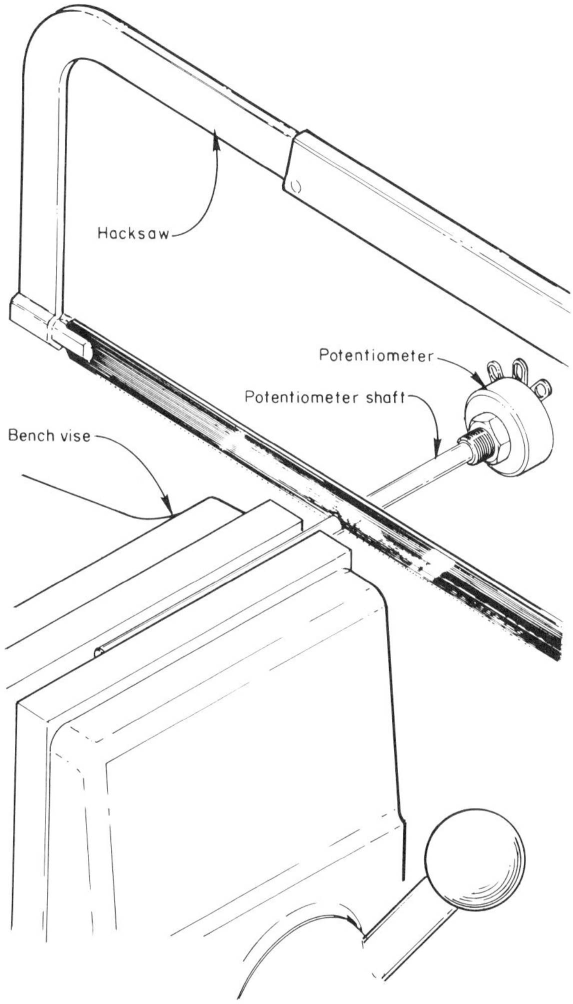

FIG. 9·16 Cutting a control shaft.

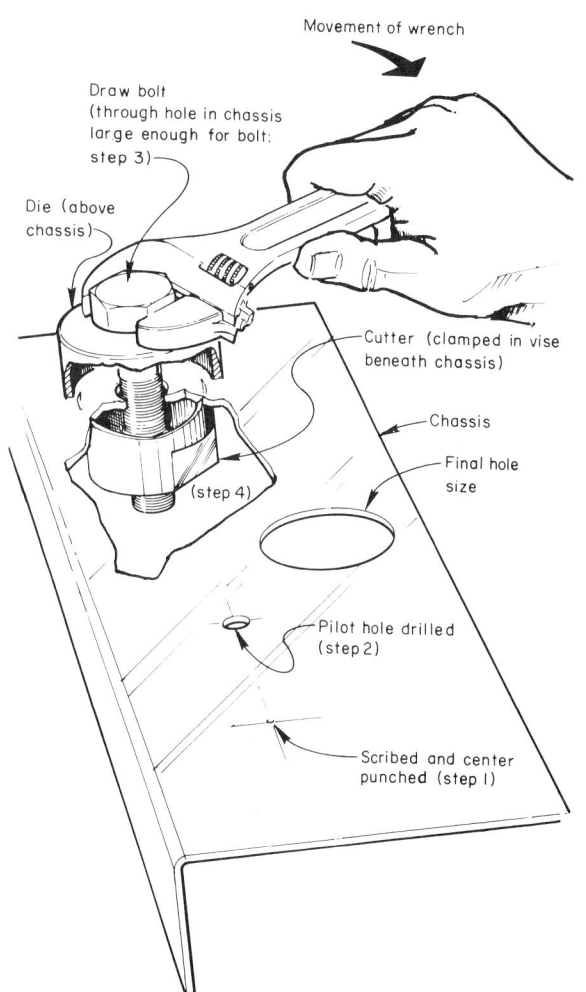

FIG. 9·17 Use of the drawpunch.

rectangular holes of various sizes. The use of drawpunches is limited to certain size holes; to certain metals; and to certain thicknesses of metal. However, drawpunches can be used on most metal chassis used in electronics fabrication.

USE OF THE DRILL PRESS

Use of the drill press (Fig. 9·18) requires the close observation of safety practices. Among these are:

1. The use of safety glasses;
2. The use of a drill-press vise or "C" clamps to secure the job;
3. Knowledge of the location of the power switch;
4. The use of hair nets when the operator has long hair;
5. The use of long-sleeved clothing;
6. The use of long pants or protective coveralls;
7. The use of fully-covered shoes;
8. Making sure that the drive belt housing is secure;

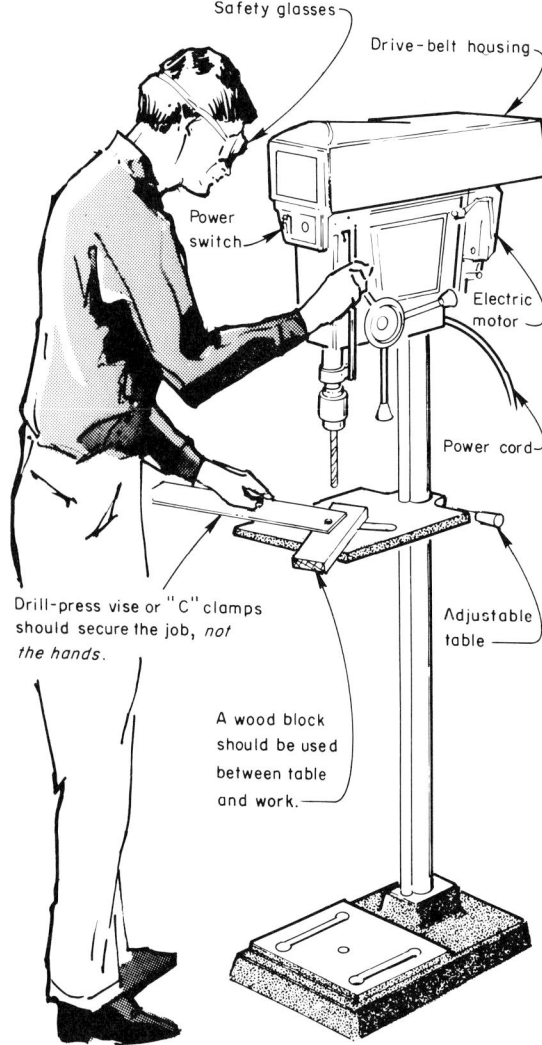

FIG. 9·18 Use of the drill press.

1. The use of good, sharp twist drills;
2. The use of a wooden block between the job and the drill-press table;
3. The proper selection of V-belt pulleys inside the drive belt housing, according to the speed of drill rotation needed;
4. The application of the right amount of pressure when drilling; and
5. The use of lubricating oil when necessary.

Usually, whenever the drilling operation causes smoke to be emitted (Fig. 9·19), it is an indication that something may be wrong. Since smoke is caused by heat, smoke indicates that the twist drill is excessively hot. A few drops of lubricating oil may be used to reduce the metal friction (which causes the heat), as well as to act as a cooling agent. However, the friction may be caused by a dull twist drill which is not cutting properly, or by the drill

9. Not allowing loose tools to remain on the drill press table;
10. Making sure that the chuck wrench has been removed from the chuck prior to turning the power on; and
11. All other safety precautions applicable to rotating machinery.

In addition, work practices that contribute to a good job are:

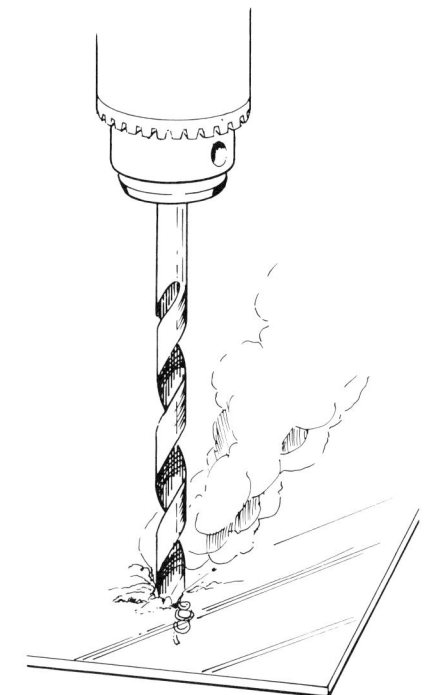

FIG. 9·19 Smoke during drilling operation is an indication that something may be wrong.

rotating at a speed not appropriate to the combination of job material and the type of steel used in the twist drill. Dull twist drills may be sharpened by using a bench grinder, such as shown illustrated in Fig. 5·24. However, sharpening drills should be done only after detailed instruction. It is not the scope of this text to cover sharpening of drills.

Regardless of what causes a twist drill to become excessively hot, this heat is damaging to the drill. A heated drill loses its temper (hardness) and becomes useless.

9·4 SUMMARY: HARDWARE AND MECHANICAL ASSEMBLY

Hardware as used in electronics work refers to parts *other than* resistors, capacitors, inductors, vacuum tubes, semiconductors, or batteries. Some of the hardware is designed primarily for physical support, while other units of hardware provide an electrical support to the circuits.

The installation and handling of hardware require the use of hand and power tools. This installation requires very careful planning, layout, and the *proper use* of the tools and equipment. All of this falls under the category of *mechanical assembly*.

The following jobs are designed to better acquaint the student with all types of hardware, the use of tools and equipment, and other routine mechanical assembly practices.

EXERCISES

JOB 9·1 How to make use of catalogs to identify hardware

OBJECTIVES
1. To become acquainted with the Radio-Electronic Master catalog
2. To become acquainted with manufacturers' catalogs
3. To learn where to get information about parts not normally found in textbooks

MATERIALS REQUIRED

Equipment
Radio-Electronic Master catalog
Assorted catalogs of hardware manufacturers or dealers

PROCEDURE
1. Look over all the catalogs supplied.
2. Inspect the index of each catalog.
3. Study the self-tapping sheet-metal screw illustrated in Fig. 9·8 in this text.
4. Look for the section on sheet-metal screws in each of the catalogs supplied.
5. Make a list of all variations in sheet-metal screws to be found.
6. Repeat steps 3 to 5 for each unit of hardware illustrated in this chapter.

QUESTIONS
1. List the areas of equipment and supplies covered in the Radio-Electronic Master catalog.
2. Look in the Radio-Electronic Master catalog for physical dimensions of a 1-watt carbon resistor.
3. If a Radio-Electronic Master catalog is not available, how can one obtain information on a particular item?

JOB 9·2 How to make a hardware display

OBJECTIVES
1. To familiarize students with the great variety of mechanical hardware available
2. To provide a means for students to actually see hardware units they may not otherwise be exposed to during training

MATERIALS REQUIRED

Supplies
Poster board, 18 × 24 inches
Rubber cement
Felt pen
Assortment of mechanical hardware, one general type

PROCEDURE

1. Various members of your class have been assigned a project such as this. Each person so assigned will make a display covering one basic type of hardware. When all of the projects are complete, the class will have a fairly complete display of all hardware normally available. This is an extra credit assignment. The following instructions *assume* that your assignment will be to make a display of *washers* available.
2. Get the assortment of washers provided to you and clean them thoroughly.
3. Familiarize yourself completely with every type of washer contained in the kit of parts given to you. You are about to become the *expert on washers* in your class.
4. If you become aware that a certain type of washer is not included in the kit given to you, do not hesitate to ask your instructor to get it for you.
5. Neatly lay the washers on the poster board supplied, arranging them to form a nice display.
6. Trace each washer with a sharp pencil.
7. Remove the washers and store them.
8. With the same sharp pencil lightly *print* the name of each washer in fairly large letters under each washer traced.
9. With a ball-point pen, briefly write under the lettering the uses for each type of washer.
10. With the felt-pen, go over the pencil-lettering of the name of the washer.
11. Get one washer at a time and coat one surface with rubber cement.
12. Place the washer at its designated location, holding it in place until the rubber cement dries.
13. Your display is now complete, assuming you have mounted all washers.
14. Be prepared to answer questions about every washer in your display.

QUESTIONS

1. Why is it important for an electronics assembler to know the various units of hardware available?
2. Could electronics exist without hardware? Explain.
3. List the various units of mechanical hardware that could possibly be used in an electronics chassis.

JOB 9·3 How to identify switches

OBJECTIVES
1. To learn to recognize switches
2. To learn to identify switches

MATERIALS REQUIRED

Equipment
Ohmmeter
Switch catalog or Radio-Electronic Master catalog
Knife switch, spst
Knife switch, dpdt
Toggle switch
Slide-lever switch
Rotary switch
Momentary-contact switch
Schematic diagram for each switch supplied

PROCEDURE

1. Study the two knife switches supplied. (Use a catalog to identify the switches if necessary.)
2. Notice that one of the knife switches has only two terminals. Further, this switch has only one arm that can complete a circuit. The switch can either open or close a single-conductor circuit. It is known as a single-pole single-throw (spst) knife switch. Compare this information with the schematic symbol supplied.
3. Make a continuity check of this switch with the ohmmeter. Connect the test leads to the terminals of the switch and make the test with the arm (pole) lifted and engaged.
4. Make the various possible continuity tests on the double-pole double-throw (dpdt) knife switch. With this switch two separate circuits can be controlled simultaneously. Compare the switch with the schematic diagram supplied for this switch.

5. Since most other types of switches are encased, it is wise to understand fully the continuity tests made above.
6. All switches follow the same principles, illustrated above. The main difference between switches is how the connecting medium is engineered. The schematic symbols and diagrams of switches vary little, as can be seen by comparing the schematic diagrams supplied for the switches you are using.
7. Practice making all the possible continuity tests on the remaining switches in your possession. Pay close attention to the schematic diagram supplied for each switch and the identification of each switch.

QUESTIONS

1. What are the limitations of an spst switch?
2. Is it possible to use a dpdt switch as an spst switch? Explain.
3. What type of switch was the rotary switch supplied?
4. Assume an spst switch on a radio receiver breaks. How can the radio receiver be operated?

JOB 9·4 How to lay out a job

OBJECTIVES

1. To become familiar with the principles of planning a mechanical assembly job
2. To encourage electronics students to plan thoroughly the installation of a switch, control, pilot lamp, etc.

MATERIALS REQUIRED

Equipment
Drafting tools
Metal scale, 6-inch

Scissors
RCA SCR-Experimenter's Manual, KM-71 or equal
RCA Solid-State Hobby Circuits Manual, HM-90 or equal

Supplies
Notebook paper and pencil
Rubber cement

PROCEDURE

1. Review the two manuals supplied to get ideas. Notice that these manuals contain information relative to planning and layout of construction of electronic devices. Included is information relative to layout and drilling on metallic chassis, and the use of templates for drilling and cutting preplanned holes in laminated circuit boards. With these ideas in mind, the combination of the two methods can be used in planning and executing construction or modification of electronics devices encountered by electronics personnel.
2. Review Sec. 9·3 of this textbook, with special emphasis on the subsection entitled *Planning and Layout*. Notice that the method of planning and layout right on the chassis is discussed. This is a well accepted practice. However, this job will acquaint you with an alternate method. You may choose the method you like best. However, both methods emphasize *planning your job before you drill*. This is the important part.
3. Using the same example given in Sec. 9·3 where there is need to install a potentiometer on a chassis, decide the general area where the potentiometer is to be installed. Then cut a piece of notebook paper to fit over the area completely, making sure that the paper reaches at least two edges of the chassis for reference purposes.
4. Having cut this paper which now represents the area of chassis to be worked on, proceed to do your planning on the paper on top of your desk. Mark the paper all you want, using drafting tools or other convenient objects.
5. When your layout is complete on the paper, turn it over and coat it with rubber cement. Then cement the paper right over the chassis, using the same edges for reference as called for in step 3 above.
6. In a few minutes, when the rubber cement is dry, you can proceed with the center punching and drilling, *right through the paper*.
7. After the holes are drilled, simply peel off the remaining paper and rub off the remaining rubber cement.

QUESTIONS
1. Why must an electronics assembler or technician know the basics of mechanical assembly?
2. Of the two methods mentioned above for planning and layout, which do you prefer? Why?
3. Why is planning the simple operation of drilling a hole so important?

JOB 9·5 How to install a potentiometer on a chassis

OBJECTIVES
1. To become familiar with good mechanical assembly practices often needed in the shop and done by nonmechanics personnel
2. To become familiar with tools and equipment needed to install physical objects

MATERIALS REQUIRED

Equipment
Hand tools for mechanical assembly
Bench vise
Wooden block, 2 × 4 inches (see Fig. 9·14)
Safety eyeglasses
Portable drill motor
Portable electric grinder
Hand tools for electrical assembly

Supplies
Potentiometer, complete with related nuts and washers
Hookup wire
Solder

PROCEDURE
1. Mark the chassis where the potentiometer is to be installed according to the discussion in Sec. 9·3, subsection entitled *Planning and Layout*, or by the method explained in Job 9·4.
2. Perform the balance of the operations explained in Sec. 9·3 of this textbook pertaining to the installation of a potentiometer on a chassis.
3. After the physical aspect of installing a potentiometer on a chassis has been completed, proceed to connect the potentiometer *electrically* to the appropriate circuit. This will, of course, require the use of hookup wire and soldering.

QUESTIONS
1. Give a thorough explanation regarding the functions of an electronics assembler working to fabricate electronics products.
2. Did you mention any functions of the electronics assembler which might be interpreted as being the function of a mechanical assembler when you answered question number one? Give your reasons.
3. Do you think an electronics technician has need to know something about mechanical assembly? Why?

JOB 9·6 How to make a heat shunt for power transistors

OBJECTIVES
1. To provide a simple project that involves planning, layout, and mechanical construction
2. To emphasize the need for heat dissipation of heavy-duty semiconductors
3. To provide the opportunity to look up transistor outlines

MATERIALS REQUIRED

Equipment
Hand tools for mechanical assembly
Safety eyeglasses
Portable drill motor
Portable electric grinder
Bench vise
Transistor manual

Supplies
Aluminum stock, right-angle shaped, 6-inch length
Power transistor, 2N2869 (complete with mounting hardware)

PROCEDURE
1. Study the heat shunt for power transistors illustrated in Fig. 9·20. You will make a similar heat shunt.
2. Plan and lay out the project. You will have to get the necessary physical dimensions of the transistor supplied from a transistor manual. Look in the *Outlines* section.
3. The same *relative size* of the heat shunt illustrated in Fig. 9·20 will be maintained.

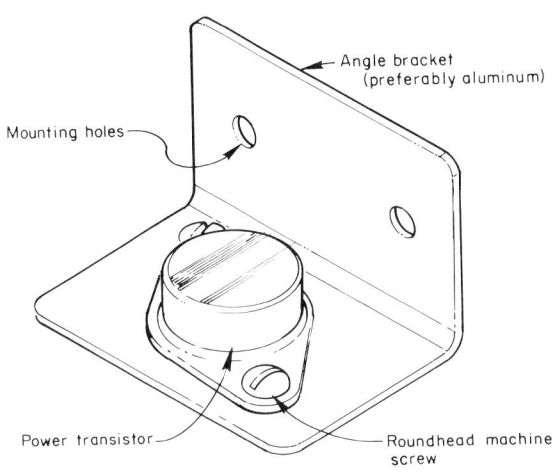

FIG. 9·20 Heat shunt for power transistors.

4. When planning and layout is complete, proceed with construction.
5. Mount the transistor on the heat shunt.
6. Submit the completed assembly to your instructor for evaluation.

QUESTIONS

1. What is the purpose of a heat shunt such as the one you have just built?
2. If aluminum is not available, can you use angle iron? Explain?
3. Were the dimensions given in the *Outlines* section of the transistor manual necessary? Explain.
4. Describe the procedure you used to construct this heat shunt.

10

INSPECTION AND QUALITY CONTROL

Since fabrication of electronic devices is normally an assembly-line progress, accomplished by teamwork of many people, there exists the possibility of an accidental oversight in construction or assembly. This oversight may come about due to an assumption that someone else had done a certain task, which in reality could have been overlooked. Also, some operations in construction may have been done poorly or not to standard. This is often the case when inexperienced workers are first employed, or it may be due to just plain human error. However, *defective workmanship cannot be allowed, regardless of the reason.* This is especially true when the finished product is in an aircraft or a spacecraft designed to carry passengers.

10·1 RELIABILITY

In order to check that the finished product has been fabricated properly, it must be *tested* in order to see if it operates properly. However, a *final test* is not enough. There must be assurance that the unit has been fabricated with high *quality* components, and that prolonged, *reliable* operation of the device is a normal occurrence. In order to have this assurance, some individuals are trained as *inspectors*. Their interest must be that of *quality control* only.

Reliable operation of the finished product can only be assured if inspectors are stationed throughout the manufacturing process. This includes inspecting all components purchased from outside sources, inspecting workmanship during assembly, the testing of subassemblies, and the final testing of the finished product. It also requires the periodic certification of all tools, test equipment, and individuals utilized in assembly and fabrication of the hardware.

10·2 NONDESTRUCTIVE TESTING

Inspection of a joint that has been soldered requires a thorough investigation of the connection. To pull the wire with force will damage or destroy the junction under test. (This is called a *pull-test*.) Although this might be done on a *sampling basis*, it does not check *every* connection. If testing every connection is mandatory, industrial *X-ray* equipment must be used. Such testing is *nondestructive*, and yet can reveal hidden flaws. Nondestructive testing utilizing magnetized particles and ultraviolet light are seldom used on electronic equipment. The units are too small for this type of testing. However, this type of testing might be done on the larger housings.

10·3 VISUAL INSPECTION

Visual inspection is the method most often used to check quality of workmanship. This type of inspection is done regularly during the manufacturing process. Inspection is accomplished with the aid of *magnifying glasses* or with *industrial microscopes*. Optical inspection devices are shown in Figs. 3·7 and 5·8. Quite often these same types of optical tools are used by the production worker during routine assembly operations. In fact, these aids are standard equipment when working in microelectronics.

10·4 CONTINUITY CHECKERS

About the simplest type of electrical test is that accomplished with the aid of *continuity checkers*. These checkers can be a simple test lamp connected in series with a small battery and two test probes. See Fig. 10·1.

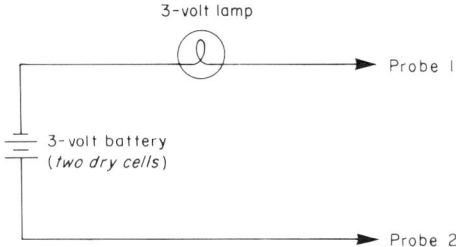

FIG. 10·1 Continuity checker schematic diagram.

The continuity checker is used to check if a complete closed circuit exists between two points. It is a "*go, no-go*" type of tester. The checker will reveal if the circuit under test is acceptable or not. This type tester finds application during simple wiring check-outs. One probe is connected to one end of a wire, the second probe is connected to the other end of the wire under test. If the lamp lights, the wiring job is acceptable. If the lamp does not light, it indicates that an "open circuit" exists. A buzzer is often used in place of the lamp. However, buzzers should not be used if the circuit under test contains semiconductor components. These active components, including integrated circuits, may be damaged by the erratic voltages associated with buzzer circuits.

OHMMETERS AS CONTINUITY CHECKERS

An *ohmmeter* can also be used to test for continuity. An ohmmeter is normally one part of a *multimeter* and is used in measuring resistance. A multimeter can also test for voltage or current. Selection of the ohmmeter function is accomplished by means of convenience switches located on the front panel of the multimeter. When the ohmmeter function is selected, this instrument can be used in the same manner as the continuity checker described above. However, deflection of the needle indicator replaces the lamp or buzzer action. Review Job 2·3 regarding how to use an ohmmeter.

10·5 TEST PANELS

During production testing, special *test panels* are designed that simplify check-out of completed sub-

or final *assemblies*. These assemblies can be a wire harness, or it can be a completely wired sub-chassis. Indicators on these test panels can either be a series of lamps, meters, or both. In fact, the test panel can include simulated operational functions, such as motors that rotate, or loudspeakers that emit sound. Test panels are used extensively where elaborate circuit testing is done by personnel lacking in-depth training in electronics testing. The panels are designed by test engineers so that it is only necessary to connect a test harness to the unit under test and then proceed through a "cookbook" testing routine, following procedures outlined by the engineer. Assuming a unit did not pass the test-panel check, it is simply set aside for further checking or troubleshooting by an *electronics technician*. See the sketch of a test panel in use, Fig. 10·2.

Testing of individual components (such as resistors and capacitors) is also possible with a test panel designed for that purpose. By using such panels, it is possible to use job-entry-level personnel to make these tests.

10·6 ENVIRONMENTAL TESTING

Use of a test panel makes possible testing of a completed assembly or subassembly under *varying conditions*. For example, the unit under test can be placed within a refrigerator while undergoing a test. In this manner, it is possible to check how the unit will function under *extreme cold* conditions. The test can also be made while the unit is in an oven to check its ability to function in *hot climates*. Or, the unit can be placed on a vibration table to see if it can continue working normally in an *unstable* environment. For example, why wait to find out if an amplifier designed to be mounted in an unheated section of an airplane will work at the cold temperatures of 30,000 feet when these cold temperatures can be *simulated* in the laboratory?

10·7 SUMMARY: INSPECTION AND QUALITY CONTROL

This chapter can be summarized with the state-

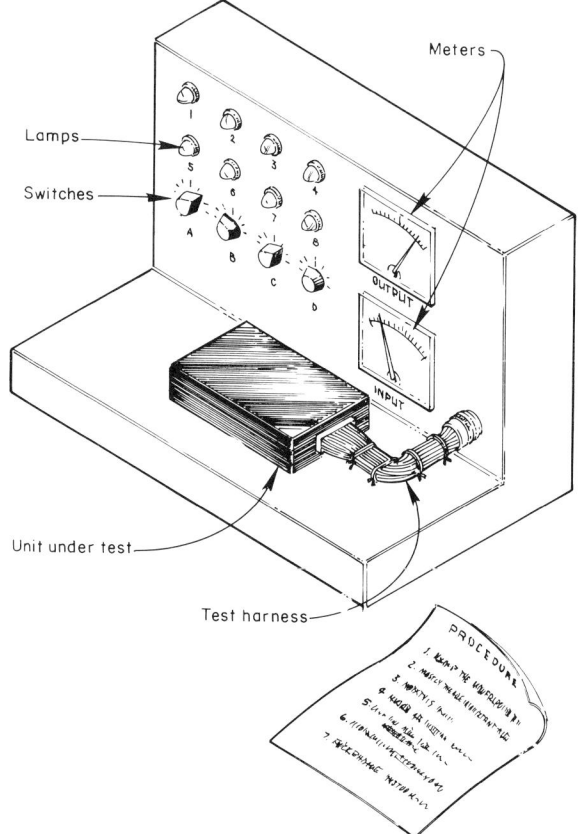

FIG. 10·2 Test panel.

ment: *In order to consider an electronics device completely fabricated, it must work and the customer must accept it.* This is possible only when the unit has been assembled properly, using good components. In order to ensure a product that will satisfy the prospective user, all parts purchased must be inspected to make sure they meet predetermined standards; a constant surveillance of workmanship must be undertaken; tools, equipment, and personnel must be certified regularly; and all completed units must pass operational tests. The summary jobs that follow are designed to better acquaint the student with inspection and quality control procedures.

EXERCISES

JOB 10·1 How to check the quality of a soldered connection

OBJECTIVES

1. To learn the principles of electrical inspection
2. To appreciate nondestructive testing and inspection
3. To learn to use optical aids for inspection

MATERIALS REQUIRED

Equipment
Components mounted on terminal board (having both solid and stranded-wire leads)
Components mounted on printed-circuit board
Magnifying glass (as in Fig. 3·7)
Industrial microscope (as in Fig. 5·8)

Miscellaneous
Well-illuminated area, preferably to include a lamp as shown in Fig. 1·2

PROCEDURE

1. Inspect the quality of soldering and installation of electrical components mounted on a terminal board. Use a magnifying glass and a microscope as necessary. Answer the questions listed below.
2. Inspect the quality of soldering and installation of electrical components mounted on a printed-circuit board. Use a magnifying glass and a microscope as necessary. Answer the questions listed below.

QUESTIONS

1. Are the wire leads pulled tight, or is the right amount of stress relief provided? (See Fig. 8·3.)
2. Are the wire leads bent smoothly, with the proper radius as illustrated in Figs. 8·3 or 8·20?
3. Are the wire leads nicked in any way?
4. If the terminal board includes stranded-wire leads, are the individual strands broken or nicked? (See Fig. 7·12.)
5. Is there sufficient clearance between the insulation (if used) and the terminal? (See Fig. 8·5.)
6. Is the insulation cut roughly?
7. Has the insulation been burnt or damaged during soldering?
8. Has the proper amount of wire wrap been allowed? And, what constitutes the *proper amount* of wire wrap?
9. In the printed-circuit board, have all the tolerances illustrated in Figs. 8·19 and 8·22 been observed?
10. How does the soldering compare to Figs. 6·11, 6·12, and 6·18?
11. Did solder flow into the stranded wire beneath the insulation? Is this good or bad? Why?
12. List other things you looked for during this inspection.
13. Did you damage or destroy any of the good soldered joints that you inspected? Is this known as nondestructive testing? Why?
14. List the things that an electronics fabrication inspector must know.

JOB 10·2 How to make a pull-test on a printed-circuit soldered joint

OBJECTIVES

1. To illustrate a method for testing the quality of a soldered connection
2. To test the student's ability to solder on printed-circuit material correctly
3. To stress the importance of paying attention to every detail in soldering

MATERIALS REQUIRED

Equipment
Hand tools for electrical assembly, including a 25-watt soldering iron
Soldering practice board (see Fig. 6·28)

Supplies
Wire solder, 60/40 alloy, No. 20 gauge, resin core
Solid copper wire, No. 24 gauge, 8 inches long

PROCEDURE

1. Remove all insulation and oxides from at least ½ inch from one end of the wire.
2. Using the method illustrated in Fig. 8·21, bend the cleaned end of the wire ¼ inch from the end. (Make a 90-degree bend.)
3. Tin the end of the wire.
4. Tin a little strip of the sheet copper mounted on the practice board, *close to the thumbtacks.* The strip tinned should be approximately ⅛ inch wide and ½ inch long. (See Fig. 10·3.)

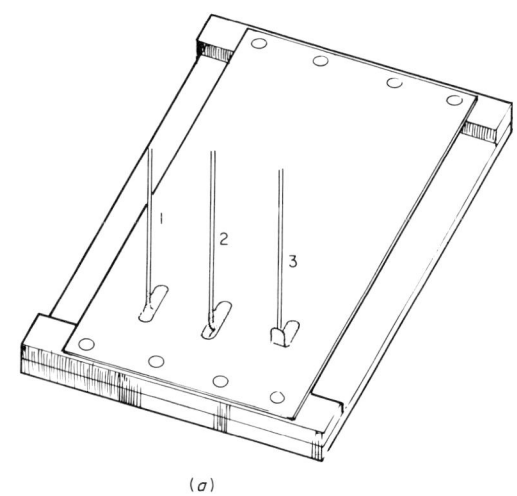

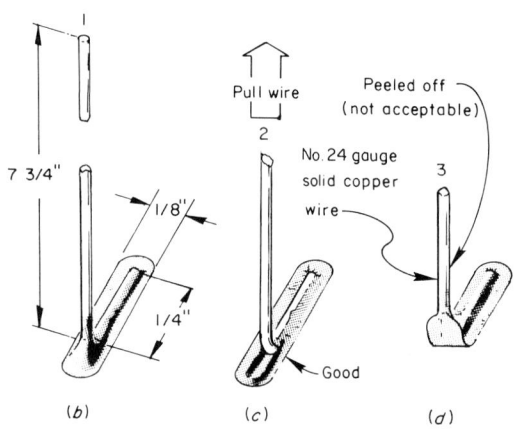

FIG. 10·3 Simulating a peel-pull test (of soldering) on printed-circuit boards.

5. Solder the tinned end of the wire to the tinned strip on the practice board. Observe contour soldering. It is important to use the minimum amount of solder. (*Note:* You may have to devise a method for supporting the wire while you solder.) The finished solder job should appear like Fig. 10·3b.
6. To test the quality of the soldering job, pull on the wire as indicated in Fig. 10·3c. The pull should be *easy, smooth,* and *steady.* If the solder joint is good, the wire will *pull through the top surface* of the solder, leaving a trough behind.
7. If the solder did not penetrate through the surface of the sheet copper, the solder will *peel off* from the sheet copper while making the pull test. This is illustrated in Fig. 10·3d.
8. The above test can also be made on the foil conductor of a printed-circuit board. In fact, when this test is conducted in industry on sample printed-circuit boards, the soldered point must withstand at least 4 pounds of pulling pressure as measured on an accurate, calibrated pull-stress machine.

QUESTIONS

1. Why was it necessary to conduct this test near the thumbtacks of the practice board?
2. Would the test be valid if you jerked the wire while making the pull? Explain.
3. Why do we consider the results *good* when the wire pulls through the top surface of the solder as in Fig. 10·3c?
4. Explain in detail why the solder peels off a soldered surface as illustrated in Fig. 10·3d.
5. Were you able to control your tinning of the sheet copper to a narrow strip as called for in step 4? Explain why you got the results that you did during this phase of the job.

JOB 10·3 How to make a continuity tester

OBJECTIVES

1. To acquaint students with the simplicity of basic testers used during inspection procedures
2. To provide a simple useful project for students

MATERIALS REQUIRED

Equipment
Hand tools for electrical assembly
Soldering iron

Supplies
Two-dry-cell flashlight, with dry cells and operating
Spool of test lead wire, rubber or plastic insulation
Solder, resin core
Electrical tape

PROCEDURE

1. Disassemble the flashlight. (You will use the parts to make your continuity tester.)
2. Cut two lengths of test lead wire to meet your needs. The lengths may be 12 inches each, or they may be 36 inches long. It all depends upon what you will be using your continuity tester for.
3. Strip the insulation from the ends of the test leads and tin. Allow about 1 inch of tinned conductor at each end.
4. Using the parts at hand, design and construct a continuity tester. Use the schematic diagram of Fig. 10·1.
5. Bundle the tester into an insulated package with the use of electrical tape.
6. Touch the two exposed, tinned ends of the test leads. The lamp should light brightly.
7. Improve on this continuity tester as you see fit.

QUESTIONS

1. What is a continuity tester used for?
2. Is there danger of getting an electrical shock with this tester? Explain.
3. Do you think you can use this tester around your automobile? Explain.
4. Could you design and build a better tester of this type? How?
5. What are the limitations of this tester?

JOB 10·4 How to make a test panel

OBJECTIVES

1. To learn the principles of production testing
2. To appreciate test panels

MATERIALS REQUIRED

Equipment
Hand tools for mechanical assembly
Portable drill motor
Bench vise
Safety eyeglasses
Hand tools for electrical assembly
Soldering iron

Supplies
1 Metal chassis, 3 × 4 × 5 inches, with cover
1 Lamp with socket, #47
2 Rotary switches, single-pole, five position
1 Toggle switch, spst
4 Dry cells, 1½-volt, size D
1 Battery holder (for 4 dry cells)
1 Cable connector, male plug (to match cable to be tested)
1 Cable connector, female jack (to match cable to be tested)
   Solder, resin core
   Miscellaneous hardware
   Hook-up wire

PROCEDURE

1. You will build a simple *mass-production tester* that is designed to test wire harnesses being fabricated in great quantity. The harness to be tested is shown in the cable wiring diagram of Fig. 7·39.
2. The schematic diagram of the tester you will build is shown in Fig. 10·4.
3. The tester you will build is to be enclosed in a portable box. However, this same tester can be mounted on the panel of a multiple production-tester as illustrated in Fig. 10·2.
4. Proceed by planning and laying out your work.
5. Construct and assemble the mechanical parts.
6. Electrically assemble the unit.
7. Identify the switches and the switch positions on the front panel of the completed unit.
8. Test the unit by turning switch S_1 to position **5**; turning switch S_2 to position **1**; and turning switch S_3 to the *calibrate* position. The lamp should now light. If the lamp does not light, the lamp itself may be defective; the dry cells may be bad; or

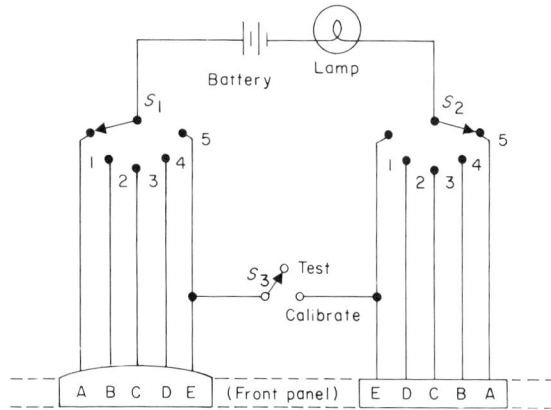

FIG. 10·4 Test panel schematic diagram.

there may be a fault with the wiring. Do not proceed further until you correct the problem.
9. To test a wire harness or cable, if assembled according to the cable diagram of Fig. 7·39, simply join the cable connectors of the harness with the corresponding connectors mounted on the test panel or box. Then, according to the diagram of Fig. 10·4, each *circuit* of the cable can be checked by rotating switch S_1 to position 1 and at the same time rotating switch S_2 to position 5; rotating switch S_1 to position 2 and switch S_2 to position 4; etc. Of course, if the lamp lights up at each set of positions, then that particular circuit can be considered GOOD. If the lamp fails to light during the test of a particular circuit, something went wrong during construction of the cable. This you must assume since the tester itself was checked in accordance with the procedure of step 8.

QUESTIONS
1. How does this tester compare with the continuity tester of Job 10·3?
2. Would it pay to build a test panel if all you were going to fabricate was three harnesses as shown in Fig. 7·39? Explain.
3. Explain the purpose for switch S_3, especially after the test panel or box has been in operation for 6 months.
4. Explain the meaning of *production testing*.

JOB 10·5 How to investigate quality control procedures

OBJECTIVES
1. To acquaint the student with the extent of quality control
2. To acquaint the student with educational opportunities in the area of quality control

MATERIALS REQUIRED

Equipment
Assortment of community college catalogs

PROCEDURE
1. Look in the catalogs supplied to you and investigate courses which are offered in the area of quality control.
2. If you experience difficulty, ask your vocational counselor for assistance. Normally, courses in quality control are taught by personnel who work, or have recently worked in the area of inspection and quality control. Therefore, their course outlines as described in school catalogs closely reflect what actually takes place in industry.
3. If possible, visit a class in quality control as offered in a community college or trade school, and talk to the instructor.
4. Having become acquainted with this area, ask your vocational counselor to help you look up job titles and descriptions dealing with quality control, inspection, and reliability.

QUESTION
1. Write a report on quality control procedures practiced in industry.

Part II

CONCEPTS OF ELECTRONICS

11

BASIC ELECTRICAL PRINCIPLES

Since the electronics assembler manufactures electronics products by following plans and construction designs prepared by engineers, he therefore needs to know the basics of oral and written communication with technical people, and he must have acquired sufficient mechanical skill in order to be able to "bring to life" the engineer's ideas. The assembler must be able to recognize electrical and electronic parts and must put them together using specialized tools and assembly methods. These are the things that the student has learned by studying the first ten chapters of this book. However, if he takes an interest in his job, and if he expects advancement, he must, in addition, acquire a basic knowledge of electronics.

Part II of this text is designed to give students basic knowledge relating to fundamental *concepts* of electronics. Although a knowledge of these basic concepts is sufficient for electronics fabrication personnel (such as assemblers, inspectors, and production-testing people), the information contained in this textbook represents *phase I* of knowledge to be acquired by electronics technicians and preengineering students.

To understand electronics, a knowledge of basic electrical principles is necessary. In fact, electronics developed out of electrical knowledge. Most electrical laws are applicable to electronics. Therefore the study of electronics begins with electricity.

The physicist tells us that electricity originates in the particles of the atom called electrons. He further tells us that all matter contains electrons. Electrons are found in paper, in wood, in metal, in glass, in rubber—in everything. Everything that we see, touch, or feel contains electrons and is electrical in nature. However, electrons are not usable in their basic state. They must be set adrift in a general direction and controlled. A general drift of electrons is known as an *electric current*.

The production of an electric current, its control, and the use we can get out of this current is known as *electricity*.

11·1 ELECTRICITY

ELECTRON MOVING FORCE

The production of an electric current requires force that will exert pressure upon the electrons to set them in motion. This force is known as an *electromotive force* (emf). The unit of measurement for this force is the *volt*.

ELECTRIC CURRENT

Although electrons are found in everything, it is easier to move the electrons contained in certain metals. Copper is one such metal. Copper is therefore said to be a good conductor, and is usually associated with heavy current flow. The unit of measurement of an electric current is the *ampere*.

CURRENT CONTROL

The control of an electric current can be exerted in two ways. One way is by providing paths for the movement of electrons, and the other involves the introduction of opposition to electron movement into the designated paths.

The paths for the electric current are composed of conductors arranged into networks known as *circuits*. The opposition offered to electron movement in its most fundamental form is known as *resistance*. All conductors exhibit a certain amount of resistance to electron movement. The unit used in the measurement of resistance is the *ohm*.

INSULATORS

Materials in which electrons find it difficult to be moved are known as *insulators*. Typical insulators are glass and rubber. A material that can be classed as a very poor conductor is normally a good insulator.

It is quite common to find a copper conductor covered by rubber insulation. The insulation helps confine the moving electrons within the conductor.

APPLICATIONS OF ELECTRIC CURRENT

An electric current is of no use unless we can derive some good from it. The following are two visible results produced by electric currents: the illumination of an electric light bulb and the rotation of an electric motor.

Electricity has also made possible such things as radio, television, remote control, automatic industrial processes, radar, space vehicles, computers, and many others.

ELECTRICAL SYMBOLS

Terms used in electricity are so numerous that abbreviations are constantly employed. For example, the letter E represents voltage, the letter I represents current, and the letter R represents resistance. Symbols in the form of illustrations are also used, to represent certain electrical meanings or conditions. To examine some of these illustrations, refer to the diagram of Fig. 11·1.

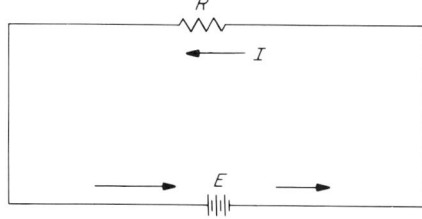

FIG. 11·1 The schematic diagram.

The symbol ⊣⊦⊦⊢ represents the source of electromotive force. The symbol ⟶ represents the flow of electrons. The symbol ⎯⋀⋀⋀⎯ represents the amount of opposition encountered by the electrons. The lines that connect the voltage source to the resistance are very good conductors, and are used to connect these two units together by providing an easy path for the electrons. The complete illustration is referred to as a *schematic diagram*.

VOLTAGE SOURCE

The schematic diagram of Fig. 11·1 shows that electrons move out of one side of the voltage source and reenter the opposite side of this same source. The side of the voltage source that is releasing electrons has an excess of electrons. The side of the voltage source through which electrons reenter is said to have a deficiency of electrons.

An example of a voltage source is a *battery*. Figure 11·2 shows a photograph of a typical automobile battery. Notice that each connecting post has a mark. One post is marked +, while the other is marked −. These posts are also identified as terminals. Therefore a battery has a positive terminal and a negative terminal. The term *polarity* is used in identifying the type of voltage at a terminal. For example, the negative terminal is said to have a negative polarity.

THE ELECTRON THEORY

Scientists have elected to identify the electron as a negative particle of electricity. Therefore a point that has an excess of electrons is said to have a large negative charge. A point that is deficient in electrons has a positive charge. A rule to be remembered is that *electrons flow from negative to positive*. This theory states that electrons flow from a point having an excess of electrons to a point that is deficient in electrons.

11·2 RESISTANCE FOR CURRENT CONTROL

In electrical circuits good conductors are used to connect the voltage source to such items as lamps. A good conductor allows a maximum electric current flow to illuminate a lamp to maximum brightness. If, by contrast, an insulator were used to connect the lamp to the voltage source, the resulting electric current would be so small that no illumination of the lamp would be possible. An illustration of this is seen in Fig. 11·3.

RESISTORS

It is not practical to change from conductors to insulators and back again every time a different amount of current is desired. In fact, there can be little variation in the control of the electric current

FIG. 11·2 Typical automobile battery.

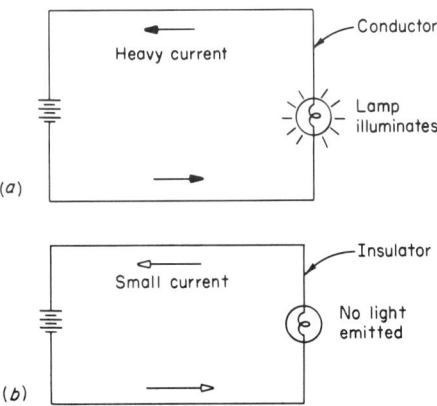

FIG. 11·3 The effects of a conductor and insulator upon the resultant electric current. (*a*) Copper wire connects lamp to battery. (*b*) Rubber band connects lamp to battery.

by the interchange of conductors and insulators, since only two conditions can exist. For these reasons the quantity of electric current is controlled by means of small devices called *resistors*. Resistors are actually poor conductors and poor insulators. However, they can be manufactured to be either better conductors or better insulators. Resistors are made in different sizes and of different materials. Figure 2·2 is a photograph of an assortment of resistors.

Resistors are rated according to their ohmic value. Either this rating in ohms is stamped on the resistor, or the resistor is color-coded. When the color code is properly interpreted, it reveals the ohmic value of the resistor. Resistors also have a wattage rating, explained in Sec. 2·3.

Some resistors are made variable. This is done so that an operator can conveniently change the value of resistance in a circuit. By thus being able to vary the resistance, an operator can indirectly vary the amount of current flowing in a circuit. The symbol for a variable resistor is —⋀⋀⋀—

Such a device is sometimes called a rheostat.

CURRENT HANDLING

Since resistors have the responsibility of controlling current flow, they are connected to the voltage source by means of copper wire—an excellent conductor. The copper wire interferes very little with the control action. In fact, for all practical purposes, the opposition to electron flow by copper wire is so small that one can actually dismiss this opposition and say that copper wire offers no resistance. The copper wire is used to *pipe* electrons from place to place, much the same way as water is piped. To elucidate this comparison, let us consider the conditions encountered when piping water.

When water is to be transported from one place to another, large pipes are used to handle large amounts of water; when small amounts of water are moved, small pipes are used. This same condition is encountered in current flow through copper wires. When heavy currents (electron movement) are to be handled, wires with large diameters are needed. When small currents are handled, small-diameter wires are satisfactory.

RESISTANCE SUMMARIZED

In summary, resistors control current flow. These resistors are connected to the voltage source by means of copper wires. Copper wires offer negligible resistance, but the amount of current flow determines the size of the wire to be used.

ELECTRICAL CIRCUIT

The interconnection of the resistor and the battery is referred to as a *closed electrical circuit*. Figure 11·4a illustrates such a completed circuit. Figure 11·4b illustrates a circuit that is not complete. This is referred to as an *open circuit*. Current cannot flow in an open circuit. In other words, if the circuit is incomplete, electrons cannot flow out of the negative terminal of the battery or into the positive terminal.

11·3 BATTERY AS A VOLTAGE SOURCE

Earlier it was established that the amount of electron movement (current) depends on voltage as well as on resistance. Hence, different values of voltages, like different values of resistors, must be available. For this reason we have many different sizes and types of batteries. Figure 2·3 is a photograph of an assortment of batteries.

Some batteries are small, while others are quite large and heavy. The physical size of the battery is no indication as to the electrical pressure (voltage) it can apply to a circuit. Many large batteries have very low voltages, and many small batteries have high voltages. However, the physical size of the battery does indicate (relatively) how much current it can handle.

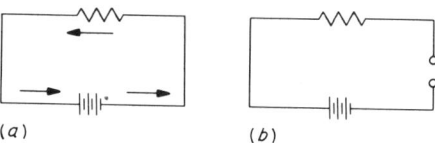

FIG. 11·4 (a) Closed circuit. (b) Open circuit.

11·4 UNITS OF QUANTITY

RESISTANCE UNITS

Resistors are rated in ohms. The Greek letter omega (Ω) is used to represent the ohm. The majority of resistors contain thousands of ohms. However, some resistors contain only a few hundred ohms, while others may have several million ohms. Because of the amount of zeros involved in the numbers dealing with thousands or millions, these two large quantities are abbreviated and have shorter designations. The letter "k" is used as a prefix to represent *thousands*, while the letters "meg" are used as a prefix to represent *millions*. The following are examples:

1000 ohms	=	1 kΩ
5000 ohms	=	5 kΩ
240,000 ohms	=	240 kΩ
1,000,000 ohms	=	1 megohm
5,000,000 ohms	=	5 megohms
10,000,000 ohms	=	10 megohms

Fractions of thousands or millions are also employed as follows:

1500 ohms	=	1.5 kΩ
7200 ohms	=	7.2 kΩ
1,100,000 ohms	=	1.1 megohms
2,200,000 ohms	=	2.2 megohms

Resistors with less than 1000 ohms of resistance are normally written like any other quantity, such as 820 ohms.

VOLTAGE UNITS

Voltages are rated in volts. Voltages encountered can be anywhere from $1/10$ of a volt to 30,000 volts. In electronic work the decimal fraction is used almost exclusively. Therefore $1/10$ of a volt will be expressed as 0.1 volt. Again, 150 volts will appear in its natural form, whereas 18,000 volts will appear as 18 KV. In the abbreviation KV, the letter "K" again represents 1000, while the letter "V" represents volts.

CURRENT UNITS

The unit of current is the ampere. The term ampere is normally abbreviated amp. In radio and electronics work the ampere is too large a unit. For this reason, the ampere is broken into 1000 parts. Each part is called a *milliampere*. A milliampere is abbreviated ma. Current quantities normally exist below 200 of these 1000 parts. It is also common to find currents low in amperes, as 3 amp, 5 amp, etc. Examples of milliampere quantities are:

0.001	amp	=	1 ma
0.01	amp	=	10 ma
0.05	amp	=	50 ma
0.15	amp	=	150 ma
1.00	amp	=	1000 ma

11·5 ELECTRICAL TEST EQUIPMENT

The quantity of voltage, current, or resistance in a circuit can be measured with electrical test equipment. Voltage is measured with a *voltmeter*, current is measured with an *ammeter*, and resistance is measured with an *ohmmeter*.

Electronics technicians make extensive use of test equipment in the pursuit of their duties. However, electronics assemblers seldom make measurements.

11·6 SUMMARY: BASIC ELECTRICAL PRINCIPLES

Electricity has its origin in the electron. Although electricity is not tangible, it can be measured and its effects observed.

A voltage will cause an electron current to flow in a closed circuit. The resultant current is dependent upon the resistance in the circuit and the voltage applied.

Conductors are materials that offer negligible resistance to current flow. Insulators, by contrast, allow very few electrons to flow as an electric current.

The pressure exerted on electrons to cause their motion is measured by a voltmeter in volt units.

The electron movement is measured by an ammeter in ampere units.

The opposition to electron movement is measured by an ohmmeter in ohm units.

The following jobs will enable the student to put into practice the basic electrical principles. These jobs not only serve as a practical review, but they are informative, since the use of test equipment is also taught.

EXERCISES

JOB 11·1 How to verify the effect of resistance

OBJECTIVES
1. To observe the effect of an electric current on a lamp
2. To observe the effect resistance has on lamp intensity
3. To prove that resistance reduces current flow

MATERIALS REQUIRED

Power Source
6.3 volts, either d-c or a-c

Supplies
#47 pilot lamp, with socket
Resistors, 22-, 47-, 68-, and 100-ohm, 1-watt

Miscellaneous
Assorted clip leads

PROCEDURE
1. With clip leads, connect the circuit of Fig. 11·5a.
2. Notice the intensity of illumination of the lamp.
3. Disconnect the power source from the lamp.
4. Connect the circuit of Fig. 11·5b.
5. Notice the intensity of illumination of the lamp.
6. Repeat steps 3, 4, and 5 using the other three resistors, one at a time.
7. Disassemble the circuit and answer the following questions.

QUESTIONS
1. Under what condition was the lamp the brightest?
2. What caused the lamp to illuminate?
3. What role did the power source play?
4. How could you tell that *electricity* was present?
5. Was there any resistance present in the circuit when the lamp was brightest? Explain.
6. Under what condition was the lamp the dimmest?
7. Draw a schematic diagram of a circuit that uses a rheostat to adjust lamp illumination.

JOB 11·2 How to interpret deflection on an ohmmeter scale

OBJECTIVES
1. To become acquainted with nonlinear calibrated scales
2. To appreciate the inverse relationship between current and resistance
3. To acquaint the student with slide-rule scales

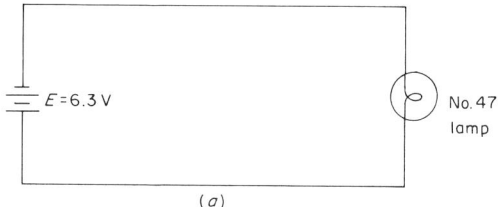

(a)

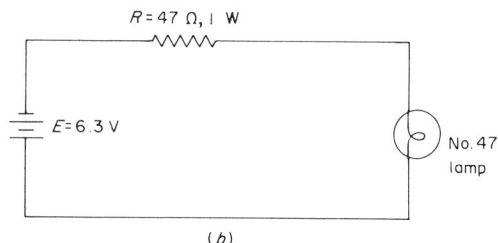

(b)

FIG. 11·5 Observing the effect of resistance on lamp illumination.

MATERIALS REQUIRED

Equipment
Vacuum-tube voltmeter with test leads
10-inch slide rule

Supplies
Resistor, 25,000 ohm, 10% tolerance
Copper wire, 1-foot long
Notebook paper and assorted colored pencils

PROCEDURE

1. Repeat Job 2•3 for review.
2. Review Job 11•1, especially the questions and answers section.
3. Compare Fig. 4•1 with Fig. 11•5b. Disregard the values of the battery and the resistor employed in each case. However, notice that in Fig. 4•1 a milliammeter is used to indicate the amount of current flowing in the circuit, whereas in Fig. 11•5b the amount of current flowing in that circuit is determined by the *relative* brightness of the illuminated lamp as compared to an earlier situation. Actually, in both examples the relative amount of *resistance* in the circuit is revealed, as well as the relative amount of *current* flowing in the circuit. Greater deflection of the meter pointer can indicate that *less resistance* is in the circuit, or that *more current* is flowing in the circuit. It all depends on how you want to *interpret* the meter deflection. It is the same situation with the lamp. A brighter lamp can mean less resistance in the circuit, or it can mean more current flowing in the circuit. Again, it all depends upon what you wish to interpret from the results.
4. Essentially, an *ohmmeter circuit* is similar to the circuit of Fig. 4•1. A meter is used to more accurately help you interpret what you want to know. In fact, the manufacturer of the meter will even provide you with a calibrated meter scale to help you further interpret the meter pointer deflection more precisely. That means that interpretation of a *meter reading* will finally rest with *your ability to read* the meter deflection. Your ability to do this accurately is developed through *practice*. (See Fig. 19•4.)
5. Compare the *D scale* on the slide rule with the ohmmeter scale on the VTVM. Notice that both scales are essentially the same. Therefore, learning to read one scale prepares you to read the other. Further, practice on one gives you practice on the other scale. (*The slide rule is a very useful tool for solving mathematical problems. The student of electronics will find much use for a slide rule during his advanced studies.*)
6. The ohmmeter (VTVM) scale has the greater flexibility of changing the multiple of its entire range by the selection of different multiplier factors through manipulation of the *range switch*. Therefore, when the meter pointer indicates **5** on the scale, it could mean **5**, or **50**, or **500**. It depends on whether the range switch is set at $R \times 1$, or $R \times 10$, or $R \times 100$, respectively. However, once again it should be noticed that the ability to read the basic scale is fundamental to the final *reading*, especially when the pointer deflects to a position between the printed numbers on the scale.
7. To further this practice of reading the fundamental scale, copy on your notebook paper that portion of the ohmmeter scale as it appears on the VTVM from **6** on the scale to **40**, including all small marks in between every printed number. Make your drawing as large as possible.
8. On your drawing, using red pencil, indicate *by each number and each mark* their respective values when the range switch is set at $R \times 1$.
9. Repeat step 8, this time using blue pencil and $R \times 100$.
10. Repeat step 8, this time using green pencil and $R \times 10K$.
11. For further-practice, close your eyes and at random set the *hairline* of the slide rule to any position. Open your eyes and read the basic value indicated on the **D** scale by the hairline. Ask your instructor for assistance as necessary. (*Remember: Practice-reading a slide rule scale gives you practice at reading an ohmmeter scale.*)

QUESTIONS

1. What is meant by the term "current is inversely proportional to resistance"?

2. Explain why the ohmmeter scale is nonlinear.
3. Describe the basic ohmmeter circuit.
4. Explain the purpose of a slide rule.

JOB 11·3 How to use a milliammeter

OBJECTIVES
1. To learn how to measure an electric current
2. To appreciate wiring and schematic diagrams

MATERIALS REQUIRED

Power source
1.5-volt No. 6 ignition dry cell

Equipment
0–1 d-c milliammeter

Supplies
Resistor, 200-ohm, 1-watt

Miscellaneous
Two clip leads

PROCEDURE
1. With the two clip leads interconnect the resistor, milliammeter, and the dry cell as illustrated in the wiring diagram of Fig. 4·1a. *Observe the polarity of the dry cell and the meter.* (See milliammeters in Fig. 11·6.)
2. Note that the milliammeter registers 0.75 ma. A small variation from this amount may be tolerable if caused by a variation in resistance or voltage.
3. Compare the schematic diagram of Fig. 4·1b with your setup. They are the same if proper results have been obtained.

QUESTIONS
1. Explain in your own words the meaning of polarity.
2. Were you satisfied that an electric current was flowing through the resistor even though you could not see the moving electrons?
3. Is 0.75 ma and ¾ amp the same amount of current?
4. Explain the difference between a wiring diagram and a schematic diagram.

FIG. 11·6 An assortment of milliammeters.

JOB 11·4 How to use a voltmeter

OBJECTIVES
1. To learn how to measure voltage
2. To observe the effects of reversed polarity

MATERIALS REQUIRED

Voltage source for testing
1.5-volt No. 6 dry cell

Equipment
Vacuum-tube voltmeter with test leads (see Fig. 19·4)

PROCEDURE
1. Apply power to the VTVM and allow at least 3 minutes for warmup.
2. Place the function switch at the d-c *plus* position.
3. Place the range switch at the 5-volt position.
4. Attach the test leads to the VTVM.
5. Switch the test probe lead to the d-c position.
6. Connect the ends of the two test leads together.
7. Adjust the ZERO knob until the indicator records 0 on the d-c scale. The d-c scale appears directly below the resistance scale.
8. Separate the ends of the test leads and apply the common test lead to the negative terminal of the dry cell.

9. With the probe test lead, touch the positive terminal of the dry cell.
10. The indicator should register 1.5 volts on the d-c scale. *Note that full-scale deflection would indicate 5 volts.*
11. Momentarily reverse the position of the test leads so that the opposite polarity exists. The pointer should attempt to move in the opposite direction.
12. Disconnect the test leads from the dry cell.
13. Change the range switch to the 1.5-volt position.
14. Repeat steps 6 through 9.
15. Note that full-scale deflection is now 1.5 volts, as indicated on the 1.5-volt scale.

QUESTIONS

1. Why do the voltages listed on the range switch indicate full-scale deflection?
2. How can reversed polarity damage a voltmeter?
3. Give a definition of voltage.
4. If you had to measure 6 volts, where would you set the range switch?

12

CIRCUITS FOR VOLTAGE AND CURRENT DISTRIBUTION

In electrical installations, voltage and current are distributed throughout the system according to need. Examples of this distribution can be found in the home, where various electrical outlets are provided throughout the house. Each of the circuits provides a different path for electron movement. Another example is the electrical system of the automobile. Here again, voltages and current are distributed to the various lamps, ignition system, and accessories according to need.

12·1 VOLTAGE DISTRIBUTION

Voltage is distributed by allocating portions of the total pressure available. The allocation is prorated according to the voltage needs of the components so that the electrons in these components can be made to move. In other words, the components requiring electromotive force from the voltage source divide this total voltage and distribute it among themselves according to their respective needs.

Whenever the electrical components in a system are allowed to dictate their voltage requirements, they are arranged in a circuit known as a *voltage divider.*

VOLTAGE DIVIDER CIRCUITS
Circuits that are considered voltage dividers have certain characteristics peculiar to them. Among

these characteristics is the fact that the electric current that flows through one of the component parts constituting such a voltage divider also flows through all the other components in this voltage divider network.

Whenever the same current flows through several components, the components are said to be connected in *series*. From this we can deduce that *a voltage divider circuit is a series circuit*. A simple resistive series circuit is illustrated in Fig. 12·1. Notice that, in order for the electrons leaving the negative side of the battery to reenter the same voltage source at the positive side, they must travel through both resistors.

To further illustrate and explain voltage division, let us examine another circuit to which actual values have been assigned (refer to Fig. 12·2). In this circuit, the interconnecting wires between the resistors and the wires connecting the battery offer a negligible amount of resistance. Therefore disregard the resistance contained in these wires.

The voltage of the battery is the total applied voltage, and is identified as E_t. Resistors R_1, R_2, R_3 have resistance values of 100, 300, and 200 ohms, respectively. The portions of voltage allocated to each resistor are indicated as E_1, E_2, and E_3. These voltage portions are identified technically as *voltage drops*.

Note especially that the greater the amount of resistance offered by a resistor, the greater will be the voltage drop *across that resistor*. Logically, the greater the resistance to electron movement, the greater will be the need for pressure to move these electrons.

In summary, voltage dividers are series circuits. The sum of all voltage drops equals the applied voltage, and the components with the most resistance will possess the greatest voltage drop. Also, in a series circuit only one current flows throughout the circuit.

LIMITATIONS OF SERIES CIRCUITS
One limitation of a series circuit is that, if a component burns out, the current to all the other components in the circuit will also cease. For this reason series circuits are not practical for home lighting.

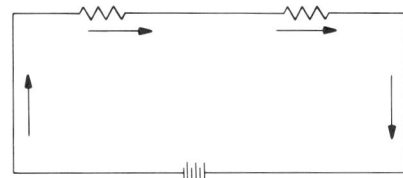

FIG. 12·1 A series circuit.

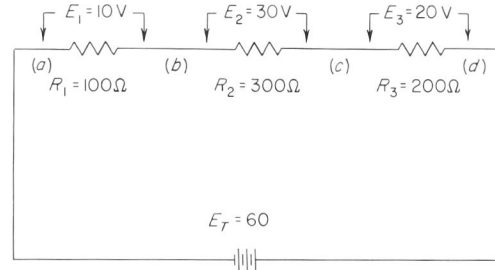

FIG. 12·2 A voltage divider circuit.

However, this limitation can be used as an asset where it would be best to stop all current if the continuation of the current could cause other components or wiring to burn out as well.

12·2 CURRENT DISTRIBUTION
Current is distributed by providing separate paths for electron movement. Each path will allow the electrons to move through it and therefore to reach any lamp or electrical appliance present in this path. Actually, each path is a separate circuit.

Circuits that provide separate paths for electron movement require only one voltage source to supply the electrical pressure for each circuit. These various circuits, each with a different current and connected in a parallel arrangement, are known as *parallel circuits*.

PARALLEL CIRCUITS
A simple parallel circuit is illustrated in Fig. 12·3. In any parallel circuit the current that flows through each branch circuit is independent of the current flowing in the other branch circuits. However, all branch currents must have their origin at

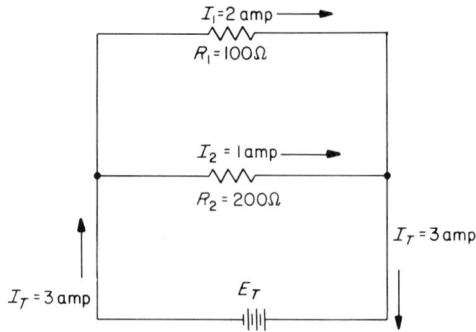

FIG. 12·3 A parallel circuit.

the voltage source, and they must also reenter the same voltage source. For this reason, the total current that originates at the voltage source will distribute itself into the various branches it feeds. The branch currents will in turn recombine to form once more the total current that will reenter the voltage source.

The distribution of the total current into the various branch circuits depends upon how much resistance each individual circuit offers to electron movement. The branches offering a greater amount of resistance to electron movement will have a corresponding smaller amount of electric current in them. Likewise, circuits with the least amount of resistance will have the greater current.

In summary, parallel circuits provide for current distribution. In a parallel circuit, a single voltage source can be applied to all branch circuits simultaneously. Also, the sum of the branch currents equals the total current.

ADVANTAGES OF PARALLEL CIRCUITS

Parallel circuits are excellent for lighting purposes. For example, in the home (where parallel circuits are used), whenever a light bulb burns out in one room, all other lamps in the house continue to illuminate. Further, each individual lamp can be turned off at will without affecting any other lamp.

12·3 SERIES-PARALLEL CIRCUIT COMBINATIONS

Combination circuits composed of series and parallel circuit arrangements are used extensively. These combination circuits are known as *series-parallel circuits*. Series-parallel circuits take advantage of the benefits offered by series circuits as well as of those made possible by parallel circuits. However, it should be understood that if a simple series circuit is needed for a given objective, then such a circuit should be used. The same is true of simple parallel circuits. Circuitry should always be as simple as practical.

SERIES DROPPING RESISTOR

An example of a series-parallel circuit is illustrated in Fig. 12·4. Two lamps that require 6 volts for proper operation are to be operated from a 12-volt source. To meet this requirement, a resistor is connected in series with the voltage source and will therefore cause a voltage drop of 6 volts across itself. When this takes place, the remaining 6 volts will be left for use by the lamps. When used in this manner, the resistor is called a *series dropping resistor*.

What has taken place is that the total circuit has become a voltage divider. However, since the lamps are connected in parallel, the voltage across each lamp is the same.

Notice that the total current flows through the resistor, but that it separates into branch currents when it reaches the junction joining the lamps. If the resistor were to burn out, thereby developing

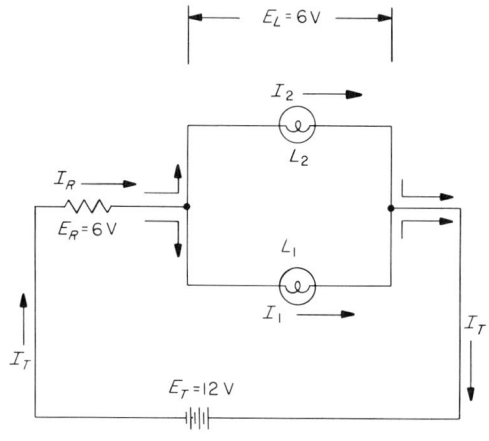

FIG. 12·4 A series-parallel circuit.

an open circuit, the two lamps would extinguish. However, if one lamp were to burn out, the other lamp would continue to illuminate.

12·4 REQUIREMENTS FOR CIRCUIT EXPERIMENTATION

A beginner in electronics should not attempt to hook up various circuit arrangements for experimentation unless he is following written instruction or is under the guidance of an experienced electronics technician. Electricity and electronics can be dangerous to human life if not handled with caution. At the very least, there can be damage to test equipment or to the component parts used in the circuits under experimentation.

For a person to experiment intelligently, he must be able to anticipate results. To anticipate results in electronic experimentation, a high degree of mathematical analysis must be made prior to the application of electrical power. A properly trained electronics technician or engineer is able to make the necessary mathematical analysis.

12·5 SUMMARY: VOLTAGE AND CURRENT DISTRIBUTION

A series circuit is used to divide voltage and to distribute this voltage according to the distribution of resistance in the same circuit. Such a circuit is known as a voltage divider. In a series circuit, the total current flows through each part of the circuit.

A parallel circuit is used to distribute a total current into branch currents according to the amount of resistance in each electrical branch. In a parallel circuit, the same voltage appears across each component that is connected in parallel.

A series-parallel circuit is a combination of circuits arranged in a manner that provides for voltage and current distribution. In a series-parallel circuit, the total current flows through the components connected in series, but distributes itself through those components arranged in parallel. The voltage in a series-parallel circuit is divided by the series-connected components, but will be the same across those components connected in parallel to each other.

The following jobs will help to summarize voltage and current distribution with the aid of simple circuits. These jobs are a practical application of the circuit fundamentals covered in this chapter. They will also serve to give an insight into the mathematical analysis involved and how test equipment is used to check results.

EXERCISES

JOB 12·1 How to test a series circuit

OBJECTIVES
1. To learn how to test resistance in a circuit
2. To learn how to measure circuit voltage

MATERIALS REQUIRED

Power source
6-volt battery

Equipment
VTVM with test leads

Supplies
Resistors, one 500-ohm and one 1000-ohm

Miscellaneous
Three clip leads

PROCEDURE
1. Set the VTVM as an ohmmeter (see Job 2·3).
2. Apply power to the VTVM and allow it to warm up.
3. Measure the resistance of each resistor and record the information.

4. With a clip lead connect the two resistors in series. *Do not connect the battery.*
5. Measure the total resistance of the combined resistors. It should be the sum of the individual resistances.
6. Using two additional clip leads, connect the battery in series with the resistors (see Fig. 12•1).
7. Set the VTVM as a d-c voltmeter. Use the 15-volt range (see Job 11•4).
8. Measure the voltage across the battery. Observe polarity.
9. Measure the voltage drop *across* each resistor. The end of the resistor nearest the negative terminal of the battery will have a negative polarity. The sum of the individual voltage drops will equal the battery voltage (review Fig. 12•2).

QUESTIONS

1. Did you calibrate the VTVM before using it as an ohmmeter and before using it as a voltmeter?
2. Why does the resistance of the two resistors add up when connected in series?
3. The 6-volt battery was not connected when resistance measurements were made. Why?
4. Why is this series circuit a voltage divider?
5. Place polarity markings on each resistor of Fig. 12•1.
6. Why must the battery be connected in order to get voltage drops across the resistors?

JOB 12•2 How to test a parallel circuit

OBJECTIVES

1. To learn about resistors connected in parallel
2. To learn about parallel voltages
3. To learn about branch currents

MATERIALS REQUIRED

Power source
12-volt battery

Equipment
VTVM with test leads, 0–1 d-c milliammeter

Supplies
Resistors, one 2200-ohm and one 3300-ohm

Miscellaneous
Assorted clip leads

PROCEDURE

1. With the clip leads connect the two resistors in parallel. *Do not connect the battery.*
2. Measure the combined resistance of the two resistors connected in parallel with the VTVM used as an ohmmeter. The total resistance should be less than 2200 ohms.
3. Remove the ohmmeter from the resistors.
4. With the resistors connected in parallel, connect the battery across the two ends of the resistors. (Use Fig. 12•3 as a guide.)
5. Set the VTVM as a voltmeter. Use the 15-volt range.
6. Measure the voltage *across* the battery.
7. Measure the voltage drop *across* each resistor. These voltage drops should *each* equal the battery voltage.
8. Connect the milliammeter in series with the battery. Maintain the resistors in the circuit connected in parallel to each other. This will give total current if polarity is correct.
9. Remove the milliammeter from the circuit and reconnect the battery to the resistors.
10. Insert the milliammeter in series with the 2200-ohm resistor and the battery. Maintain the 3300-ohm resistor connected to the battery direct. This will give the current through the 2200-ohm resistor only.
11. Repeat step 10, but exchange the resistors in order to read the current flowing through the 3300-ohm resistor.
12. The sum of the separate resistor currents should equal the total current indicated in step 8.

QUESTIONS

1. Explain why the total resistance of resistors connected in parallel is less than the smallest resistance used.
2. Explain why the voltage drop across each resistor was the same as the battery voltage.
3. Explain why branch currents in a parallel circuit combine to equal the total current.

JOB 12·3 How to test a series-parallel circuit

OBJECTIVES
1. To summarize voltage and current distribution
2. To stimulate thinking on current paths

MATERIALS REQUIRED

Power source
12-volt battery

Equipment
VTVM with test leads, 0–1 d-c milliammeter

Supplies
Resistors, 1000-, 2200-, and 3300-ohm

Miscellaneous
Assorted clip leads

PROCEDURE
1. Connect a 1000- and a 3300-ohm resistor in parallel.
2. Add to this combination a 2200-ohm resistor in series.
3. Apply the negative terminal of a 12-volt battery to the loose end of the 2200-ohm resistor.
4. Apply the positive terminal of the battery to the free junction of the parallel resistors. (Use Fig. 12·4 as a guide.)
5. Place the VTVM as a voltmeter across the 2200-ohm resistor and record the voltage.
6. Subtract this voltage drop from the battery voltage. The result should be the voltage across the parallel resistors.
7. Verify the answer of step 6 by measuring the voltage across the 1000-ohm resistor.
8. Measure the current flowing through each resistor by alternately connecting the milliammeter in series with each resistor. However, maintain the series-parallel arrangement of the resistors at all times. Record these current measurements. The sum of the branch currents should equal the current flowing through the 2200-ohm resistor.
9. Draw a schematic diagram of the circuit arrangement used in step 8.
10. On the schematic diagram show how the current flows. Use arrows to indicate electron movement.

QUESTIONS
1. What is meant by voltage distribution?
2. What is meant by current distribution?
3. How can the *actual resistance* of the 2200-ohm resistor be determined? Give two methods.

13

ALTERNATING-CURRENT FUNDAMENTALS

An electric current that alters its direction of flow periodically is referred to as an *alternating current*. When the flow of the electric current is always in a given direction, as is the case when a battery provides the source of voltage, *direct-current* electricity is in existence.

Alternating-current electricity and direct-current electricity respond to the same principles as covered in Chaps. 11 and 12. However, in a-c electricity, conditions that are not common to d-c electricity are highly evident. This chapter will acquaint the student with some of the more popular manifestations peculiar to a-c electricity.

13·1 FREQUENCY OF ALTERNATION

The direction of current flow alters because the polarity of the voltage source also changes. Take as an example the illustration of Fig. 13·1. Assume the battery to be mounted on a turntable. A set of contacts allows the battery to be connected to the circuit periodically.

During condition *a*, no current flows through the resistor. During condition *b*, an electric current flows through the resistor toward the reference point. We can call this a positive current for purposes of identification. When the battery is rotated on the turntable to condition *c*, again no current

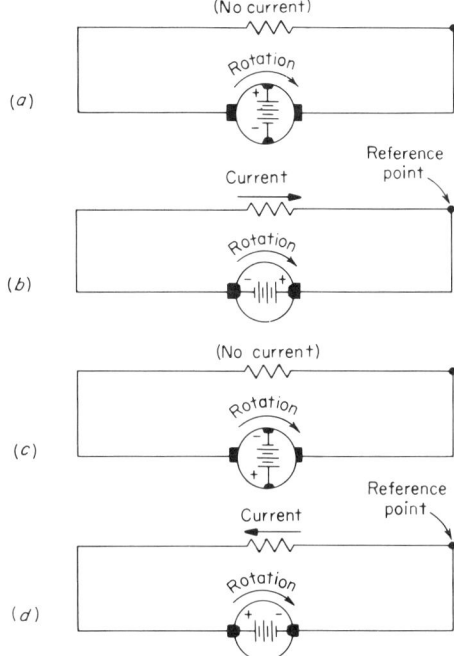

FIG. 13·1 Altering the direction of current flow.

will flow. However, by rotating the turntable clockwise by 90°, the battery is connected to the circuit once again, but this time as shown in condition d. Here the current flows through the resistor, but the direction of current flow is away from the reference point, or in a reversed direction. This current can be identified as a negative current. As can be seen, when the flow of current is present, it is either positive or negative with respect to the reference point.

Had we timed the position of the turntable so that it rested in each condition for 1 second of time, the complete cycle would have taken 4 seconds. This means that it would have been possible to complete 15 cycles in 1 minute, or 30 alternations of current in 1 minute. The frequency of repetition is therefore 15 cycles per minute.

PRACTICAL FREQUENCIES

For most practical applications a frequency of 15 cycles per minute is too slow. In electricity everything happens very fast; therefore the standard unit for time measurement is the second.

The voltage source available in the home is alternating in polarity. The frequency of the resulting alternating currents is 60 *cycles per second* (cps). In radio receivers, frequencies in the millions of cycles per second are encountered, while in television receivers one finds frequencies in the hundreds of millions of cycles per second. Cycles per second are also identified as *Hertz*. The letters Hz are used to represent Hertz.

PRODUCTION OF ALTERNATING CURRENTS

Alternating currents can be produced electrically with rotating machinery, or they can be produced electronically with vacuum tubes or transistors.

When alternating currents are required as a source of power for lighting, or to operate appliances in the home or business, rotating motor-generator units as seen in Fig. 13·2 are used. The generator is the device that creates the alternating current. The purpose of the motor is to cause a mechanical rotation within the generator.

The dynamotors shown in Fig. 13·2 are also motor-generator units having a common housing.

Motor-generator units as seen in Fig. 13·2 can produce frequencies up to 800 cps (800 Hz) without too much trouble. Beyond that, mechanical problems limit their practicability.

When the power requirements are not too great, or the frequency requirement is high, electronic circuits can produce alternating currents. These circuits are known as *oscillators*, and they can employ either vacuum tubes or transistors. More will be said about these circuits in later chapters. For now, it is sufficient to say that, regardless of how an alternating current is produced, a-c principles are the same.

13·2 GRAPHICAL REPRESENTATIONS

Relative conditions of alternating currents or voltages, with respect to magnitude, direction, and time, are often plotted on graphs. Such a graph appears in Fig. 13·3. The sequence of events is plotted along the horizontal axis and is related to

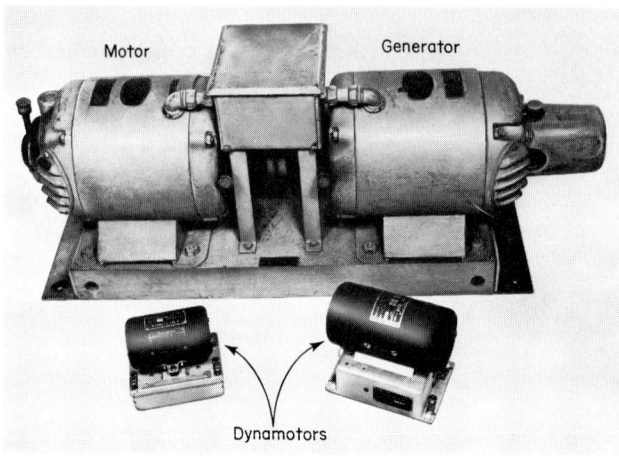

FIG. 13·2 Rotating electric generators.

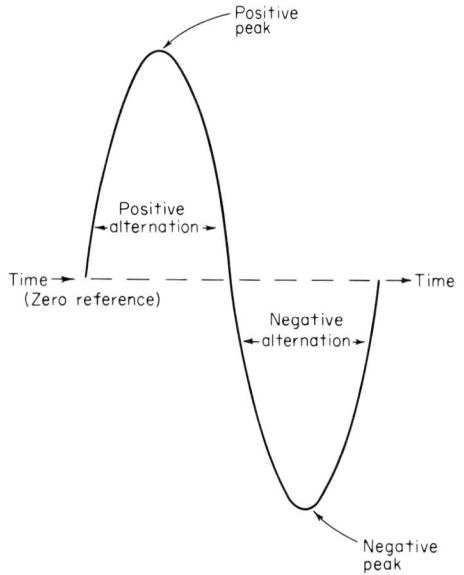

FIG. 13·3 Sine-wave relations.

SINE WAVES

The curve illustrated in Fig. 13·3 is identified as a sine wave. A sine wave is the result of plotting the output of an electrical generator as it makes one complete revolution. Since one complete revolution represents circular travel of 360°, it is proper to designate a sine wave as taking 360° for completion. However, a sine wave is said to be composed of 360 *electrical degrees*, since often the design of the generator will not tolerate a 1:1 ratio in regard to degrees.

ELECTRICAL DEGREES

Attention is called to Fig. 13·4, where the various electrical-degree points are illustrated in regard to a sine wave. Note that the positive peak appears at the 90° point of the sine wave. The negative peak is located at the 270° point. There are 180° *between* the positive and negative peaks. No current will flow at the 0, 180, and 360° positions.

SINE-WAVE PERIOD

The length of time that it takes a sine wave to complete 360 electrical degrees is known as a *period*. For an a-c frequency of 60 cps, the period of 1 cycle (or sine wave) would be $1/60$ of a second. In other words, a sine wave would travel through 360 electrical degrees in $1/60$ of a second.

It can also be said that there are 60 periods in an a-c frequency of 60 cps (60 Hz).

elapsed time. Magnitudes, or quantities, are plotted on the vertical axis, using the time base line as the zero reference. The direction of current movement (or polarity of voltage) is represented relatively with respect to the zero reference line, taking into account whether a given point on the graph appears above or below the line.

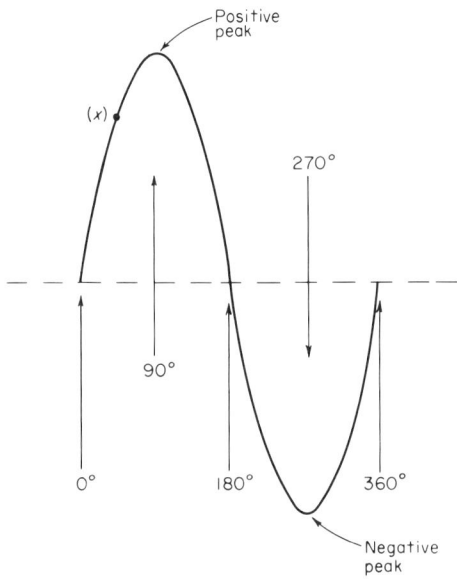

FIG. 13·4 Electrical degrees.

APPLICATION OF SINE WAVES

A sine wave plotted on a graph can be used to determine current or voltage conditions at a given time. Figure 13·4 illustrates this point.

Let us assume we are interested in finding out what current conditions exist at the 45° point of a given sine wave. It can be seen, as indicated by the X on the curve, that the current is positive and has reached a level of 70 percent of the total current possible.

13·3 PEAK AND EFFECTIVE VALUES

A direct current flowing through a resistor produces a constant amount of heat in the resistor. An alternating current flowing through a resistor produces a heat in the resistor that fluctuates with the varying current. However, the resistor cannot respond with its heat variation at the same rate that the current varies and reverses. Therefore the resistor will heat to a level with alternating currents that is less than the heat level it will reach with an equal maximum direct current.

It has been found that direct current having 70 percent of the peak value of an alternating current produces the same *heating effect* in a resistor as the heat produced with the greater alternating current (refer to Fig. 13·5).

In Fig. 13·5 it is seen that to produce a given heat in a resistor with alternating current requires a maximum (peak) of 10 volts. Yet, to produce the same heat, a direct current requires only 7 volts. The a-c voltage in this case can also be assigned the value of 7 volts, *effective*. Therefore a-c voltages carry two identifications, peak values and effective values. *An effective value is 70 percent of the peak value.*

13·4 VOLTAGE AND CURRENT RELATIONS

In the illustration of Sec. 13·3, current and voltage were used interchangeably, since an increase in voltage will cause an increase in current through a resistor. However, when dealing with alternating currents, often the resultant current will not respond immediately to voltage changes. For this reason, among others, it is important to indicate

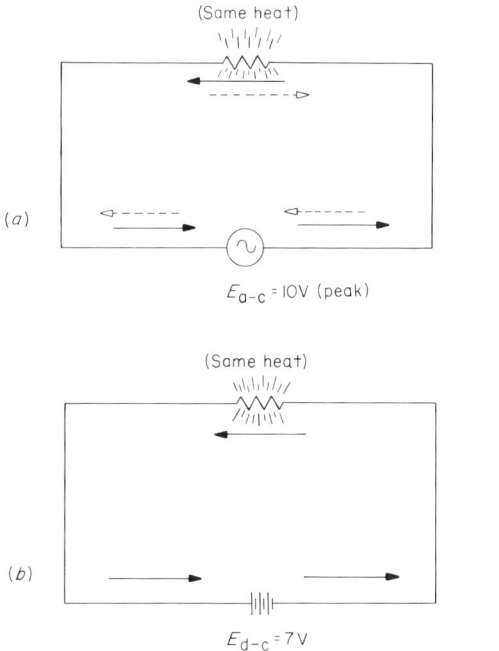

FIG. 13·5 Heating effect with different types of currents. (a) Alternating current. (b) Direct current.

on any diagram whether the voltage source will produce a direct current or an alternating current.

The symbol used to designate an a-c voltage source is a circle with a small sine wave within it (see Fig. 13·5a). This symbol does not indicate whether the alternating currents (or voltages) are produced electrically by rotary generators or electronically by vacuum tubes or transistors. It indicates only that the voltage source will produce alternating currents.

The abbreviation d-c serves to indicate that d-c conditions exist; a-c indicates that a-c conditions are to be observed.

13·5 MAGNETISM AND INDUCTORS

When a direct current flows through a wire, the current creates a steady magnetic field around the wire. When an alternating current flows through the wire, the magnetic field produced will alter in intensity, and, periodically, reverse its polarity. The magnetic field is able to keep in step with current variations. Because of this relation between a current and a magnetic field, the magnetic fields produced find many applications in electronics. In fact, magnetism is responsible for the development of many components associated with a-c electricity.

INDUCTORS

Inductors are components made of wire. These components are manufactured for the sole purpose of making use of the magnetic field that surrounds the wires when an electric current flows through them. An assortment of inductors is seen in Fig. 2·5.

An inductor actually is a coil of wire. These coils are used to choke, or restrict, alternating currents, and may be known as *chokes*. The opposition to alternating currents developed by coils of wire is known as *inductive reactance*. The ability of a coil to create this opposition is referred to as *inductance*.

INDUCTIVE REACTANCE

Inductive reactance is measured in ohms, since it too restricts the movement of electrons. This opposition is dependent upon the physical dimensions of the coil, the frequency of the alternating current in the coil, and the mathematical constant 2π.

INDUCTANCE

The ability of a coil to create inductive reactance, called inductance, is determined by the physical dimensions of the coil. Inductance is rated in a unit known as the *henry*.

The advantages of inductors lie in the fact that the magnetic field that surrounds a current-carrying wire can be harnessed and concentrated around the coil of wire. The total magnetic field concentrated around a current-carrying coil is illustrated in Fig. 13·6.

The core of the coil can either be vacant or can contain soft iron. When an iron core is used, the inductor is designed to operate with low-frequency alternating currents or with direct currents. If the core is vacant, containing air only, the coil is designed to operate with very high a-c frequencies, in the order of millions of cycles per second (MHz).

TRANSFORMERS

Two coils of wire, when adjacent to each other, can transfer electrical energy from one to the other by using the magnetic field as the coupling agent. Such a two-coil device is called a *transformer*.

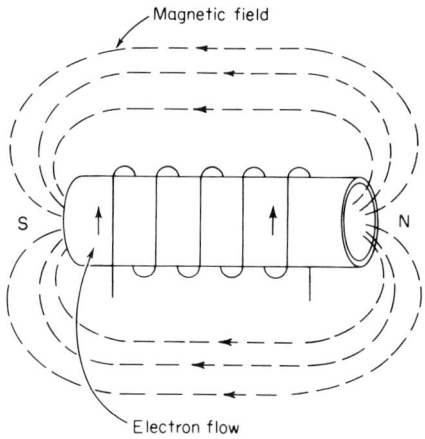

FIG. 13·6 Current in a coil of wire produces a magnetic field.

Many transformers contain more than two coils. In all transformers, one coil receives the original electrical energy from a voltage source by normal wire hookup. This coil is known as the *primary* coil. The primary coil then proceeds to set up a magnetic field around itself. Any other coils in close proximity will acquire a voltage from the magnetic field. This is said to be *induced* voltage. The coils acquiring this induced voltage are called *secondary* coils.

If the secondary coils of a transformer are part of a closed circuit, a current will flow. This current flowing in the secondary coils is referred to as an *induced current*. An induced current will always flow opposite in direction to the original, or primary, current. Figure 13·7 is a schematic diagram of a low-frequency transformer and the related primary and secondary circuits and currents. The two vertical lines between the coils indicate that this transformer is using an iron core.

13·6 ELECTROSTATICS AND CAPACITORS

Scientists have determined that the electron has an electric charge and have proceeded to designate this charge as negative. It has also been established that like charges repel each other. Therefore electrons push each other away and never come in contact with one another.

To counterbalance the negative charge, scientists have established that the absence of electrons means that a positive electric charge exists. Also established is the fact that unlike charges attract each other. It is therefore possible for the negative electrons to be attracted to a positive point that has a deficiency or lack of electrons.

The forces acting to repel or attract electrons are called *electrostatic forces*. Electrostatics plays important roles in both d-c and a-c electricity. A component used widely with alternating currents that applies the principles of electrostatics is the *capacitor*.

CAPACITORS

A capacitor is a device used to minimize voltage variations. It can also store electrostatic charges. Figure 2·4 shows an assortment of capacitors. As can be seen, capacitors are found in a variety of sizes and shapes. Some capacitors have the flexibility of being variable.

A capacitor is defined as being *two conductors separated by an insulator*. By such a device, one of the conductors can have an excess of electrons while the other can have a deficiency of electrons. The insulator prevents a movement of electrons from the negatively charged conductor to the positively charged conductor. The insulating material separating the conductors is very thin. Therefore a strong electrostatic force exists between the charged conductors. In this way, a capacitor can store electrostatic energy. Since voltage is the pressure that causes electrons to move, the capacitor is also subject to a voltage across the insulating material (see Fig. 13·8).

CAPACITANCE

The ability of a capacitor to store an electrostatic charge is termed *capacitance* and is rated in *farads*. Capacitance is determined by the physical dimensions of the conductors and insulator separating them.

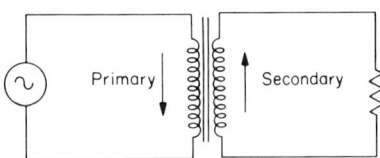

FIG. 13·7 Current flow in a transformer.

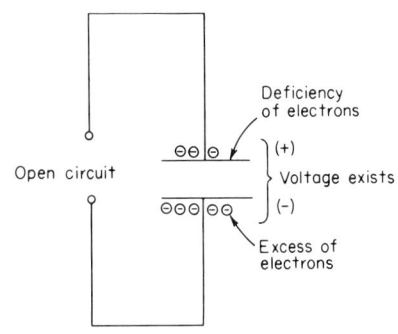

FIG. 13·8 Conditions of a charged capacitor.

Capacitors are also rated according to the amount of voltage they can withstand when charged. This rating is known as the *breakdown voltage rating.*

CAPACITIVE REACTANCE

A capacitor will oppose any further charge or discharge from a given state. This opposition is known as *capacitive reactance.* Since a movement of electrons is necessary to change the charge condition of a capacitor, this opposition to charge change is an opposition to electron movement, and can be rated in ohms. Capacitive reactance is therefore rated in ohms.

When a capacitor is used with alternating currents, capacitive reactance is continuously in evidence. In fact, it is not possible to have capacitive reactance with pure direct currents since no changes ever take place during operation.

Capacitive reactance is determined by the capacitance of a capacitor, the frequency of alternating currents involved, and the mathematical constant 2π.

13·7 IMPEDANCE

In circuits using alternating currents, the total opposition to electron movement is known as *impedance.* Impedance is the combination of resistance and reactance. Resistance by itself is due to friction. By contrast, a reactance is an opposition that is developed either magnetically or electrostatically.

Transformers are often rated according to impedance since the wire in the coils contains resistance and the coils develop inductive reactance.

13·8 SUMMARY: ALTERNATING-CURRENT FUNDAMENTALS

Alternating-current electricity is identified as such because the direction of electron movement is altered periodically. By contrast, in d-c electricity the general movement of electrons is constant and in a given direction at all times.

The frequency of current reversal is expressed in cycles per second, and Hertz. Alternating currents can be produced electrically with rotating generators, or they can be produced electronically by the utilization of vacuum tubes or transistors.

The voltage and current conditions with respect to time, polarity, and magnitude are expressed with graphs. One such graphical presentation is the sine wave.

The magnitude of an alternating current or voltage can be expressed by its peak value or by its effective value.

Two groups of components designed to operate primarily with alternating currents are inductors and capacitors. Inductors utilize the magnetic fields that surround current-carrying wires. Chokes and transformers are samples of inductors. Capacitors work on the principle of electrostatics, and are defined as two conductors separated by an insulator.

Inductors and capacitors have the ability to create an opposition to alternating currents. This opposition is termed reactance and is expressed in ohms. Chokes are used to minimize current variations. Capacitors are used to minimize voltage variations.

A transformer is an inductor that has the ability to transfer electrical energy from one circuit to another. Low-frequency transformers and chokes use an iron core. High-frequency transformers and chokes use an air core.

The total opposition to electron movement offered by resistance and reactance is termed impedance. Impedance is also measured in ohms. Transformers are often rated according to their impedance values.

The following jobs will better acquaint the student with the components used in connection with alternating currents. These jobs will also give practical experience in the testing of capacitors and inductors.

EXERCISES

JOB 13·1 How to use an oscilloscope to display a-c waveforms

OBJECTIVES

1. To become familiar with waveform display
2. To better understand the meaning of *alternating* currents
3. To learn to use an oscilloscope

MATERIALS REQUIRED

Power source
117-volt, 60-cycle alternating current

Equipment
Oscilloscope, complete with test leads having clips
Operation manual for the oscilloscope
Transformer, filament type (117-volt input, 6.3-volt output)

PROCEDURE

1. Carefully examine the front panel of the oscilloscope (see Fig. 19·2).
2. Refer to the operation manual for the explanation of each control and terminal.
3. Attach the power cord to the line voltage source.
4. Turn the a-c power switch on. Allow a few minutes for warmup.
5. Rotate the *intensity* control until a spot or a horizontal line of light appears on the face of the screen. Adjust the intensity control so that it produces a moderate amount of light. The light produced should not be too bright.
6. Center the spot or line of light on the screen by adjusting the vertical and horizontal centering controls.
7. Adjust the *focus* control to produce a sharp, definite spot or line of light.
8. If the light produced is in the form of a spot, set the *sweep* selector at any of the numbered positions. A horizontal line should now appear instead of the spot. (*Note:* If a spot of light is allowed to continue on the screen, the screen will be damaged.)
9. Adjust the length of the horizontal line with the *horizontal gain* control.
10. Apply power to the transformer.
11. Connect one output terminal of the transformer to the *vertical input* ground terminal on the oscilloscope. Use a clip lead.
12. Connect the second output terminal of the transformer to the "hot" vertical input terminal of the oscilloscope with another clip lead.
13. Set the *sync* selector at the *internal* position. The polarity is not important at this time.
14. Set the horizontal frequency *sweep* selector at a number close to the input frequency (60 Hz in this case).
15. Adjust the *sweep vernier* control until it is possible to see one or a few sine waves as shown in Fig. 13·3. The horizontal time base line will not be seen as illustrated in Fig. 13·3.
16. Adjust the vertical peak-to-peak presentation with the *vertical gain* or *calibration* control on the oscilloscope.
17. If the sine wave drifts horizontally, adjust the *sync adjust* control to stop it. To stop the sine wave, it may also be necessary to readjust the sweep vernier control.

QUESTIONS

1. Did the waveform obtained on the oscilloscope compare favorably with the sine wave of Fig. 13·3?
2. Explain in your own words what the sine wave of Fig. 13·3 represents.
3. Describe an oscilloscope.
4. What is the intensity control on an oscilloscope used for?
5. What does focus mean?
6. What does sync stand for?
7. What does vernier stand for?
8. Why is it necessary to set the sweep selector on an oscilloscope close to the frequency of the signal that one wishes to view?

9. Should one ever leave a spot of light on the screen of an oscilloscope for a prolonged period of time? Why?
10. Why is an oscilloscope necessary to *view* alternating-current representations?

JOB 13·2 How to test capacitors

OBJECTIVES
1. To appreciate that a nonelectrolytic capacitor is an open circuit to direct current
2. To observe the charging action of an electrolytic capacitor
3. To appreciate a condenser checker

MATERIALS REQUIRED

Power source
12-volt battery

Equipment
VTVM with test leads, 0–1 d-c milliammeter

Supplies
0.01-mfd 150-volt capacitor
8-mfd 450-volt capacitor
25,000-ohm resistor

Miscellaneous
Assortment of clip leads

PROCEDURE
1. Connect the battery, 0.01-mfd capacitor, and the milliammeter in series. Note that no current flows continuously through the meter. A small flicker of the meter pointer may be observed upon completion of the circuit. This is normal.
2. Replace the capacitor with the 25,000-ohm resistor. Note that a current is present.
3. Set the VTVM as an ohmmeter on the $R \times 1000$ range.
4. Remove the 0.01-mfd capacitor from the circuit and test its resistance. It should be infinity, indicating that it is an open circuit to direct current.
5. Now test the resistance of the electrolytic capacitor with the ohmmeter. Observe polarity. The common lead of the VTVM is negative. The pointer of the meter should gradually climb. The climb indicates that the large value of capacity is slowly reaching a charged condition. This charging current in an electrolytic capacitor is identified as a *leakage current* since, while charging an electrolytic capacitor, electrons traverse the electrolyte.
6. The capacitance of a capacitor must be measured with a condenser or capacitor checker. Manufacturers of capacity testers provide adequate operating instructions. In addition to testing for capacitance, a condenser checker is able to test a capacitor for *short circuits*.

QUESTIONS
1. What stops electron movement through a capacitor?
2. In step 2, what did the use of the resistor prove?
3. What was the ultimate resistance of the electrolytic capacitor?
4. Can the capacitance of a capacitor be measured with an ohmmeter?
5. Can an ohmmeter check to see if a capacitor has a *short*? Explain.
6. Can an ohmmeter check to see if an electrolytic capacitor has excessive *leakage*? Explain.

JOB 13·3 How to test transformers

OBJECTIVES
1. To learn about continuity testing
2. To learn about short circuits
3. To appreciate impedance
4. To recognize transformers

MATERIALS REQUIRED

Equipment
Ohmmeter

Supplies
Power output transformer

Miscellaneous
One clip lead

PROCEDURE
1. Compare the power output transformer in your possession with the transformer shown in Fig. 2·5. (The transformer in the photograph has 7000 OHMS stamped on it.)

2. Note that the transformer has four leads. One pair of leads is for the primary coil, while the second pair is for the secondary coil.
3. Measure and record the resistance of the primary coil.
4. Measure and record the resistance of the secondary coil.
5. If an indication of resistance is obtained, the coil can be considered continuous and therefore as not being *open*. This simple test is called *continuity testing*. In continuity testing the exact value of resistance is not important.
6. Make a continuity test between the primary coil and the secondary coil of the transformer. An open circuit should exist since insulation separates the two coils.
7. Connect a clip lead across the primary coil. These coil leads are heavily insulated.
8. With the clip lead in place repeat step 3. Zero ohm of resistance should now be obtained because the primary coil is *shorted* by the clip lead.
9. The value of resistance obtained in step 3 (without the short) represents the resistance of the wire in the coil. In the case of the transformer shown in Fig. 2·5, the 7000 OHMS stamped indicates the *impedance* offered by the primary coil to alternating currents. An ohmmeter cannot measure impedance.

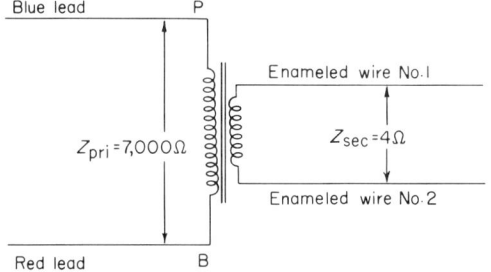

FIG. 13·9 Impedance and lead identification of an audio output transformer.

FIG. 13·10 The impedance bridge.

10. The schematic diagram for a typical power output transformer is shown in Fig. 13·9. Compare the values given with your results.
11. Study the *impedance bridge* shown in Fig. 13·10. This instrument is used to measure reactance and impedance.

QUESTIONS
1. What does a continuity test indicate?
2. Can a continuity test be made on a capacitor?
3. Explain an open circuit.
4. Explain a short circuit.
5. What is impedance?

14

SOLID-STATE ELECTRONICS

The term *solid state* normally refers to active components in the semiconductor family. This includes the various types of diodes, transistors, and integrated circuits. The handling and utilization of these components is covered in various sections of this book; therefore, this chapter will concentrate mainly on the basic theory of how these units work and what makes them work. In keeping with the objectives of the text, coverage of the subject will be nonmathematical in nature. Further, in the interest of well-rounded coverage, some comparison of semiconductor devices to electron tubes will be made. This is both useful and necessary since the history of electronics development has its roots deeply imbedded in electron tubes. Also many products are still manufactured which use electron tubes, and many more are out in the field that need service and repair. In fact, it is recommended that the reader refer to Appendix I which covers the basics of electron tubes.

Principles of solid-state diodes and transistors commonly found as discrete components are emphasized in order to provide flexibility for the instructor or student who wishes to experiment beyond the scope of this text. Regardless of packaging techniques and specific applications, the basics of semiconductor components are rather uniform.

14·1 SOLID-STATE DIODES

A semiconductor is constructed as a solid, unified device. No vacancy or gap exists between electrodes, such as the space between the cathode and the plate of a vacuum-tube diode. Although several electrodes are found within semiconductors, they are fused together to maintain a solid feature.

Within the family of semiconductors is found the *solid-state diode*. The solid-state diode is comparable with the vacuum-tube diode. The solid-state diode and the vacuum-tube diode both allow an electron current to flow from a cathode to an anode. (*The plate of a vacuum tube can also be referred to as an anode.*) Therefore both diodes can perform many of the same functions.

Several types of solid-state diodes are used in electronic work. Some of these diodes are required to handle heavy currents, while others may be designed to carry only a small current. Figure 2·28 shows an assortment of semiconductor, or solid-state, diodes. The symbol for a solid-state diode is shown in Fig. 14·1a. Figure 14·1b compares the electrodes of a solid-state diode with those of the vacuum-tube diode. Notice that the cathode of the solid-state diode carries a *plus* sign, the standard mark that is used to distinguish the cathode of a solid-state diode. An explanation for this mark will be found in Chap. 15, under rectification.

Solid-state diodes have several advantages over electron-tube diodes. Three of the most important are that (1) solid-state diodes *do not require a heater or filament*; (2) solid-state diodes are *much smaller*, physically, than electron-tube diodes; (3) solid-state diodes *can withstand vibration* levels that would shatter ordinary electron tubes.

Solid-state diodes normally are soldered into a circuit. By contrast, electron-tube diodes use a socket for plug-in purposes. This is done because solid-state diodes have a longer life expectancy than electron tubes.

DIODE THEORY

Generally, the purpose of any diode is to allow electrons to flow from cathode to anode, but *not from anode to cathode*. The vacuum-tube diode accomplishes this quite well, since the cathode emits electrons when heated and the plate (or anode) attracts and collects these electrons.

In the solid-state diode, many electrons are allowed to travel toward the anode from the cathode (within the diode). However, in this type of diode, a few electrons are also able to travel from the anode toward the cathode, *within the diode.*

As in any circuit, a voltage or pressure is needed in order to have an electric current. Figure 14·2 illustrates the conditions for electron flow within a solid-state diode. When the electron movement is from cathode to anode, as in Fig. 14·2a, the diode is said to be *forward-biased*. During forward-bias conditions (notice the polarity of the battery), a *heavy forward current is normal*. However, when the polarity of the battery is reversed, as in Fig. 14·2b, a *small back current* will flow. The circuit of Fig. 14·2b is said to be *reversed-biased*.

To determine the quality of a solid-state diode the *ratio of the forward current to the back current* must first be determined. This can be done with the use of a milliammeter. However, since current and resistance are only inversely proportional, an ohmmeter can be used as well. When an ohmmeter is used to test a solid-state diode, the resistance of

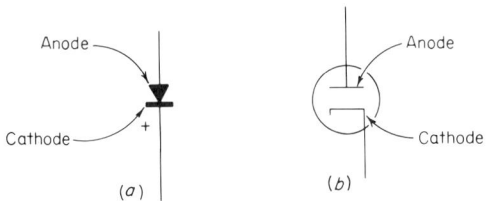

FIG. 14·1 Comparing the semiconductor diode with the vacuum-tube diode. (*a*) Semiconductor diode schematic symbol. (*b*) Vacuum-tube diode symbol.

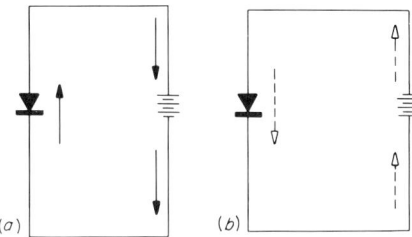

FIG. 14·2 Results of bias polarity on electron flow in a semiconductor diode circuit. (*a*) Forward bias results in a heavy forward current. (*b*) Reverse bias results in a small back current.

the diode is measured both forward and reversed. The results are evaluated in terms of the *front-to-back ratio.*

Front-to-back ratios usually exceed 1000:1, lower ratios indicating poor diodes and higher ratios indicating better diodes.

14·2 PN MATERIALS

With the above explanation, a practical concept can be developed on solid-state diodes for testing purposes. However, a more complete technical approach is necessary since the study of solid-state diodes precedes the study of transistors.

A solid-state diode is often made of either germanium or silicon that has been modified. Germanium, for example, can be modified with arsenic. It can also be modified with gallium. Arsenic and gallium are referred to as *impurities* in the study of semiconductors. Other impurities are available as well. When impurities are added to pure germanium, thus modifying the germanium, the germanium is said to be *doped.* The amount of impurities used is extremely small. In fact, it is quite minute compared with the amount of germanium used. For this reason, such a diode would simply be referred to as a germanium diode. The same is true of silicon diodes.

N-TYPE GERMANIUM

A germanium crystal that is doped with arsenic gains electrons. This crystal therefore has an excess of electrons as compared with the pure germanium. Germanium under these conditions is said to be of an N type, where N stands for negative. In other words, many extra electrons exist in N-type germanium because of the addition of arsenic.

P-TYPE GERMANIUM

A germanium crystal that is doped with gallium loses electrons. This crystal therefore has a deficiency of electrons as compared with the pure germanium. Germanium so doped is referred to as a P-type germanium, where P stands for positive. Since a positive polarity implies a deficiency of electrons, the term *hole* is used in the study of semiconductors to indicate that an electron is absent. P-type germanium is therefore said to contain many holes. These holes are considered *positive electric charges.*

14·3 PN JUNCTIONS

In semiconductors, N- and P-type germanium are both used at the same time. In fact, they are placed side by side and united. The surface at which each type makes contact with the other is called a *junction.* The electrons and holes in semiconductors combine at these junctions. The combination of electrons and holes at the junctions of semiconductors allows for electron flow in the external circuits of such semiconductors.

To investigate the practical application of semiconductors, a few illustrations will be helpful. Referring to the circuits of Fig. 14·2, note that they can also be drawn as shown in Fig. 14·3. By use of the latter method of illustration, the N and P nomenclature can be employed. Carefully compare Fig. 14·3 with 14·2. Note the heavy forward current and the small back current in each case. Also notice the battery polarity in both cases.

With forward bias, the N-type germanium is connected to the negative terminal of the battery. With reverse bias, the N-type germanium is connected to the *positive* terminal of the battery.

In the illustrations, the cathode is the N-type germanium while the anode is the P-type germanium. Although a germanium diode was used in the

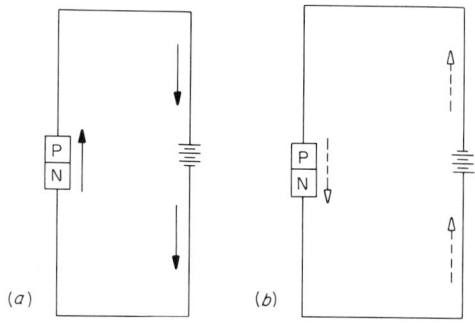

FIG. 14·3 (*a*) Forward-biased (heavy forward current). (*b*) Reversed-biased (small back current).

illustration, a discussion of silicon diodes would be much the same.

HOLE CURRENT

A very peculiar analysis of *relative conditions* takes place in the study of semiconductors. This analysis deals with the presence or absence of extra electrons.

In electronics, an electric current is generally said to be the movement of electrons from a point having an excess of electrons to a point having a deficiency of electrons. This is referred to as electron flow. This explanation for an electric current is maintained when studying semiconductor devices and is defined as an *electron current*. However, in the study of semiconductor devices, the student will also find that the absence of an electron, or hole, must be considered important.

Figure 14·4 and the following explanation will clarify the significance of a hole.

If one ball is available to fill two cavities appearing in a piece of wood, then only one *hole* will be vacant, as in Fig. 14·4a. When the ball is taken from cavity 2 and moved from right to left to fill in cavity 1, as in Fig. 14·4b, then the hole will appear as cavity 2. In effect, in moving the ball from right to left, the empty *hole was moved* from left to right. It can therefore be said that the ball movement was from right to left and that the hole movement was from left to right.

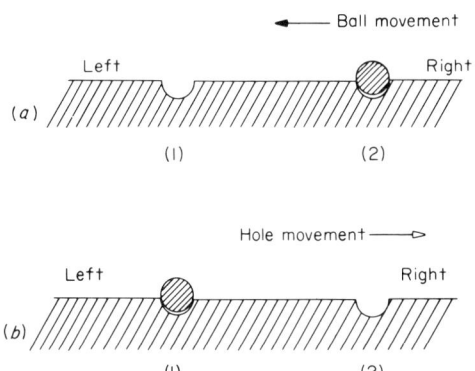

FIG. 14·4 Hole movement.

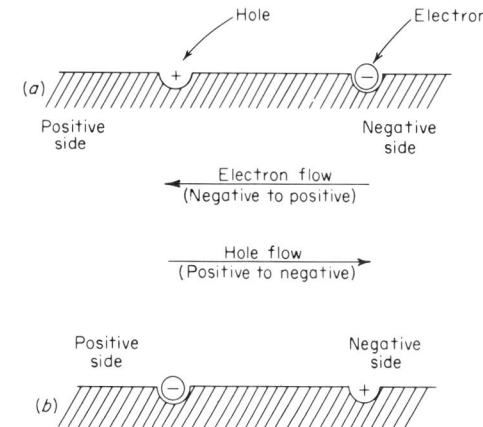

FIG. 14·5 Hole flow and polarity.

With this in mind, and remembering that electrons have a negative charge and that holes have a positive charge, study Fig. 14·5. Notice that, while the *electron flow* is from the negative side to the positive side, the *hole flow* is from the positive side to the negative side. Figure 14·6 is a more technical illustration, showing the relation of electron flow versus hole flow in a forward-biased diode.

The practical and experienced electronics technician is able to work with the familiar electron-flow theory during most of his experience with semiconductors. However, the detailed study of transistors and solid-state diodes requires of necessity acknowledgment of the presence of holes and their behavior. This is especially true since a combination of electrons and holes takes place at the junction of the P and N germanium materials.

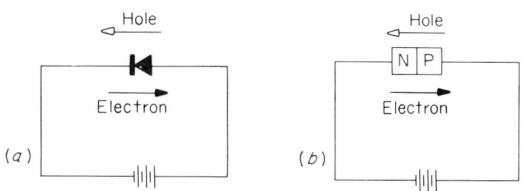

FIG. 14·6 Electron and hole flow under forward biasing.

14·4 TRANSISTORS

The function of a transistor is similar to that of a vacuum-tube triode; that is, both provide a means to control the amount of electric current that flows through them. The main difference between the two is how this is accomplished.

A vacuum-tube triode exerts control on the current by varying the input voltages. In contrast, the transistor accomplishes the control on the comparable current by varying the input currents. In other words, *a vacuum tube uses voltage to control current, whereas the transistor uses current to control current.* An exception to this rule is the field-effect transistor (FET) which is discussed later in this chapter.

To study basic transistor circuitry, one needs to apply the knowledge gained from the study of solid-state diodes. It is said that a transistor is nothing more than two diodes placed back to back. Previous knowledge of vacuum-tube triodes is also advisable. These two basic requirements will be discussed in this text in order to provide the student with an understanding of transistors.

TRANSISTOR SYMBOLS

As with most components used in electronics, the construction of the transistor somewhat dictates the form the symbol takes. As stated earlier, the study of semiconductor diodes precedes the study of transistors. With this in mind, let us briefly review the construction of these diodes. A solid-state diode is made of two sections of modified germanium or of silicon. While one section takes the form of an N material, the other section will be of a P material. These two sections are joined together. At this junction, the electrons of the N-type material combine with the holes of the P-type material.

In a transistor two similar junctions are employed, with three different sections of doped germanium or silicon used. One *thin* section of P-type germanium is sandwiched in between two thicker sections of N-type germanium. This produces what is known as an NPN transistor (see Fig. 14·7a). As indicated, the center thin section is called the *base*. One outer section is referred to as the *emitter*, while the other section is called the *collector*. Two types

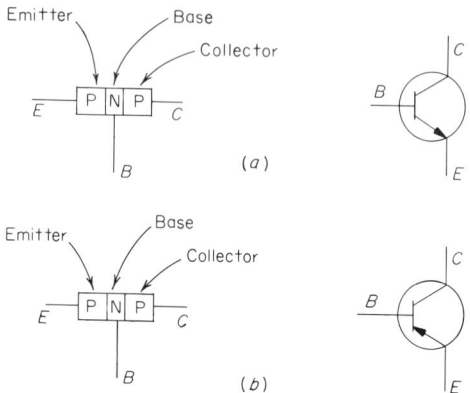

FIG. 14·7 (a) NPN transistor symbols. (b) PNP transistor symbols.

of symbols are used to represent an NPN transistor. They are both shown in Fig. 14·7a.

The symbols in Fig. 14·7b apply to transistors made with the N-type germanium sandwiched in between two P-type germanium sections. This type of transistor is known as a PNP transistor. Again note that the base section is thin, compared with the emitter and collector sections. Further, note that, in the symbols that employ an arrow in the emitter, the *arrow points out* for an *NPN transistor*, but *points in* for a *PNP transistor*.

TRANSISTOR BIAS

A transistor can be considered as being two diodes placed back to back, with each diode section having an external applied voltage for bias purposes. In applying voltages to a transistor, *the emitter must be forward-biased* and *the collector must be reversed-biased.* See Fig. 14·8a and b for the proper biasing of an NPN transistor. Also study the biasing arrangement for PNP transistors as shown in Fig. 14·9a and b.

TRANSISTOR CIRCUITS

Although in theory and construction the transistor varies greatly from the familiar vacuum tube, comparison of the circuit and function of each facilitates the practical understanding of transistors. For this reason, to simplify the explanation of transistor circuits, most textbooks and other publications use

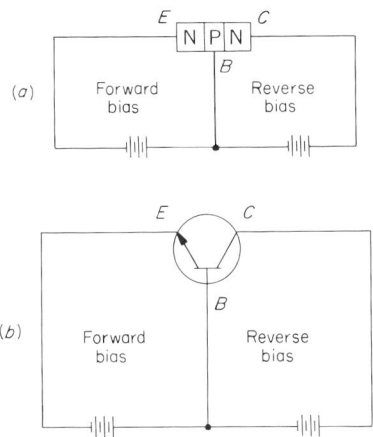

FIG. 14·8 Fundamental circuit conditions for NPN transistors. (a) NPN transistor block diagram. (b) NPN transistor schematic diagram.

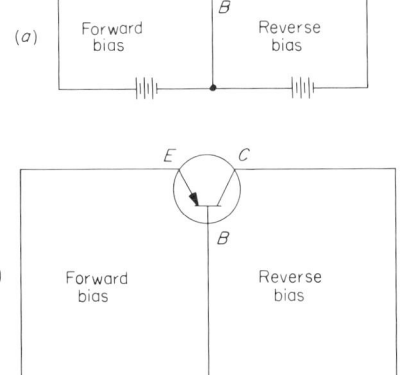

FIG. 14·9 Fundamental circuit conditions for PNP transistors. (a) PNP transistor block diagram. (b) PNP transistor schematic diagram.

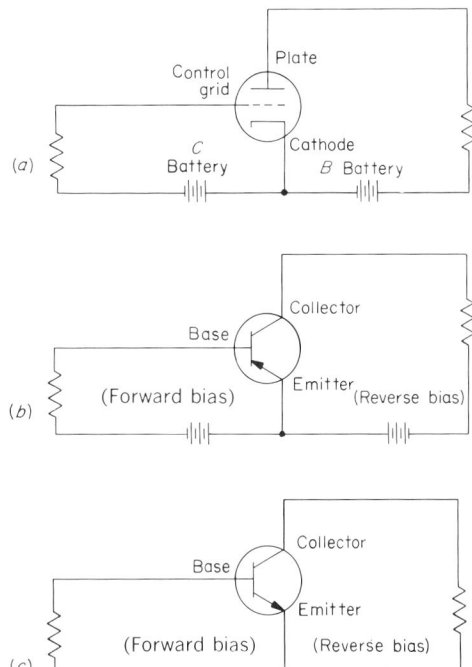

FIG. 14·10 (a) Basic vacuum-tube triode circuit. (b) Basic PNP transistor triode circuit. (c) Basic NPN transistor triode circuit.

simple vacuum-tube circuits in the discussion. This practice works out very satisfactorily, and is used in this text.

To begin, Fig. 14·10a is helpful for a quick review of a simple vacuum-tube triode circuit arrangement. Now compare the other two circuits in Fig. 14·10b and c. Generally speaking, the emitter is comparable with the cathode, the collector is comparable with the plate, and the base serves the same function as the control grid. Note that the emitter of the PNP transistor (Fig. 14·10b) is positive with respect to the base for proper forward bias. Also to be noted is the fact that the collector is reversed-biased, being connected to the negative side of the battery through the load resistor. Further examination reveals that all the polarities of the NPN transistor circuit (Fig. 14·10c) are reversed, because of the difference in transistors.

Although transistors can be compared with vacuum tubes in many ways, the two are radically different in construction and design. Figure 2·6 shows a typical vacuum tube surrounded by an assortment of transistors. Note the size variation. One of the advantages of transistors is that they are quite small.

TRANSISTOR CURRENTS

In a vacuum-tube triode circuit, the plate current (output) is controlled by the voltage (input) on the control grid. In the transistor, the collector current (output) is controlled by the base current (input). From this, a vacuum tube is said to be a voltage-controlled device, while the transistor is said to be a current-controlled device. (Again, the FET is the exception. See below.)

The three circuits in Fig. 14·11 show the path of the various currents involved. That the output currents of the two transistor circuits are quite heavy, even though a reverse bias exists, is due to the fact that the base is extremely thin and also to the combining of holes with the electrons within the transistor. It is not within the scope of this text to explain all these details. However, the student should see that, in the circuits outside of the transistor, all circuits function in a normal manner as concerns voltage, resistance, and electron flow.

In the vacuum-tube circuit of Fig. 14·11a, the plate current can be changed by varying the grid voltage at its source. In the transistor circuits of Fig. 14·11b and c, the collector currents can be changed by varying the rheostat in the base circuit, thereby causing a corresponding change in base current. (A rheostat is a variable resistor used to vary, and therefore control, current flow directly.)

14·5 FIELD-EFFECT TRANSISTORS

A special transistor that encompasses the principles of both transistors *and* vacuum tubes as discussed above is known as a *field-effect transistor*. The letters **FET** are a popular designation for this special transistor. The basic operation of a field-effect transistor as it regards current flow parallels the discussion on transistor currents above, with some minor variations. The FET transistor is composed of PN materials joined solidly to each other, and therefore both *electron* and *hole currents* exist. Biasing polarities must be considered, depending upon the arrangement of the PN materials, and front-to-back ratios are important considerations in some tests. However, where in the type of transistor discussed in Sec. 14·4 there was an interaction of the two types of currents (electrons and holes, also called *charge carriers*), in the field-effect transistor only one type of *active* current (charge carrier) is

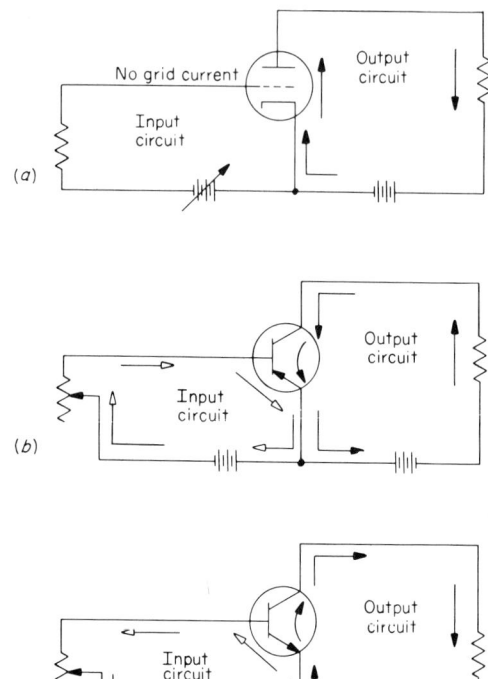

FIG. 14·11 (a) The grid voltage has an effect upon the plate current in a vacuum-tube circuit. (b) The base current has an effect upon the collector current in a PNP transistor circuit. (c) The base current has an effect upon the collector current in an NPN transistor circuit.

normally present. Therefore, the type of transistor discussed in Sec. 14·4 is known as a *bipolar* device (two charge carriers), and the field-effect transistor is known as a *unipolar* device (one charge carrier). Control of whatever type carrier is used is by means of an electrode, known as a *gate*, which exerts an electrostatic influence over the output current in a manner that parallels the method of control used by the *control grid* in a vacuum tube.

Two types of field-effect transistors exist: the semiconductor *junction* type, and the *metal-oxide-semiconductor* type. Of the two, the metal-oxide-semiconductor type, also known as the MOS-FET is the most popular. Further, the MOS-FET most closely resembles the vacuum tube in operation and characteristics.

METAL-OXIDE SEMICONDUCTOR FIELD EFFECT TRANSISTORS

The MOS-FET looks like most other small transistors, with the exception that one should expect to see *four* leads. There are four possible schematic diagram symbols for MOS-FET devices as shown in Figs. 14·12 and 14·13. The variation is due to the type of PN material used as a path, or *channel*, for the charge carriers and to the internal arrangement or form that this channel takes.

An MOS-FET is built upon a *substrate* PN material. This substrate constitutes the *bulk* of the transistor, and is assigned the letter *B* for designation. The bulk (substrate) in turn supports another PN material which is opposite to itself in type. This other PN material will be of the N-type if the bulk is of the P-type, and vice versa. This second PN material is identified as the *channel*. The channel is drawn as a solid line in the center of the symbols in Fig. 14·12, and as a broken line in the symbols of Fig. 14·13. When the arrowhead that is on the bulk (*B*) line points in to the channel, the channel is made of N-type material and the charge carriers will be electrons. When the arrowhead points away from the channel, the channel is made of P-type material and the charge carriers will be holes. This method of indicating the type of material the channel is made of applies to the symbols in both Figs. 14·12 and 14·13.

Actually, the channel provides the path to complete the circuit of an external interconnection of resistors, wires, and battery as illustrated in Fig. 14·14; such a path allows an electric current to flow. Electrons enter the channel through a *source* (S) electrode, and are pulled out of the channel through a *drain* (D) electrode. The flow of current through the channel is in turn affected by an *electrostatic charge* created by a *voltage* on an insulated control electrode known as a *gate* (G). Now this latter statement is very important to comprehend. What all of this means is that in an MOS-FET, *voltage controls current*; just like in a vacuum tube.

When the line representing the channel is solid, it means that a current will normally be flowing, as illustrated in Fig. 14·14; even when no voltage is applied to the gate. This condition is said to be *normally ON*. This condition exists because the channel is physically built as a continuous path. Application of a *reverse gate voltage* serves to *choke* the channel, thereby reducing the amount of current flowing through the channel. This is called *depletion* of the charge carriers.

When the line representing the channel is broken, it means that normally a current will not flow through the channel. This condition is said to be *normally OFF*. This condition exists because the channel is physically built in two sections, separated by a small section of the bulk (substrate) material. In order to effectively join the two sections of the channel so that current will flow, it is necessary to apply a *forward-biased gate voltage* to the gate electrode, thereby completing or closing the circuit.

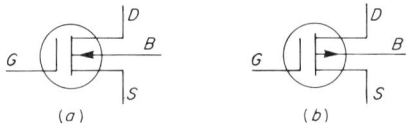

FIG. 14·12 Symbols for depletion-type MOS–field-effect transistors. (*a*) N-channel. (*b*) P-channel.

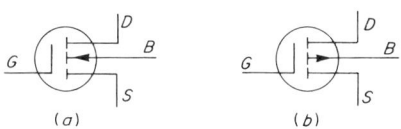

FIG. 14·13 Symbols for enhancement-type MOS–field-effect transistors. (*a*) N-channel. (*b*) P-channel.

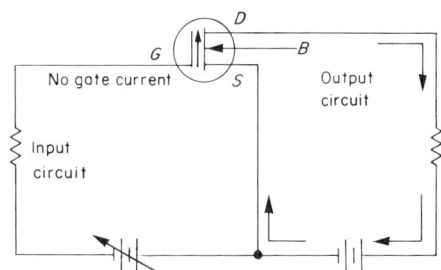

FIG. 14·14 The gate voltage has an effect upon the drain current in an MOS-FET circuit.

A forward-biased gate voltage in this case serves to draw or produce active charge carriers in the section of bulk separating the channel sections. Again, this is done electrostatically. This method of closing the *gap* is called *enhancement*.

The bulk electrode B can be used to extract an output signal, or it can be used as a ground or reference point. At any rate, its connection can be considered secondary in principle. Regarding connection of the other three leads and the resulting current, the student should compare closely Figs. 14·11a and 14·14.

MOS-FET CIRCUITS

MOS field-effect transistors find application in amplifiers, oscillators, and switching circuits. Although discussion on this type of circuitry is covered in later chapters, two simple examples of basic amplifiers circuits that utilize MOS-FET devices are shown in Figs. 14·15 and 14·16 in order to give the student an insight as to the use of these transistors. Notice that in each circuit, the depletion, N-channel type of MOS-FET is used. While this type of MOS-FET device is excellent for amplifier applications, the enhancement type of MOS-FET is best for switching circuits.

HANDLING MOS-FET DEVICES

Standard semiconductor handling precautions apply to MOS-FET devices. However, the nature of this component demands additional precautions. For example, since an MOS-FET device operates on an electrostatic charge principle, static electric-

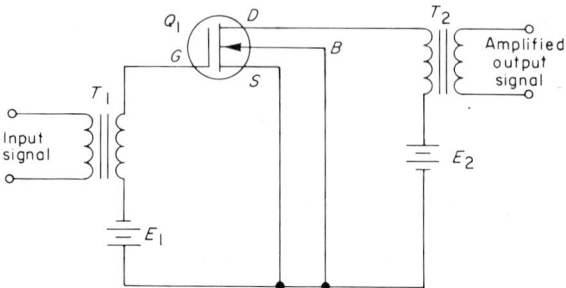

FIG. 14·16 Transformer-coupled MOS–field-effect transistor amplifier (N-channel, depletion-type FET).

ity can be extremely troublesome and hazardous to this type of transistor. A metal-oxide-semiconductor field-effect transistor that is allowed to slide freely in a plastic container will build up a large electrostatic charge quite rapidly. If this happens, the charge may become sufficient to pierce through the thin insulation that separates the gate from the channel. This will render the transistor useless. An electrostatic charge can also be built by allowing the leads to brush against clothing made of silk or nylon; and the charge can also be transferred to the transistor by the person handling this delicate device. In order to avoid or minimize this invisible hazard, MOS-FET devices should be kept in their boxes, their leads should be kept wrapped in aluminum foil or some similar conductive material, and the person handling these units should devise ways to keep himself electrically grounded while working.

14·6 SPECIAL-APPLICATION DIODES

There are three types of special-purpose diodes. These are the *zener diode*, the *tunnel diode*, and the *silicon-controlled rectifier*.

THE ZENER DIODE

The zener diode is used in circuits where it is necessary to maintain a given voltage between two established limits under all operating conditions. It can therefore be identified as a *voltage regulator*. One can expect to find a voltage regulator at the output of some power supplies. Zener diodes are

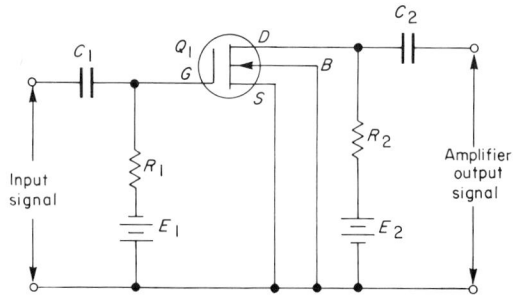

FIG. 14·15 RC-coupled MOS–field-effect transistor amplifier (N-channel, depletion-type FET).

also used to protect voltage sensitive components such as meters in test instruments. The schematic symbol for a zener diode may be seen in Figure 2•38b.

The zener diode compares with the gas-filled electron-tube diode. Both of these units operate on the principle that they will begin to conduct heavily and suddenly whenever a voltage that exceeds their established limit is placed across their cathode and anode terminals. These diodes are designed to withstand any voltage up to that limit. In other words, a zener diode (as well as a gas-filled diode) will by design *break down* and cause a short circuit across the device it is supposed to protect whenever the applied voltage exceeds the maximum tolerance. Specifically, a zener diode will conduct heavily when the reversed bias is pushed beyond a critical point.

THE TUNNEL DIODE

A *tunnel diode* is one that has had its basic semiconductor material (such as germanium) extensively modified with impurities. Because of this high concentration of impurities, the charge carriers are able to *sneak* across the PN junction with ease. This condition is known as *tunneling*. Because of this tunneling action, the amount of current that travels through this type of diode is not consistent in responding in a normal manner to changes of the applied bias voltage. That is to say, the tunnel diode conducts heavily for all reverse bias voltages, and it will also conduct heavily for some of the forward bias voltages. However, there is a wide area of forward bias that conduction actually *diminishes* with an increasing forward bias. This in effect makes a tunnel diode a *nonlinear* device. Other nonlinear devices are the vacuum tube and the transistor. Therefore, a tunnel diode can play the role of an amplifier, just as vacuum tubes and transistors. In some cases, the tunnel diode is actually a better amplifier. Since amplifiers are used in oscillator and switching circuits, the tunnel diode works nicely here too as the basic active component. Figure 2•38c is the schematic symbol of a tunnel diode.

SILICON CONTROLLED RECTIFIERS

A *silicon controlled rectifier*, normally identified simply as an SCR, is a heavy-duty diode that has the added feature of being able to be turned ON by a remote signal. In a way, it resembles a transistor in that it too is built on the *back-to-back diode principle*. It has a *cathode* and an *anode* like all other diodes, but it also has a control electrode called a *gate*. The symbol for an SCR is shown in Fig. 2•38d. An SCR differs from a transistor in that a transistor's output can gradually be varied by the corresponding variation of an input signal to its *base* control electrode, whereas in an SCR the output current can only be turned ON by the *gate* electrode. In order to turn OFF the output current, the applied voltage across the cathode and anode terminals must be temporarily removed. An alternate way to turn OFF the output current is to temporarily reverse the polarity of the applied cathode-to-anode voltage (reverse bias), as when an alternating current is applied. For this reason, SCRs find wide application in the control of a-c operated electrical equipment. Actually, an SCR is a heavy-duty, fast-acting switching device that can control heavy current equipment with a small weak signal. It is not considered an amplifier. The electron-tube counterpart of an SCR is the gas-filled *thyratron* tube. Physically, an SCR resembles a power transistor and it too must be mounted on a permanent type heat sink since large amounts of heat are generated within the SCR. Failure to use a heat sink will damage or ruin the SCR.

14•7 INTEGRATED CIRCUITS

Semiconductor technology has expanded beyond the manufacture of discrete components such as diodes and transistors. Engineering and research has led to the development of manufacturing processes whereby the same principles and techniques used to manufacture transistors can be used to manufacture resistors and capacitors. However, by using these techniques the resistors and capacitors are manufactured *simultaneously* with the manufacture of semiconductor diodes and transistors. Not only that, but while the above-mentioned components are being manufactured, they are also *interconnected* to each other into predesigned useful circuits. These total component arrays are known as *integrated circuits*.

As indicated in Sec. 2·16, there are two basic classifications of integrated circuits, *monolithic* and *hybrid*. The simultaneous manufacture of components and related interconnections produces the monolithic integrated circuit. If, however, only part of the circuit is produced by the method described above, and it is necessary to solder in added capacitors or resistors, or if it is necessary to interconnect two or more integrated circuits and yet consider it one complete package, the total overall arrangement is then known as a *hybrid integrated circuit*. Figure 14·17 compares various miniature electronic components and circuits with a coin piece, including a comparison of integrated-type circuits. Integrated-circuit packaging illustrated in Fig. 2·35 is in widespread use also.

TYPICAL MONOLITHIC INTEGRATED CIRCUIT

An integrated circuit is often identified by the letters **IC**. The general symbol for an integrated circuit is seen in Fig. 2·38*j*. The numbering of the leads, the quantity of leads indicated, and the location of the leads around the triangle will vary according to the complexity of the circuitry involved. For a typical example of a commercially available IC, see Figs. 14·18 and 14·19. Figure 14·18 shows how the integrated circuit number CA3020 is connected to serve as an audio amplifier without the need of transformers. Figure 14·19 shows in schematic diagram form what is *inside* the triangle of the CA3020. The circuit is monolithic, and notice that it contains *eleven* resistors, *three* diodes, and *seven* transistors. The CA3020 is enclosed in a TO-5, 12-lead package similar to that illustrated in Fig. 2·35*b*.

Actually, the CA3020 is a multipurpose IC. The basic package can serve in several applications.

FIG. 14·17 Miniature electronic components and circuits: (*a*) Ceramic printed circuit (example of hybrid approach). (*b*) Transistor (TO-5 package). (*c*) Monolithic integrated circuit (packaged). (*d*) U.S. ten-cent coin.

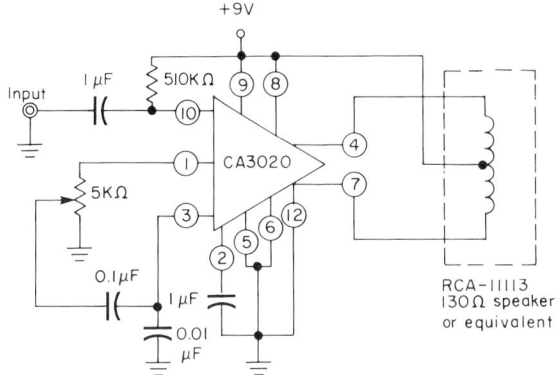

FIG. 14·18 An IC audio amplifier without power output transformer, driving a high impedance loudspeaker. (*Radio Corporation of America, Electronic Components Division, Harrison, New Jersey.*)

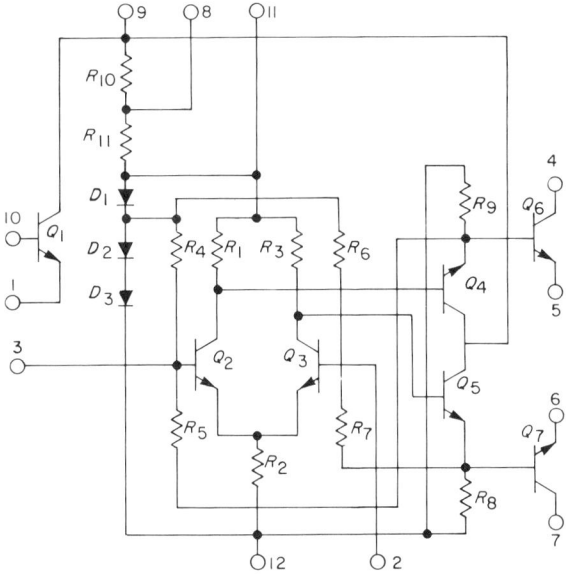

FIG. 14·19 Schematic diagram of the CA 3020 integrated-circuit, multipurpose wide-band amplifier. (*Radio Corporation of America, Electronic Components Division, Harrison, New Jersey.*)

The connections and added components one uses around it determine its function. For example, if one wishes to use a low impedance loudspeaker with the CA3020 serving as an audio amplifier, it is necessary to use an external impedance matching transformer and make a few minor circuit modifications around the outside of the IC as shown in Fig. 14·20. Compare Figs. 14·18 and 14·20.

14·8 SEMICONDUCTOR LIMITATIONS

Although the life expectancy of solid-state diodes and transistors is quite long, their longevity can be shortened considerably by improper techniques or practices used by electronics assemblers and technicians. Therefore, a few words of caution are included here.

1. Exercise care not to overheat a semiconductor when soldering. Heat sinks are used to minimize this heat (see Fig. 6·14).
2. Do not allow a semiconductor to get wet or damp. (Water is a conductor of electricity.)
3. Keep semiconductors away from any hot item or area.
4. The construction of a glass-encapsulated semiconductor device is quite fragile. Therefore avoid all mechanical shocks when handling this instrument.
5. Be sure to check all battery polarities before applying power to semiconductor circuits.

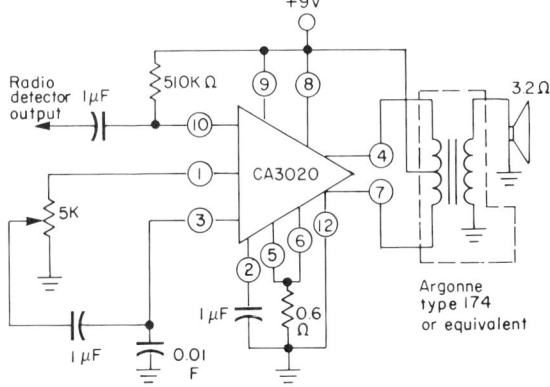

FIG. 14·20 An IC audio amplifier using a power output transformer to match the impedance of a loudspeaker. (*Radio Corporation of America, Electronic Components Division, Harrison, New Jersey.*)

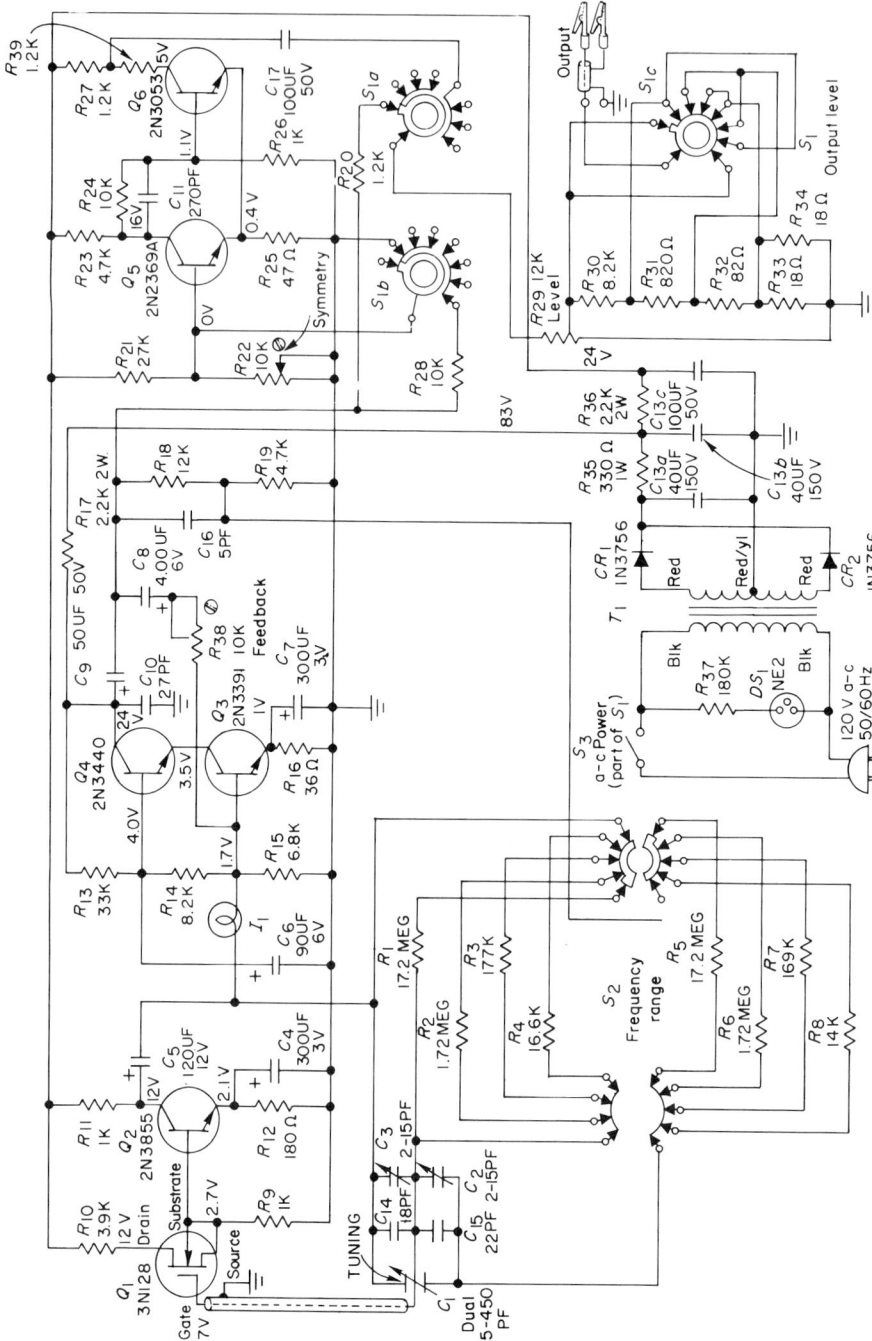

FIG. 14·21 A solid-state-circuit audio signal generator schematic diagram. (*Radio Corporation of America, Electronic Components Division, Harrison, New Jersey.*)

6. Pay close attention to biasing conditions.
7. Keep the current through semiconductors to a minimum.
8. If a semiconductor is designed with a stud for mounting, do not operate the circuit unless the diode or transistor is mounted.

14·9 SUMMARY: SOLID-STATE ELECTRONICS

Solid-state electronics is centered around the semiconductor family of components. However, closely related are the resultant manufacturing techniques for these devices which have made possible the fabrication of *microminiature* circuits as a whole, including resistors and capacitors. These minute arrays of components which are manufactured as a unit are known as *integrated circuits*.

Since the semiconductor is the basis for all of this technology, it is important to become quite familiar with the basic semiconductor units, the *diode* and the *transistor*. Further, one must not forget that for the most part semiconductors were developed as replacement for the familiar electron tube. However, semiconductor electronics has not completely eliminated the need for electron tubes, nor have integrated circuits replaced discrete components wired individually to each other. For example, Fig. 14·21 is the schematic of a solid-state audio signal generator where separate diodes, transistors (including MOS field-effect transistor), and passive components (all as separate units) are *wired* together to produce the finished product. And although this signal generator utilizes semiconductor devices as the active components, electron tubes could easily do the same thing, and in the past have done just that.

In summary, semiconductors perform many of the functions performed by electron tubes. Therefore a prior knowledge of electron tubes helps in the understanding of solid-state diodes and transistors. Also, the study of solid-state diodes precedes the study of transistors.

In the study of semiconductors, two currents are considered: first, the regular electron flow, and second, the flow of holes, or hole current. Where electrons flow from negative to positive, holes flow from positive to negative. Holes are considered a positive electric charge.

Transistors are current-controlled devices. The base current controls the collector current. For proper operation of a semiconductor, proper biasing is important. During forward-bias conditions, heavy currents flow. During reverse bias, small currents are the normal result. In transistors, the emitter is forward-biased and the collector is reversed-biased.

Care in working with semiconductors ensures a normal long life for them. Semiconductors are sensitive to heat. Heat can damage a semiconductor.

The jobs that follow are designed to give the student a simple but *practical* understanding of semiconductors. However, close attention should be paid to the cautions listed in Sec. 14·8.

EXERCISES

JOB 14·1 How to check front-to-back ratios in semiconductors

OBJECTIVE
1. To learn about front-to-back ratios in semiconductors
2. To learn to use a voltohmmeter
3. To appreciate a semiconductor tester

MATERIALS REQUIRED.

Supplies

A good semiconductor diode
A 10-inch length of copper wire
Voltohmmeter, 20,000 ohms per volt, with test leads

PROCEDURE

1. Compare the voltohmmeter in your possession with that shown in Fig. 14·22.
2. Compare the voltohmmeter in your possession with the VTVM used in previous jobs.
3. The 20,000 ohms per volt rating of a voltohmmeter is an indication of the sensitivity of the instrument. This particular instrument, presently in use, is quite adequate for most electronics work.
4. Note that the meter scales and switches on both the VOM (voltohmmeter) and the VTVM (vacuum-tube voltmeter) are almost identical. Some differences are encountered, but they are minor.
5. The use of a VOM is very similar to the use of a VTVM. However, the VOM does not require external power or a warmup period.
6. Apply your knowledge of resistance measurements with a VTVM when using the VOM as an *ohmmeter*. While a ZERO ADJUST knob will be found on both instruments, the VOM will not have an OHMS ADJUST knob.
7. Insert the test leads in the COMMON and OHMS jacks, respectively.
8. Set the range switch in the $R \times 10,000$ position.
9. Measure the resistance of the length of wire supplied.
10. Reverse the polarity of the test leads and again measure the resistance of the same length of wire. In both cases the resistance should be zero.
11. Remove the length of wire and proceed to measure the resistance of the semiconductor diode supplied. *Maintain the range switch at the $R \times 10,000$ position.* This is necessary in order to protect the diode.
12. Reverse the polarity of the test leads and once more measure the resistance of the semiconductor diode. Note that, when the polarity of the test leads is reversed, the resistance measured is much different.
13. Calculate how many times greater one resistance measurement is over the smaller reading. This *ratio* of resistance measurements can indicate the quality of the diode. The higher this ratio, the better the diode under test. Since an ohmmeter in reality causes a current to flow through the component under test, a measure of resistance is an *inverse* indication of current flow. The resistance ratio and the current ratio will therefore be the same, except that they are *inversely proportional*. This analysis is known as the front-to-back *ratio* of a semiconductor. It can refer to either resistance or current.
14. Front-to-back ratios (resistance measurements) determined between any two electrodes of a transistor can quickly reveal the *relative* condition of the transistor. Ratios should be *at least 10:1*.
15. Semiconductor testers, similar to that shown in Fig. 14·23, sometimes also provide for simple front-to-back ratio measurements. However, the meter indication may show either GOOD or BAD for quick analysis. A more complete, technical tester is shown in Fig. 14·24. However, it too provides for front-to-back ratio testing of diodes.

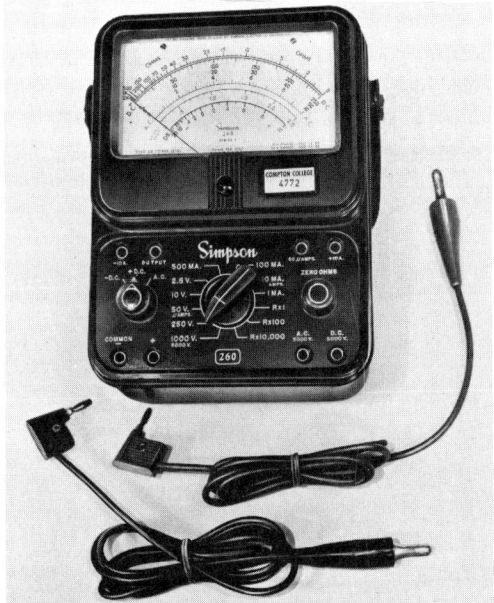

FIG. 14·22 A voltohmmeter (VOM).

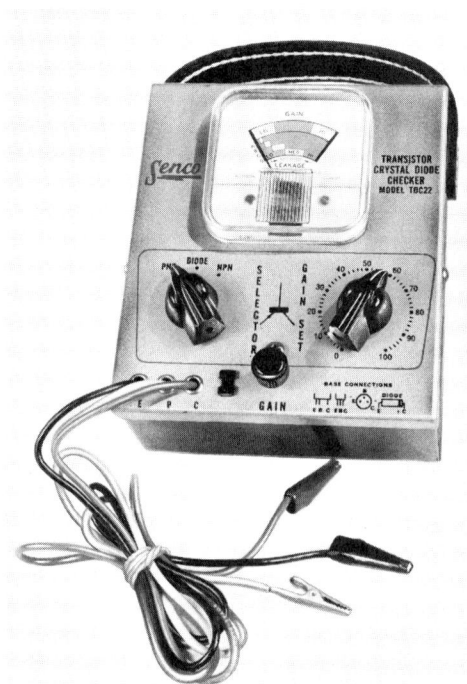

FIG. 14·23 A simple transistor and diode tester.

FIG. 14·24 A technical transistor tester. (*Radio Corporation of America, Electronic Components Division, Harrison, New Jersey.*)

QUESTIONS

1. List all differences between a VOM and a VTVM.
2. Will more or less current flow through a semiconductor under test when the range switch of an ohmmeter is set at $R \times 1000$ than when it is at the $R \times 10,000$ position? Explain.
3. If the front-to-back ratio of a semiconductor diode is 100:1, is the diode satisfactory for use? Explain.
4. Were holes flowing through the ohmmeter as a current while the semiconductor was undergoing its front-to-back ratio test? Explain.
5. Was it necessary to know about PN junctions in order to understand the tests made on front-to-back ratios? Explain.
6. Explain the term inversely proportional.
7. Was forward and reverse bias involved during testing of the front-to-back ratios? Explain.
8. Does a low-resistance measurement indicate that a heavy current will flow when a voltage is applied to that resistance? Explain.

JOB 14·2 How to investigate basic circuit arrangements of transistors (including testing considerations)

OBJECTIVES

1. To acquaint the student with common-base, common-emitter, and common-collector circuit configurations
2. To lay the foundation for examination of biasing circuits
3. To lay the foundation for testing of transistors

MATERIALS REQUIRED

Equipment

Transistor tester (see Fig. 14·24), with instruction manual
"RCA Transistor Manual, Technical Series SC-13" or equal

Supplies

Assortment of transistors
Notebook paper and pencil

PROCEDURE

1. There are three basic circuit arrangements within which transistors can be connected: *common-base, common-emitter,* and *common-collector*. A person must become familiar with these three *circuit configurations* before further practice can become meaningful.
2. The common-base circuit configuration is shown in its most basic state in Figs. 14•8 and 14•9. Study these circuits closely. Notice that the *base* electrode is the common element to both *biasing* circuits (input and output circuits).
3. The common-emitter circuit configuration is shown in Figs. 14•10 and 14•11. Study these circuits closely. Notice that the *emitter* electrode is the common element to both the input and output circuits. In addition, notice the comparison of the transistor circuits with the vacuum-tube circuit in each case. The vacuum tube circuit illustrated in these two figures is known as the *common-cathode* circuit configuration. The circuit configuration shown in Figs. 14•10 and 14•11 are the most popular circuit arrangements found around the triode-type of active components in electronics. Since these circuit configurations are popular, most transistor and tube testers use this type of configuration to evaluate performance of these active devices.
4. Compare the MOS field-effect transistor circuit configuration of Fig. 14•14 with the circuits of Fig. 14•11. The MOS-FET circuit is a *common-source* circuit configuration.
5. Notice that in Fig. 14•11, the output currents of both transistor circuits are affected by the input currents. The ratio of the current in the output circuit to the current in the input circuit is useful in determining the quality of a transistor. This ratio is known as the *forward current-transfer ratio*, and is often called *beta*. Since the collector current (output) is always greater than the base current (input), expect to encounter *beta* values anywhere from 10 to several hundred. Although values of beta normally will not change with transistor age, they will differ drastically from one type transistor to another. Therefore, the use of a transistor manual is essential in order to evaluate measured results.
6. Refer to Fig. 14•10. Notice that the collector circuit is reverse-biased. That means that if no base current is available (which causes a heavy collector current to flow, as in Fig. 14•11), then only a small *back-current* will flow in the collector circuit. This back-current is known as *collector leakage current*. A measure of this back current will also help to evaluate the condition of a transistor. Another term given this back-current is *collector-cutoff current*.
7. With the aid of the transistor tester manual, and with the guidance towards evaluation given by the transistor manual, proceed to test the assortment of transistors provided to you. However, before commencing the actual testing operation, thoroughly study the transistor tester manual. Also, turn to the front pages of the transistor manual and further research the areas of transistor circuit configuration, beta, and collector-cutoff current.
8. Since it is not the scope of this text to delve too deeply in any subject relating to electronics theory, do not hesitate to ask your instructor for assistance.

QUESTIONS

1. Draw the schematic diagram of a common-collector with a NPN transistor circuit configuration.
2. Using the transistor manual, list the normal applications for each of the three transistor circuit configurations.
3. With the aid of the transistor-tester manual and the transistor manual, give the letter abbreviations used for *beta* and for *collector-cutoff current*.
4. Draw a simplified sketch of the transistor-tester scales and explain what each scale indicates. (See Fig. 14•24.)
5. List each transistor supplied and indicate its condition as you determined by testing it with a transistor tester.

JOB 14·3 How to investigate basic biasing circuits for transistors

OBJECTIVES

1. To acquaint the student with bias circuits designed to provide a transistor with proper polarities.
2. To acquaint the student with bias circuits designed to provide a transistor with stable applied d-c voltages

MATERIALS REQUIRED

Equipment
"RCA Transistor Manual, Technical Series SC-13" or equal
Transistor theory reference material supplied by instructor

PROCEDURE

1. Study Figs. 14·8 and 14·9. Notice the circuit arrangement regarding the relation of the battery polarities and the individual electrodes within the transistors. Notice the terms *forward* and *reverse bias* in each circuit.
2. Study Fig. 14·10 in the same manner.
3. Although these circuits provide the transistors involved with the proper polarity of the basic operating d-c voltages to satisfy minimum requirements, in an actual operating situation when a signal is applied for amplification, the *levels* of these d-c voltages will vary slightly. This is not acceptable since variable applied d-c voltages cause an unstable foundation upon which the signal to be amplified must rely. To minimize this instability, the basic transistor circuits of Fig. 14·10 are *varied slightly*. This variation can include the simple addition of one resistor between the collector and base terminals of the transistor, or it can include the addition of at least two resistors and a capacitor to the basic circuit. Further, the points at which the bias voltages are applied may alter from one circuit to the next. Essentially, these voltage stabilizing circuits are nothing more than voltage dividers and filtering circuits.
4. An example of a voltage divider network used to apply the proper bias to a transistor can be seen in Fig. 17·5. Here, resistors R_4 and R_5 combine to divide the applied –6 volts so that only a portion of this voltage is presented to the base electrode. This contributes to increased stability.
5. An example of resistor-and-capacitor combination used to further increase circuit stability is also seen in Fig. 17·5. In this case emitter resistor R_6 and capacitor C_7 serve this purpose.
6. To further acquaint yourself with this subject, research the matter of *biasing* and *bias stability* in the theory section (front pages) of the transistor manual supplied to you for this purpose. To supplement this material, also refer to the various reference materials on transistor theory supplied by your instructor.

QUESTIONS

1. Define *bias* as applied to transistor circuits.
2. Define *bias stability* as it regards transistor circuits.
3. Is bias stability a problem with MOS field-effect transistors? Explain.
4. Redraw the NPN transistor circuit of Fig. 14·10 to include bias stability circuits. Give three possible circuit variations to accomplish bias stability.
5. Explain how in Fig. 17·5 resistors R_4 and R_5 serve to divide the applied voltage for presentation to the transistor.
6. Explain how in Fig. 17·5 resistor R_6 and capacitor C_7 serve to increase circuit stability.

JOB 14·4 How to investigate special applications for diodes

OBJECTIVES

1. To become acquainted with special purpose diodes
2. To become acquainted with experimenter's and hobby-type projects that can serve to motivate the student in the area of semiconductor applications

MATERIALS REQUIRED

Equipment

"RCA Silicon Controlled Rectifier Experimenter's Manual, Technical Series KM-71" or equal

"RCA Solid-State Hobby Circuits Manual, Technical Series HM-90" or equal

"RCA Transistor Manual, Technical Series SC-13" or equal

Supplies

Notebook paper and pencil

PROCEDURE

1. Review Sec. 14·6 entitled *Special Application Diodes*.
2. Using the manuals supplied, study in more detail about *Zener diodes*.
3. Using the manuals supplied, select an application for Zener diodes that interests you personally. Write a report, including schematic or other diagrams explaining this particular application for Zener diodes. If you use additional reference material, include this information also in your report.
4. Repeat steps 2 and 3, but this time select the area of *tunnel diodes*.
5. Repeat steps 2 and 3, but this time select the area of *silicon-controlled rectifiers*.
6. Repeat steps 2 and 3, but this time select an area of special purpose diodes not covered above.

QUESTIONS

1. In your opinion, what is the main application for Zener diodes?
2. In your opinion, what is the main application for tunnel diodes?
3. In your opinion, what is the most useful application for silicon-controlled rectifiers around the home?
4. Using the manuals supplied above, copy the schematic diagram of a single circuit that used as many different type of diodes as possible. Explain the purpose of such circuit. (*It is all right to include transistors, but the assignment is centered around the diodes used.*)

JOB 14·5 How to get acquainted with integrated circuits

OBJECTIVES

1. To provide for a student to get more information on integrated circuits
2. To motivate the student to experiment with integrated circuits

MATERIALS REQUIRED

Equipment

"RCA Linear Integrated Circuits (manual), Technical Series IC-41" or equal

"RCA Solid-State Hobby Circuits Manual, Technical Series HM-90" or equal

Supplies

Notebook paper and pencil

PROCEDURE

1. Look for a simple application of integrated circuits, such as in a code-practice oscillator. Look in the Hobby Circuits Manual supplied.
2. Write a report as to what supplies and equipment would be needed to construct such an oscillator. Also include in your report the schematic diagram that the builder would be concerned with. Finally, draw a sketch of what the finished product would look like.
3. Review Sec. 14·7 entitled *Integrated Circuits*, in this textbook.
4. Look over the Integrated Circuits manual supplied. Locate in this manual the circuits and diagrams that most closely resemble those in Figs. 14·18, 14·19, and 14·20 found in this textbook. Write a report on the *various applications* for the integrated circuit used.
5. With the aid of the above manuals, answer the questions listed below. (*Use other sources of information if they are available.*)

QUESTIONS

1. What is the difference between a discrete and integrated component?

2. What type of amplifier is considered the optimum configuration for general-purpose, linear integrated circuits? Why?
3. Write a brief statement regarding the limitations of supply-voltage to integrated circuits.
4. Describe the three most popular packages housing integrated circuits.
5. In working with integrated circuits, what in your opinion are the three most important precautions to observe?

15

ACTIVE CIRCUITS: DESCRIPTIONS

In manufacturing or maintenance of electronic equipment one encounters a great variety of circuits. However, numerous as these circuits are, it is possible to fit them into one of four categories. These are *rectification, amplification, oscillation,* and *switching*.

15·1 RECTIFICATION
Rectification is the process of changing an alternating current into a direct current. Circuits used to accomplish this are known as *rectifier circuits*.

POWER RECTIFIERS
Rectifier circuits are found in *power supplies*. A power supply is used to substitute for the batteries needed by vacuum-tube circuits or transistor circuits, eliminating the problem of frequent battery replacement. An a-c source of voltage is used to supply electrical energy to the power supply.

Rectifier circuits evolve from diodes. These diodes may be either of the vacuum-tube type or of the semiconductor variety. Two simple power rectifier circuits are shown in Fig. 15·1. Figure 15·1a employs a vacuum-tube diode for rectification; the circuit in Fig. 15·1b employs a solid-state semiconductor diode for rectification. In both circuits the output current flows from the cathode to the anode. Since the polarity of alternating current is continuously changing, current can flow only

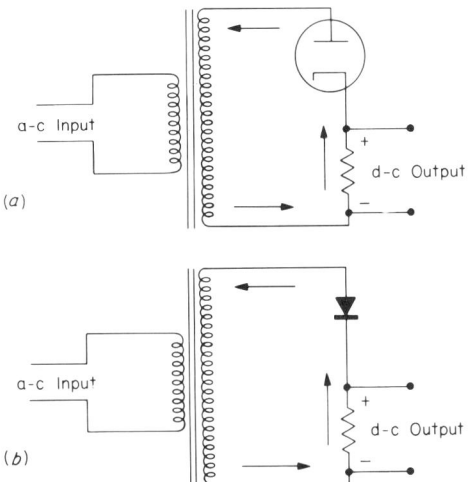

FIG. 15·1 (a) Vacuum-tube diode used as a power rectifier. (b) Semiconductor diode used as a power rectifier.

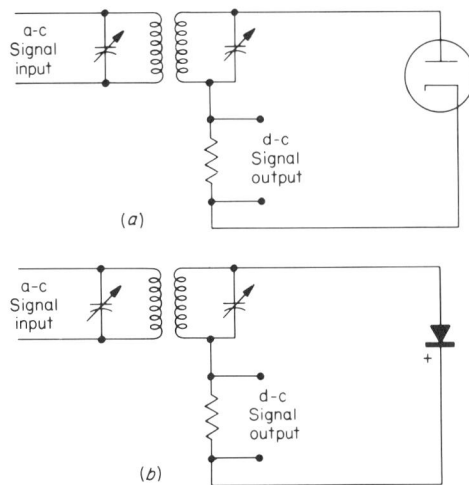

FIG. 15·2 (a) Vacuum-tube signal rectifier. (b) Semiconductor-diode signal rectifier.

through the load resistor during one-half of the applied a-c cycle. Chapter 16 covers in greater detail the rectification process. Note that the positive mark on the cathode of each diode represents the *rectified polarity*. The cathode of a solid-state semiconductor diode is often identified with a plus mark.

Very often power rectifier circuits will not use transformers at all. When transformers are used, however, an iron core is needed, because frequencies encountered by power supplies are at the low end of the audio-frequency band. Typical power-supply frequencies are 60, 115, 400, and 800 cps.

SIGNAL RECTIFIERS

Not all rectifiers are concerned with the elimination of batteries. Another application for diodes is to rectify electrical *signals*. Knowledge of power rectifiers helps in understanding signal rectifiers. Figure 15·2 illustrates an a-c signal being changed into a d-c signal. In this case it is common to find the frequency of the signals involved in the thousands or even millions of cycles per second. When the frequency of the signal is in megacycles, no iron core is used and resonant circuits are employed. Once again, the output is *developed* across a resistor. Signal rectifier circuits are also known as *detec-*

tors and *demodulators*. The current requirements of these circuits are quite low.

15·2 AMPLIFICATION

Amplification is the process of building up a weak electrical signal into a much stronger electrical signal. Vacuum-tube triodes or transistors are used as the core of a circuit constructed to accomplish this task.

TYPES OF AMPLIFIER CIRCUITS

Numerous circuit arrangements are made into amplifiers. However, two amplifier categories make up the bulk of electrical signal amplifiers. An amplifier is either a *voltage amplifier* or a *power amplifier*.

A voltage amplifier is concerned primarily with increasing the strength of an electrical signal for future use.

A power amplifier, also known as a current amplifier, has the task of building up a signal for immediate use or work.

Under normal conditions, voltage amplifiers obtain a very weak electrical signal and build it up to *drive* a power amplifier. The power amplifier in turn operates working devices such as the loudspeaker in a radio receiver, or it may operate an electric motor or relay.

COUPLING CIRCUITS

Amplifiers are also identified by the circuit network between stages. These networks are arrangements of transformers, capacitors, resistors, or simple inductors. The networks are called coupling circuits. Two of the most common coupling circuits are *transformer coupling* and *RC coupling*, the latter being associated with resistors and capacitors.

A great deal of the amplifying process is a result of the circuit conditions in these coupling circuits. Vacuum tubes and transistors are not alone responsible for amplification.

Two transformer-coupled amplifier circuits are shown in Fig. 15·3. Figure 15·3a employs a vacuum-tube triode, while the circuit in Fig. 15·3b makes use of a PNP transistor. Transformer T_1 couples a weak signal into these amplifiers, and transformer T_2 couples the strengthened signal out of the amplifier.

The amplifier circuits employed in Fig. 15·4 use the RC type of coupling. Once more a vacuum-tube triode is used in Fig. 15·4a. However, an

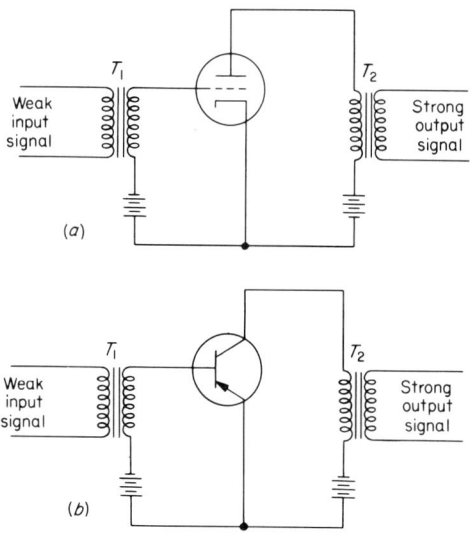

FIG. 15·3 (a) Transformer-coupled, vacuum-tube amplifier. (b) Transformer-coupled, transistor amplifier.

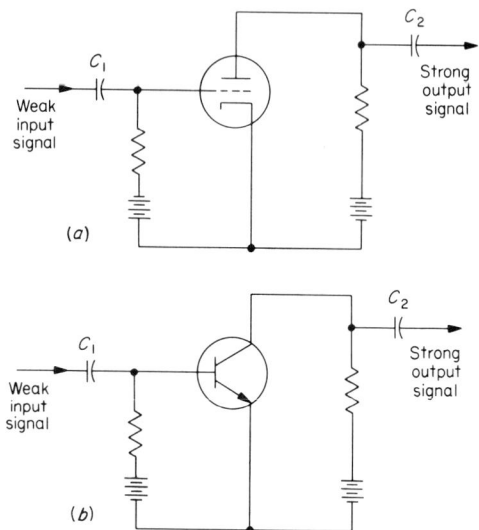

FIG. 15·4 (a) RC-coupled, vacuum-tube amplifier. (b) RC-coupled, transistor amplifier.

NPN transistor is used in Fig. 15·4b. In these circuits, capacitor C_1 couples the weak signal into the amplifier, while capacitor C_2 couples the strengthened signal out of the amplifier.

POPULAR AMPLIFIER CIRCUITS

Amplifier circuits are widely used. For example, a typical radio receiver has two rectifier stages, one for power and one for signal. It uses only one oscillator circuit, but four or five amplifier stages. One of these stages is a power amplifier, and the rest are voltage amplifiers. Detailed information about amplifiers is contained in several chapters, where the study of amplifiers is associated with their application.

15·3 OSCILLATION

Oscillation, as applied to electronics, is the process of generating a-c signals with the aid of amplifier circuits. Any amplifier can be made to *oscillate* by making a few minor circuit modifications.

REGENERATION

An oscillator circuit is also identified as a *regenerative amplifier*, because when the normal output of

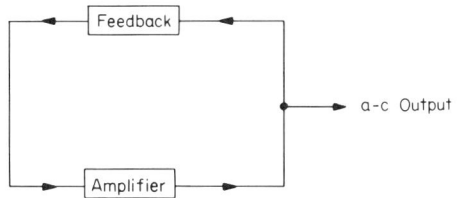

FIG. 15·5 Block diagram of a regenerative amplifier that oscillates.

an amplifier is properly fed back into its own input circuit, the amplifier will oscillate. This oscillation is caused by the *reamplification* of its own output signal. See Fig. 15·5 for a block diagram of an oscillator, or regenerative amplifier.

Two simplified schematic diagrams of oscillator circuits are shown in Fig. 15·6. One circuit employs a vacuum tube, while the second makes use of a PNP transistor. As shown in the illustration of Fig. 15·6a, a transformer couples the plate signal back into the grid circuit for regeneration. This will cause the tube amplifier circuit to oscillate. A small third winding is employed by the coupling

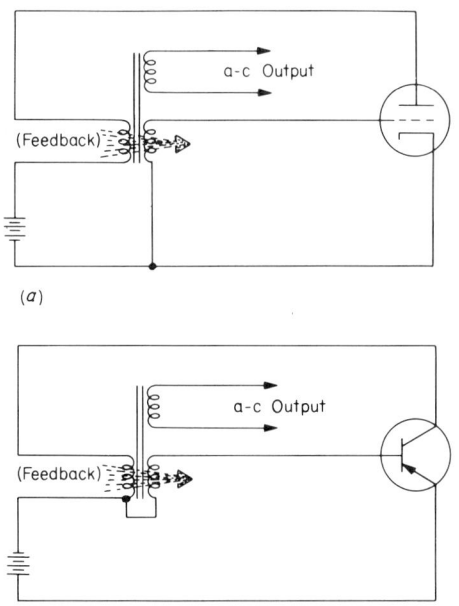

FIG. 15·6 (a) Simplified vacuum-tube oscillator. (b) Simplified transistor oscillator.

transformer so that an output can be extracted. Note that the explanation given above for the vacuum-tube oscillator also applies to the transistor oscillator shown in Fig. 15·6b. Additional and more detailed information on oscillators will be found in Chap. 17.

15·4 SWITCHING

Switching circuits are really an application of rectification, amplification, and oscillation principles. They are used for *pulse gates, wave shaping,* and *pulse timing.* Gating circuits are used for fast switching of electrical pulses from one circuit to another. Switching circuits are found in great quantity in computers and computer-type instruments. In fact, a given computer may utilize *thousands* of these circuits. Because of this, semiconductors are used almost exclusively since they are much smaller than electron tubes and require less power to operate.

THE AND GATE

Figure 15·7 shows a series of diagrams that illustrate the principles of a representative switching circuit, the *AND* gate. Figure 15·7a illustrates the *fundamental* principles of the gate. As can be seen, switch *A and* switch *B* must be closed in order for the lamp to function (light). Figure 15·7b uses diodes that contribute to the selection of the proper combination of signal inputs in order to get a desired output. Notice that in this illustration, the output voltage at point *f* is +10 volts when either diode is connected to a +10-volt signal. However, when both input *A* and input *B* are connected to a 50-volt signal (matching the power supply voltage), the output voltage (*f*) will be 50 volts. Therefore, this circuit gives only a high voltage output signal when *both* inputs have a high voltage applied. This type of circuit is also classified as a *logic circuit.* The *AND* gate normally appears in the format shown in Fig. 15·7c. To simplify matters still more, the symbolic diagram of a two-input *AND* gate is shown in Fig. 15·7d.

THE LOGICAL FLIP-FLOP

Another representative switching circuit is the *logi-*

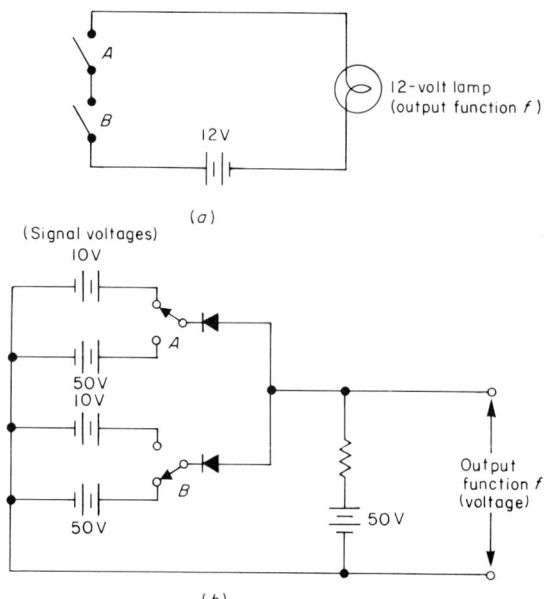

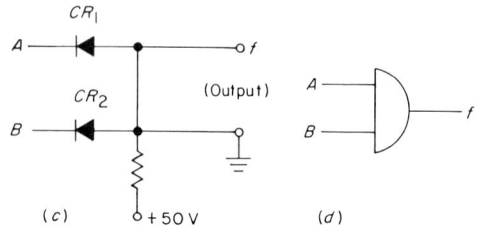

FIG. 15·7 Principles of the AND gate.

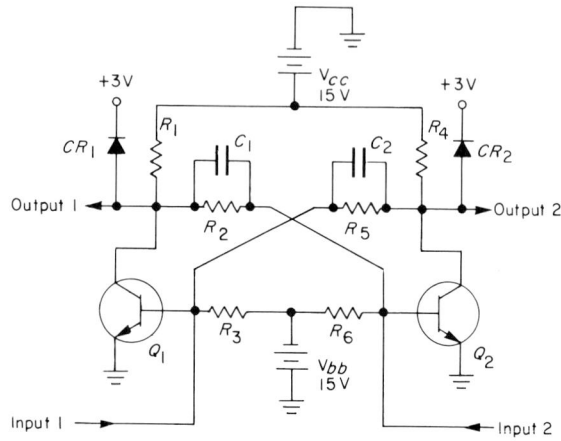

FIG. 15·8 The *logical* flip-flop circuit (bistable multivibrator).

cal flip-flop, shown schematically in Fig. 15·8. This circuit belongs to the oscillator family, and is also known as a *bistable multivibrator*. The circuit is used extensively in *digital devices*, such as computers and computer-type instruments. Its purpose is to store *binary information* electronically. Binary information refers to the system of logic where only *two* conditions exist: either ON or OFF; light or dark; plus or minus; high or low; north or south; one or zero. The logic involved is based on the *binary numbering system*.

15·5 SUMMARY: ACTIVE CIRCUITS

In summary, active circuits are centered around the use of electron tubes or semiconductors. These circuits can *act upon* a signal voltage or current, either to generate it or to modify its level or waveshape. Resistors, capacitors, and inductors by themselves cannot accomplish this. Active circuits can be categorized as rectifiers, amplifiers, oscillators, or switching circuits.

The following jobs are designed to guide the student into quick recognition of these four classifications of circuits.

EXERCISES

JOB 15·1 How to identify a rectifier circuit

OBJECTIVES
1. To become acquainted with rectifier circuits
2. To appreciate the requirements for electron flow

MATERIALS REQUIRED

Supplies

Paper and pencil

PROCEDURE

1. Refer to the rectifier circuits in Fig. 15·1.
2. The arrows indicate that an electric current exists in the secondary circuit of the transformer. Since these arrows all point in a counterclockwise direction only, they imply that the *electron movement is a direct current*.
3. The current flowing up through the resistor develops a voltage drop across the resistor with the corresponding polarity as marked. Since the current flows in only one direction, the voltage drop produced is d-c.
4. Electrons are able to flow as a current as shown because the circuit is complete. The vacuum tube should be considered to have a hot cathode, although the heater is not illustrated.
5. In a rectifier circuit electrons flow only in the direction shown because electrons flow readily from a cathode to an anode (plate) but not from an anode to a cathode.
6. In the primary winding of the transformer, electrons alternately travel clockwise and counterclockwise.
7. On your paper, copy the rectifying circuits of Fig. 15·2.
8. On the diagrams you have made, show with arrows the paths taken by the electrons as they move in the completed circuits. Remember that electrons cannot travel through a capacitor and that they must be able to return to the point from which they started.
9. Have your instructor check your work.

QUESTIONS

1. Define an electron.
2. If electrons were not able to move in the circuits of Fig. 15·1, would the minus and plus marks remain by the resistor?
3. How is it possible to indicate with arrows that an alternating current is flowing?
4. What purpose do the circuits of Fig. 15·1 serve?

JOB 15·2 How to identify an amplifier circuit

OBJECTIVES

1. To become acquainted with amplifier circuits
2. To become acquainted with coupling circuits
3. To recognize the limitations of electron flow in amplifier circuits

MATERIALS REQUIRED

Supplies
Paper and pencil

PROCEDURE

1. On your paper, copy the vacuum-tube amplifier of Fig. 15·3a.
2. Prepare to draw in the arrows that will indicate electron movement.
3. Electrons will leave the negative terminal of the B battery and travel through the vacuum tube from the cathode to the plate. They will leave the plate and travel down through the primary winding of transformer T_1 and reenter the B battery at the positive terminal. Refer to Fig. A1·11, illustrating the *plate current*.
4. There is no grid current flowing through the secondary winding of coupling transformer T_1 since the control grid is not able to emit electrons.
5. On your paper, copy the transistor amplifier of Fig. 15·3b.
6. Draw in the arrows that indicate electron current.
7. Electrons flow clockwise in the base-emitter circuit. They leave the negative terminal, flow up through the secondary winding of transformer T_1, to the base, against the arrow of the emitter, down the emitter lead, and left to the positive terminal of the battery they left. Refer to Fig. 14·11b, illustrating the *base current* as well as the *input circuit current*.
8. Electrons travel counterclockwise in the collector circuit. They leave the negative terminal of the collector battery and travel up through the primary winding, of coupling transformer T_2 and into the collector. These electrons join the emitter electrons and proceed down through the emitter lead. The two groups of electrons separate upon reaching the wire junction below, the collector electrons returning back to the positive terminal of the battery they originally left. Refer to Fig. 4·11b, illustrating the *collector current* as well as the *output circuit current*.

9. On your paper, copy the vacuum-tube amplifier of Fig. 15·4a.
10. Draw in the arrows indicating electron flow. Remember that electrons cannot travel through a capacitor. Refer to Fig. 4·11a.
11. On your paper, copy the transistor amplifier of Fig. 15·4b.
12. Draw in the arrows indicating electron flow. Refer to Fig. 4·11c.
13. Submit all diagrams to your instructor for approval.

QUESTIONS

1. What purpose do the circuits of Figs. 15·3 and 15·4 serve?
2. What type of coupling is used in the circuits of Fig. 15·3?
3. What type of coupling is used in the circuits of Fig. 15·4?
4. What type of vacuum tube is used in Fig. 15·3a?
5. What type of transistor is used in Fig. 15·3b?
6. What type of transistor is used in Fig. 15·4b?
7. What is the meaning of coupling?
8. Explain why a control grid cannot emit electrons.

JOB 15·3   How to identify an oscillator circuit

OBJECTIVES

1. To become acquainted with oscillator circuit principles
2. To appreciate the meaning of coupling
3. To recognize an application for amplifiers
4. To test ability to trace electron flow

MATERIALS REQUIRED

Supplies
Paper and pencil

PROCEDURE

1. Refer to Fig. 15·6.
2. Notice that the feedback is coupled from the output circuit to the input circuit through the transformer.
3. Study the two circuits and notice that they are transformer-coupled amplifiers similar to those shown schematically in Fig. 15·3, although only one battery is used in each case.
4. On your paper copy the two *oscillator circuits* of Fig. 15·6.
5. Show how electrons travel in these oscillators. Use arrows to indicate electron flow. Refer to Job 15·2 as necessary.
6. Have your instructor check your finished diagrams.

QUESTIONS

1. What purpose does an oscillator serve?
2. How is regeneration illustrated in Fig. 15·6?
3. Explain how the feedback signal is coupled from the output circuit to the input circuit in both oscillators of Fig. 15·6.
4. Explain what corrections were necessary for the mistakes you made, if any, in step 5 above.

JOB 15·4   How to identify a switching circuit

OBJECTIVES

1. To become acquainted with types of switching circuits
2. To become acquainted with applications for switching circuits

MATERIALS REQUIRED

Equipment
"RCA Transistor Manual, Technical Series SC-13" or equal
"RCA Solid-State Hobby Circuits Manual, Technical Series HM-90" or equal

Supplies
Notebook paper and pencil

PROCEDURE

1. Study the *gate* circuit of Figure 15·7c. Notice that two simultaneous inputs to this circuit are possible, as well as the possibility of a normal one-input signal. However, of major importance here for purposes of recognition is the use of the two diodes for signal-input paths. These diodes, together with the resistor, serve as electronic gates that only allow a certain combination of signals

through. Being a gate, either the proper signal combination *is allowed* through, or it *is not allowed* through. This then, is just like the action of a switch. In an electrical circuit that uses a switch, current either gets through or it does not. For this reason, an electronic gate, such as shown schematically in Figure 15•7c, is classified as a *switching circuit*. Another term associated with gates and switching circuits is *digital circuits*.

2. Study the *flip-flop* circuit of Fig. 15•8. A flip-flop circuit is a form of *oscillator circuit*. Specifically, the flip-flop falls in the family of *multivibrator* circuits. As with all oscillators, there must be a feedback of an output signal for it to be regenerated. A predominately recognizable trait of a multivibrator circuit is the feedback link. Because of the way this circuit is usually drawn, this feedback link appears as a large **X** right in the middle of the schematic diagram of a multivibrator. In the circuit of Fig. 15•8, the **X** is formed by the lead that joins R_2 with R_6, and with the lead that joins R_3 with R_5. The multivibrator type of oscillator will cause each transistor (or vacuum tube) to rapidly and suddenly *switch* from a fully conducting state to a cutoff state. Multivibrator circuits are therefore also identified as *switching* or *digital* circuits.

3. With the above information as a guide, look in the two manuals supplied under *switch* from a fully conducting state to a cutoff state. Multivibrator circuits are therefore also identified as *switching* or *digital* circuits.

4. With the above information as a guide, look in the two manuals supplied under *switching circuits* and see how many of these digital circuits you are able to recognize. Make a list of the various special switching circuits you encounter. Be sure to use the correct designations as given in these manuals.

5. For more complete acquaintance with this classification of active circuits, read the entire section on switching circuits in these two manuals. Pay close attention to applications mentioned for these circuits.

QUESTIONS

1. Give your definition of a switching circuit.
2. What does the term *digital* mean as applied to switching circuits?
3. How many types of electronic gates did you encounter in the manuals supplied to you? Name them.
4. How many types of multivibrator circuits did you encounter in the manuals supplied to you? Name them.
5. List the various possible applications for switching circuits as listed in these manuals.

16

POWER SUPPLIES

Previous chapters have laid the foundation necessary for the study and understanding of electronics technology. While the electronics assembler may have gained enough information thus far to be successful in his trade, electronics technician trainees must continue to master the more technical aspects of their chosen field. This chapter will commence an explanation of how the various components used in electronics unite to achieve the wonderful results obtained from electronic devices.

16·1 THE NEED FOR A POWER SUPPLY

Electron tubes and semiconductors used as amplifiers or oscillators require direct currents and voltages for their operation. Batteries can, of course, meet this requirement and are the simplest form of power supply used in electronics. Batteries allow for complete portability of the electronic device using them as a power source. However, quite often the power requirements of large electronic units are far greater than can be supplied by a practical number of batteries. In order to improve the practicability in supplying power to the various electron tubes or semiconductors, a circuit composed of resistors, capacitors, inductors, and electron tubes (or semiconductors) is used to develop the necessary voltages. Such a circuit is called a *power supply*. The power-supply section found in some radio or television receivers is illustrated in Fig. 16·1. The

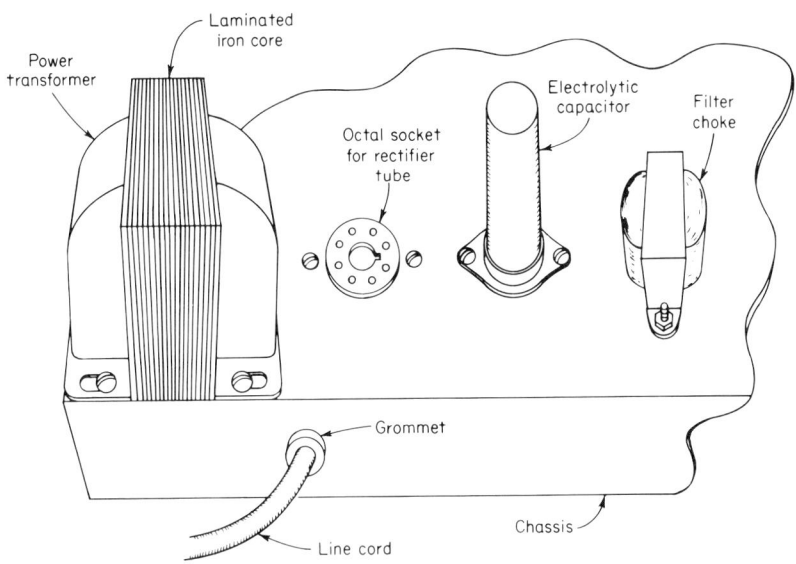

FIG. 16·1 The electronic power supply.

schematic diagram of the power supply illustrated is shown in Fig. 16·15.

The electronics repairman or experimenter often finds it necessary to have a convenient source of power for his work. To meet this need various power supplies are manufactured as separate units. Fig. 16·2 is an example of this type of power supply.

ORIGINAL POWER SOURCE

It is not possible to obtain an output from any device without a corresponding input. Energy may be transferred from one form to another, but it can never be created or destroyed. Therefore, although the power supplies mentioned above are expected to deliver power, they also require a power input from an external source. Batteries obtain power through charging and store it electrically by chemical means. Power supplies utilizing electronic circuits are continuously fed power from an external source. In effect, an electronic power supply transfers only electrical energy. It transfers voltages from one level to another and usually rectifies an alternating current so as to produce a direct current. However, the input to a power supply may be either an alternating current or a direct current (see Fig. 16·3).

POWER SUPPLY OUTPUTS

Power supplies that are to provide power to electron tubes have a minimum of two outputs. One output supplies the heaters or filaments, while the second supplies the plates and screen grids. Some power supplies also develop bias voltages to be used by the control grids. The output for the heater or filaments is termed the A supply. The output supplying the plate and screen grids is identified as the B supply. Bias voltages for the control grids are obtained from the output known as the C supply. The C supply is not widely used since bias voltages are often developed by the electron tube circuits requiring them.

The output of a power supply to be used in conjunction with transistors is normally a low-level d-c voltage. Transistorized devices often make use of small batteries as their power supply.

TYPES OF POWER SUPPLIES

Electronic power supplies are normally identified according to their electrical input or to the compo-

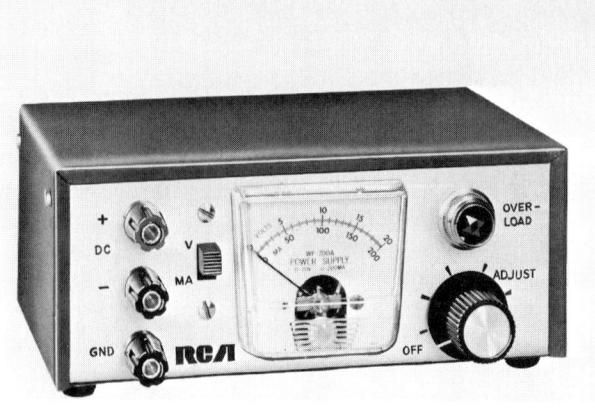

FIG. 16·2 The bench-type power supply. (*Radio Corporation of America, Electron Tube Division, Harrison, New Jersey.*)

FIG. 16·3 The power supply.

nents used in their input circuit. An *a-c power supply* requires an a-c input. An *a-c/d-c power supply* may operate with either an alternating current or a direct current for an input. A *transistorized power supply* makes use of a transistor in its input circuit to invert periodically a direct current to a form of alternating current. Note that pure direct current in the input circuit of a power supply is not allowed. It must be a pulsating direct current or an alternating current.

16·2 FUNCTIONS OF A POWER SUPPLY

A power supply must be able to produce alternating or direct currents for the outputs. These a-c and d-c outputs are also required to be at various voltage levels. All this must be done with only one input voltage of either alternating or direct current.

In order to transfer a voltage from one level to another, a transformer must be used. However, a transformer requires a changing magnetic field for its operation. Alternating currents provide this changing magnetic field very nicely. In order for a direct current to produce a change in the magnetic field, the direct current itself must be interrupted. This will give the effect of an alternating current. Regardless of the type of electricity used as an input for the transformer, the output of the transformer will be an alternating current. For the A supply output of the power supply this may be all right. The B and C supply outputs must be direct current. To have a direct current for an output, the alternating current must be rectified with either an electron tube or with a semiconductor device. However, rectification alone is not sufficient to produce a nearly perfect direct current. Filtering with electrolytic capacitors and choke coils or resistors must also be employed.

THE POWER TRANSFORMER

The transformer utilized in power supplies is known as a *power transformer*. Figure 16·4 shows two power transformers. This item is the largest and heaviest of all the components of an electronics chassis. Power transformers are manufactured with various windings (coils), depending on the power supply circuit in which they are to be used. A schematic diagram of a power transformer is shown in Fig. 2·26h.

Transformers are often encapsulated, or *potted* in square, metallic or plastic cube forms to the extent that they lose their normal identity. This is especially true in the area of miniature components, designed in packages that are to be mounted on printed-circuit boards. The student is cautioned about being misled by outward appearances. Figure 16·5 illustrates an encapsulated transformer. The case in Fig. 16·6 also contains a transformer: an *isolation transformer*.

FIG. 16·4 Typical power transformers.

FIG. 16·6 The isolation transformer. (*Radio Corporation of America, Electron Tube Division, Harrison, New Jersey.*)

THE RECTIFICATION PROCESS

An introduction to rectification was presented in Chap. 15. Here a more complete coverage of this process is necessary for a better understanding of power supplies.

The rectifying circuit consists of a diode and a resistor connected in series. When these two items are connected to an a-c source, an electron current will flow through the resistor when the plate or anode of the rectifier has applied to it a positive potential (see Fig. 16·7). The current through the resistor develops a voltage drop across the resistor. Notice that *two* positive potentials exist. One positive potential is applied from the voltage source, while the second is developed by the resistor.

During the half cycle of the alternating current when the plate or anode of the rectifier has a negative potential applied, no electron current will exist. (Actually, in the semiconductor a small current will flow, but it is negligible for this discussion.) In Fig. 16·8, notice that no voltage is developed across the resistor.

Figure 16·9 is a summary of the foregoing explanation. When an a-c voltage is applied to the diode-resistor combination, current will flow during one-half of the cycle only. Therefore the voltage developed across the resistor will always have the same polarity. This is pulsating direct current, since the flow of current through the resistor is interrupted, but not reversed. Note that the plate of

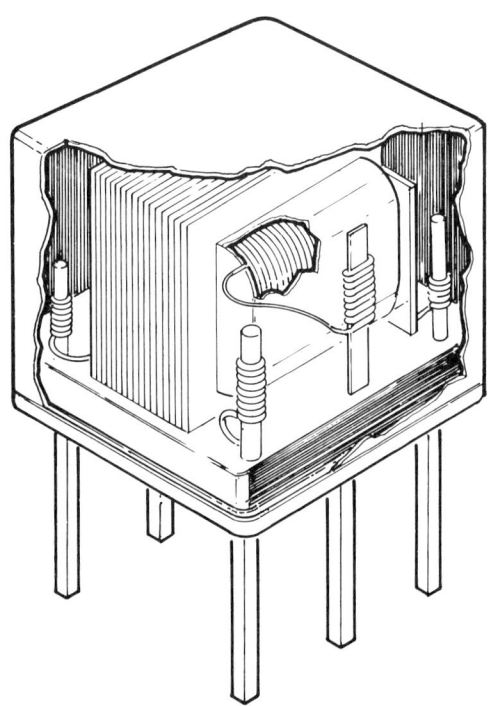

FIG. 16·5 The encapsulated transformer.

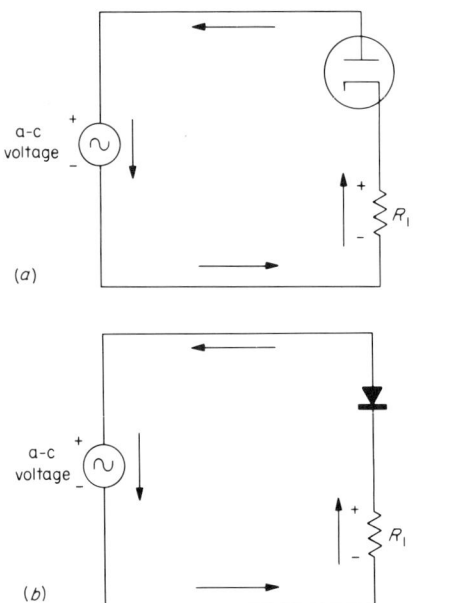

FIG. 16·7 The conducting rectifier. (a) Electron tube conducting. (b) Semiconductor diode conducting.

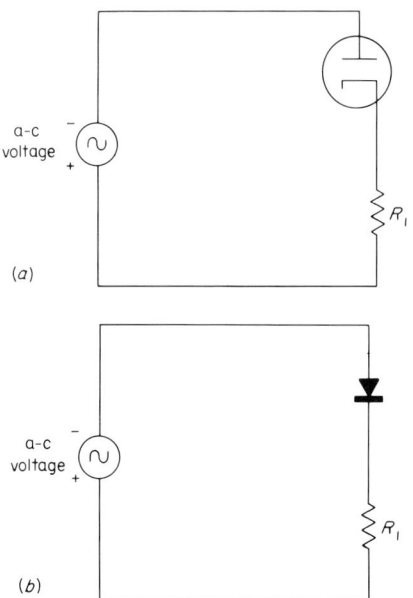

FIG. 16·8 Nonconducting rectifiers. Notice that the polarity of the a-c voltage source has reversed.

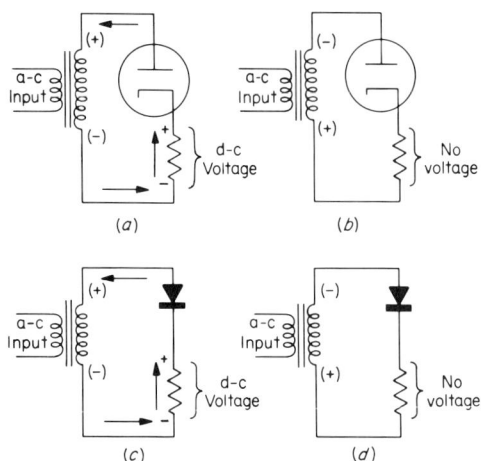

FIG. 16·9 Diode conduction and nonconduction conditions. (a) Compare with diagram (b). (c) Compare with diagram (d).

the rectifier is alternately positive and negative. However, the end of the resistor next to the cathode is always positive when a polarity exists. This is the rectified positive. For this reason, the cathode of many semiconductor diodes is marked positive.

The circuit described above is known as a *half-wave rectifier*.

FILTERING THE RECTIFIED OUTPUT

A pulsating direct current as an output from a power supply is not suitable. In a true direct current there can be no variation of the current flow. To minimize the variation in the current flowing through the resistor of Fig. 16·9, an electrolytic capacitor is connected across it (see Fig. 16·10).

The object of the electrolytic capacitor is to store electrical energy while the rectifier is conducting. This energy is then released during the time that the rectifier does not conduct.

Figure 16·10a shows that, during conduction, electrons flow through R_1 in the usual manner. During conduction also some of the electrons will charge the electrolytic capacitor as shown. What actually happens is that some electrons go into the negative plate of the capacitor while other electrons are removed from the positive plate. Now referring to Fig. 16·10b during the nonconduction

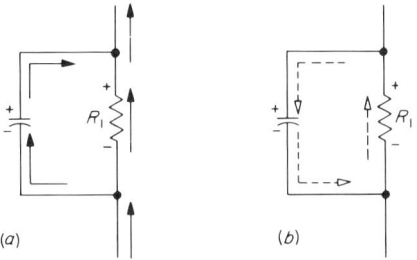

FIG. 16·10 Capacitive filtering. Notice that an electrolytic capacitor is used. (a) Charge current. (b) Discharge current.

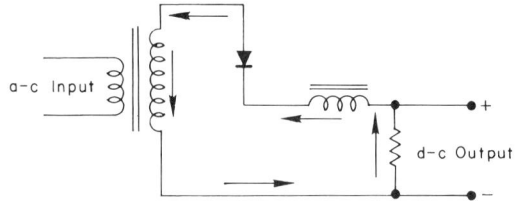

FIG. 16·12 Half-wave rectifier with inductive filter.

period, the electrons that were stored in the negative plate of the electrolytic capacitor move through the resistor while on their way to replace the electrons previously lost at the positive plate of the capacitor. This discharging current flowing through the resistor will produce a voltage drop across the resistor *during the nonconduction period*. It can now be seen that an electron current will flow through the resistor R_1 at all times. This will in turn cause a continuous voltage drop across the resistor, thereby creating a better d-c voltage. Since the capacitor cannot offer a constant supply of electrons, the voltage drop across the resistor will fluctuate a small amount. A schematic diagram of a half-wave rectifier with a capacitive filter is shown in Fig. 16·11.

While a capacitor used as described above can minimize voltage variations, a choke coil when placed in a circuit in series can minimize current variations (see Fig. 16·12). By minimizing the current variation through R_1, the choke coil indirectly reduces the voltage drop variation across the resistor. However, a choke coil used alone as shown will not reduce the current fluctuation appreciably.

To be more effective, it must be used in conjunction with electrolytic capacitors, as shown in Fig. 16·13. The extra capacitor C_1 aids in better filtering. It too charges during conduction and discharges through the resistor (and the choke coil) during the nonconduction period.

16·3 THE FULL-WAVE RECTIFIER

The output of the B supply section of a power supply must be as pure direct current as possible. To meet this requirement, improvements to the circuits described above are constantly under design. Perhaps the greatest improvement to come out of this research is the simultaneous operation of two half-wave rectifiers (see Fig. 16·14). The circuit shown allows both halves of the a-c waveform to be used. If one diode is not conducting, the other is. The circuit is known as a *full-wave rectifier*.

In a full-wave rectifier the current that flows through both rectifiers (alternately) flows through the resistor. In this case the capacitors charge to the peak value during heavy conduction of each rectifier and discharge through the resistor during the lesser values of conduction. A full-wave rectifier utilizing a dual-plate vacuum tube is shown in Fig. 16·15. To simplify the drawing of the sche-

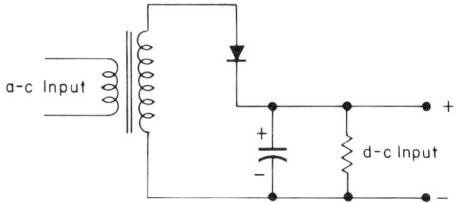

FIG. 16·11 Half-wave rectifier with capacitive filter.

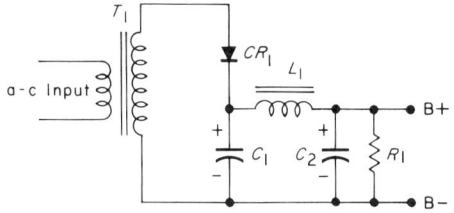

FIG. 16·13 The B supply.

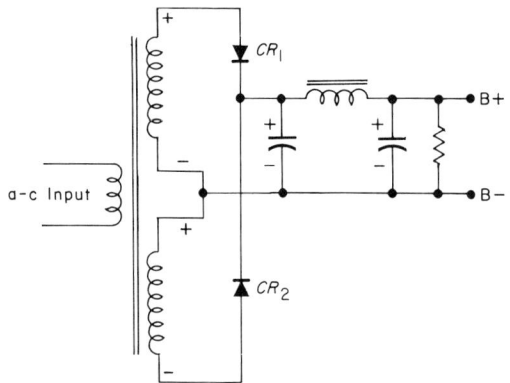

FIG. 16·14 A two-diode, full-wave rectifier.

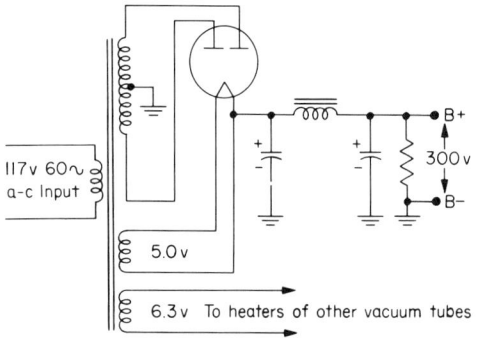

FIG. 16·15 A complete full-wave power supply that utilizes a directly heated cathode.

matic diagram, ground symbols are used. This means that all points having a ground symbol are connected together. In this case, ground is also B−.

The values of voltages indicated in Fig. 16·15 are typical. A higher value of B potential than the potential appearing as input to the power supply is possible because, in the power transformer, the high-voltage secondary windings have a greater number of turns than the primary coil. (All a-c voltages are *effective*, also known as *rms*, voltages. The B supply output is d-c.)

The full-wave power supply discussed above, or any other power supply utilizing a power transformer, is classed as an a-c power supply since the primary of the transformer requires alternating currents.

16·4 THE TRANSISTORIZED POWER SUPPLY

Power supplies discussed thus far have obtained the input power from an alternating-current source. The power transformer has been used to change the level of input voltage to whatever level of output voltage was needed. The ability of a transformer to change voltage levels depends upon a changing or altering input voltage. However, sometimes the input is a direct current; such as when a device is to operate from a battery. To solve the problem, it is necessary to convert the d-c to a-c, or at least to cause the d-c to be interrupted periodically. The transistorized power supply has been designed to accomplish this task.

Essentially, all that is needed is an electronic switching circuit that can interrupt the d-c input voltage and current so that the transformer can step-up or step-down the voltage to whatever level is needed. Then through rectification and filtering, a d-c output can again be obtained; this time at a different voltage level than the input. The switching circuit used is an *oscillator*; specifically a multivibrator circuit. Figure 16·16 shows a simplified transistorized power supply.

16·5 THE BLEEDER RESISTOR

The resistor across which the B supply has been developed in all of the power supplies discussed above (see Fig. 16·16), is known as the *bleeder* resistor. Most good power supplies employ such a resistor. However, very often this resistor is eliminated from a power supply. When a bleeder resistor is not used, the current flow is through the conducting tubes or transistors which operate from the same power supply.

16·6 SUMMARY: POWER SUPPLIES

In summary, a power supply can be considered a battery eliminator. Power supplies remove the burden of having to maintain, store, and handle large and heavy batteries that consume much space. However, the elimination of all batteries is not desirable. Portable equipment requires the use of batteries.

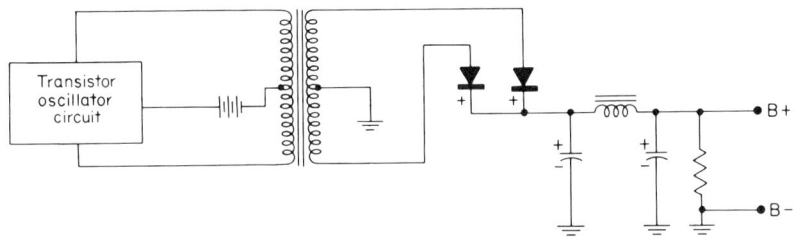

FIG. 16·16 A simplified transistorized power supply.

The replacement of batteries with electronic power supplies does not eliminate the need for an external power source. Power supplies commonly make use of alternating currents for input power. This alternating current must then be rectified and filtered in order to produce a d-c output. With alternating currents, a transformer can be used to change the voltage level to any desired value.

The following jobs are designed to give the student a practical understanding of power supplies.

EXERCISES

JOB 16·1 How to identify components in an a-c power supply

OBJECTIVES
1. To learn to recognize power-supply components
2. To appreciate the physical size of power-supply components

MATERIALS REQUIRED

Equipment
Radio receiver or audio amplifier with transformer-type power supply utilizing vacuum tubes

PROCEDURE
1. Compare the radio receiver (or audio amplifier) supplied with the chassis illustrated in Fig. 16·1.
2. Locate on the radio receiver supplied all the components indicated on the chassis of Fig. 16·1.
3. If needed, help in identifying the electrolytic capacitor may be obtained by referring to Fig. 2·17.
4. If the filter choke is missing, it may be that a resistor is used in its place. This resistor will be relatively large and will be found connected to the electrolytic capacitor.

QUESTIONS
1. Why can one expect the power-supply components to be the largest parts used in a chassis?
2. How many electrolytic capacitors were found? Were any of them dual capacitors?
3. Were all the power-supply components close to each other?

JOB 16·2 How to identify the leads of a power transformer

OBJECTIVES
1. To become better-acquainted with power transformers
2. To appreciate resistance variations in the windings of a transformer
3. To learn the color code for leads of a power transformer

MATERIALS REQUIRED

Power source
117-volt, 60-cycle alternating current

Equipment
VTVM

Supplies

Power transformer with color-coded leads and a line cord attached to the primary winding
Schematic diagram of the transformer

PROCEDURE

1. Inspect the condition of the color in the leads of the transformer supplied.
2. If the colors are faded or impregnated with varnish, it may be necessary to scrape lightly the surface of the insulation to reveal the true color of the leads.
3. Use the schematic diagram of the power transformer in Fig. 16·15 as a guide, unless a schematic diagram is supplied with your transformer.
4. Do not apply power to the transformer until told to do so.
5. Note that the *black* leads are attached to the line cord. These leads belong to the primary winding.
6. The *green* leads apply 6.3 volts alternating current to the heaters of all vacuum tubes except the rectifier.
7. The *yellow* leads apply 5.0 volts alternating current to the filament of the rectifier tube.
8. The *red* leads are connected to the plates of the rectifier tube and therefore apply alternating current to these plates. This voltage will have a magnitude of several hundred volts.
9. The *red-and-yellow* lead is the center tap of the high-voltage secondary winding.
10. Measure the resistance of all windings and record the results.
11. The primary winding should have approximately 15 ohms.
12. Both the heater and filament windings should indicate approximately zero ohms since they are wound with a few turns of heavy wire.
13. The resistance between the two red leads will be on the order of 100 ohms.
14. The resistance from either red lead to the red-yellow lead should be approximately one-half of the resistance obtained in step 13.
15. Apply 117-volt, 60-cycle a-c power to the primary winding of the transformer.
16. Measure the a-c voltage of the secondary windings with the VTVM. Record your results. Be sure that the VTVM is set up for measuring a-c volts.
17. The heater and filament voltages will be approximately the values indicated in Fig. 16·15.
18. The voltage between one red lead and the red-yellow lead will be approximately 300 volts a-c.
19. The voltage between the two red leads will be twice the voltage obtained in step 18.
20. Disconnect the power from the power transformer.

QUESTIONS

1. What was the resistance in the primary winding in your transformer?
2. How much a-c voltage will the transformer you tested present to one plate of the rectifier tube?
3. Did the transformer supplied to you have any additional leads?
4. Make a schematic diagram of the transformer you tested and enter on the diagram all the information you recorded. Present it to your instructor for approval.
5. Is an a-c voltage of 100 volts, effective, the same as an rms voltage of 100 volts? Explain.

JOB 16·3 How to locate the B supply outputs

OBJECTIVES

1. To learn to locate terminals in an actual circuit
2. To recognize a B+ and B− terminal

MATERIALS REQUIRED

Equipment
Radio receiver
Schematic diagram of the receiver

PROCEDURE

1. Note on the following schematic diagrams of power supplies that B+ and B− always appear on opposite ends of the same electrolytic capacitor in the filter circuit: Figs. 16·13 to 16·16.
2. To locate the B+ and B− terminals, first locate the electrolytic capacitor or capacitors used in the filter circuit of the power supply.
3. With the aid of a schematic diagram of the power supply involved, determine which electrolytic ca-

pacitor is the *output filter capacitor*, such as C_2 in Fig. 16·13.
4. The positive terminal of this capacitor is B+.
5. The negative terminal of the same capacitor is B−.
6. Locate the B+ and B− terminals on the radio receiver supplied.
7. Show your instructor their location, for approval.

QUESTIONS
1. Define B+.
2. Define B−.
3. Why are electrolytic capacitors necessary in a power supply?
4. Can a filter circuit be used in a power supply without a rectifying device?

JOB 16·4 How to investigate various power supply circuits

OBJECTIVES
1. To become familiar with power supply circuits as found in complete schematic diagrams of electronic devices.
2. To build confidence in the student that he is able to concentrate on a portion of a total schematic diagram.

MATERIALS REQUIRED

Supplies
Notebook paper and pencil

PROCEDURE
1. List the parts that constitute the power supply in Fig. 4·2.
2. List the parts that constitute the power supply in Fig. 4·3.
3. In Fig. 4·3, what is the complete name and function of C_{21}?
4. List the parts that constitute the power supply in Fig. 14·21.
5. List the parts that constitute the power supply in Fig. 17·10.

QUESTIONS
1. Define a power supply.
2. Define a rectifier.
3. Define a filter.
4. Can a simple battery be considered a power supply? Explain.
5. What is missing from the power supply capacitors in Fig. 14·21? Explain.

JOB 16·5 How to use a bench-type power supply

OBJECTIVES
1. To learn to use the bench-type power supply as a battery eliminator
2. To appreciate the role of the power supply in electronic devices

MATERIALS REQUIRED

Equipment
Bench-type power supply (see Fig. 16·2)
Transistor radio receiver (see Fig. 17·10)
Hand tools for electronics work
Pair of test leads

PROCEDURE
1. Remove the batteries from the radio receiver.
2. Determine the total voltage supplied to the radio receiver by the battery or combination of cells. Make a note of this.
3. Determine the polarity of the leads that go to the radio receiver from the batteries. Make a note of this.
4. Apply power to the bench-type power supply.
5. Adjust the bench-type power supply to the voltage you determined in step 2.
6. With the test leads, connect from the output terminals of the bench-type power supply to the battery leads protruding from the radio receiver. *Be sure to observe the polarities as noted in step 3. Failure to do this normally will destroy the transistors in the radio receiver.*
7. Turn ON the radio receiver. The receiver should play normally, *even without the batteries.* The reason being that the bench-type power supply is serving as a *battery eliminator.* Compact little power supplies are often used as *power packs* to *save* on batteries. These power packs normally come equipped with small plugs, ready to be in-

serted in an adapter jack such as part J_2 in Fig. 17·10.

8. As you continue to play the radio receiver, realize that the receiver is no longer portable. Further, realize that the bench-type power supply has become a part of the receiver. To prove this, turn the bench-type power supply off and you will see that the receiver stops playing.

9. Disconnect the bench-type power supply from the receiver and reinsert the batteries in the holders of the radio receiver.

QUESTIONS

1. List the possible uses for a bench-type power supply.
2. What is the main difference between a portable radio receiver and one that is not portable?
3. Explain how a bench-type power supply could be used in servicing a nonportable tube-type radio receiver.

17

LOW-FREQUENCY AMPLIFIERS AND OSCILLATORS

In electronics there is no device as popular as the amplifier. An amplifier is a network of circuits working together around an electron tube (or transistor) for the purpose of strengthening an electrical signal. Actually, the birth of electronics was centered around the vacuum tube. In 1903 Dr. Lee De Forest proved that a control grid placed between a cathode and an anode (plate) could interrupt the flow of electrons as they traversed the space between the cathode and the anode. In so doing Dr. De Forest invented the triode vacuum tube. A transistor also accomplishes the task of current control, and can therefore be used in an amplifying circuit.

A review of Chapter 14 is advisable before continuing with this chapter.

17·1 TYPES OF AMPLIFIERS

Amplifiers may be one-stage units or a composite of smaller amplifiers. For example, a public-address amplifier is composed of several individual amplifier stages. Individual amplifier circuits are designed to amplify either voltage variations which appear as signals or the current variations which are also components of the signal. Since these voltage or current variations represent the signal, a signal can be evaluated in terms of the rate of voltage or current change. This rate of change is better expressed in terms of the *frequency* of variations in cycles per second. Because of this, amplifiers are also rated according to the frequency of the signal they best amplify.

From the above it can be seen that there are *voltage amplifiers, current amplifiers, low-frequency amplifiers*, and *high-frequency amplifiers*. Actually, combinations of these amplifiers are used. One may encounter a low-frequency voltage amplifier or a low-frequency current amplifier. It all depends on what is needed to perform a given task. Figure 17·1 shows the basic block diagram of an audio-frequency amplifier. It is a low-frequency amplifier composed of two individual amplifying stages. The *first audio-frequency amplifier* is a voltage amplifier, while the *power output stage* is a current amplifier.

17·2 APPLICATIONS OF LOW-FREQUENCY AMPLIFIERS

Low-frequency amplifiers are used in radio receivers, television receivers, public-address systems, phonographs, tape recorders, automatic machine controls utilizing electronics, remote-control equipment, and in many other civilian and military electronics equipment.

The human ear can respond only to frequencies below 20,000 cycles per second (cps or Hz). For this reason frequencies below this amount are called *audio* frequencies. Any device that makes use of a loudspeaker or microphone must therefore use at least one audio-amplifying stage. The audio-amplifier block diagram of Fig. 17·1 represents a fundamental application of low-frequency amplifiers. It is widely used and is easily accessible for investigation, testing, and experimentation.

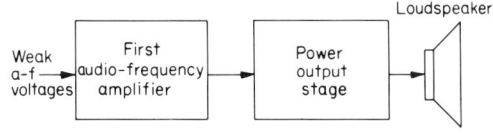

FIG. 17·1 Block diagram of an audio amplifier.

FUNCTION OF AN AUDIO AMPLIFIER

The audio amplifier of Fig. 17·1 is found in radio receivers, television receivers, phonographs (record players), tape recorders, and in any other electronic unit that utilizes a loudspeaker. The object of the total amplifier is to amplify or strengthen weak audio-frequency voltages obtained from a pickup device, such as in a record player. This amplified *signal* voltage is then used to *drive*, or control, a current amplifier which delivers the necessary power to operate the loudspeaker. The current amplifier thus used is known as a *power output stage*.

THE LOUDSPEAKER

While the phonograph pickup converts mechanical vibrations to electrical signal voltages, a loudspeaker converts electrical signal currents to mechanical vibrations. A loudspeaker is shown in Fig. 17·2. Vibrations in a loudspeaker are produced in the paper cone. By vibrating the paper cone, the surrounding air is made to move in unison. In so doing *sound* is generated. This sound is a reproduction of the pickup-needle vibrations. The loudspeaker is therefore a *reproducer*.

In summary, a phonograph pickup converts mechanical needle vibrations to weak voltages, and a loudspeaker converts varying electric currents to sound information. One schematic symbol for a loudspeaker is shown in Fig. 17·3. Its operation is explained below in Sec. 17·4 (The Voice Coil).

AUDIO SIGNALS FROM RADIO RECEIVERS

Audio signals are obtained from radio receivers after a stage known as the *detector*. Prior to the detector the frequencies involved are very high and are beyond the hearing range of the human ear. These high frequencies are known as radio frequencies because they carry the radio wave. High-frequency amplifiers are discussed in Chap. 18. Detectors are discussed in Sec. 15·1.

17·3 THE FIRST AUDIO-FREQUENCY AMPLIFIER

The first audio-frequency amplifier as indicated in Fig. 17·1 is a voltage amplifier. Basically, its objective is to build up the weak signal voltages fed to

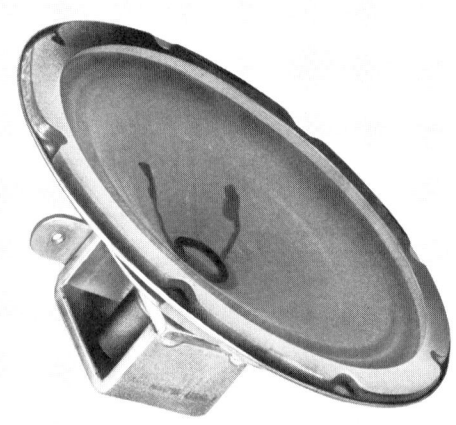

FIG. 17·2 The loudspeaker.

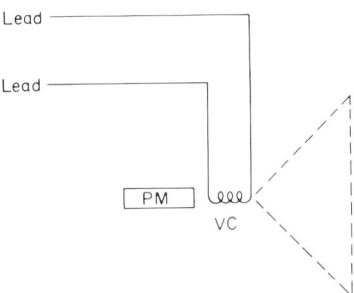

FIG. 17·3 Schematic symbol for a loudspeaker.

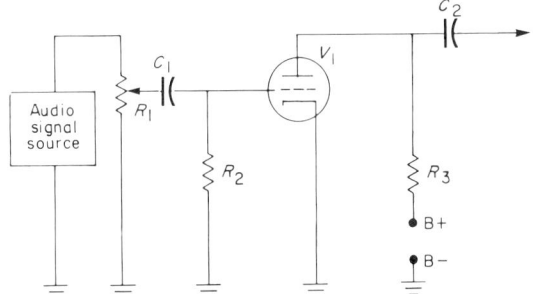

FIG. 17·4 Schematic diagram of a first audio-frequency amplifier stage that uses a vacuum tube.

it as input so that they may be strong enough to drive the power output stage. Representative schematic diagrams of the first audio-frequency amplifier are drawn in Figs. 17·4 and 17·5.

The input signal in Fig. 17·4 is applied across potentiometer R_1, which acts as a *volume control*. A potentiometer is seen in Fig. 2·13. The potentiometer shown in the photograph also contains a switch mounted on its back. The switch is not illustrated in Fig. 17·4. Output is obtained from the potentiometer through its *wiper* terminal. The wiper is represented by the arrow in the center of R_1 in Fig. 17·4. The purpose of the potentiometer is to regulate the amount of voltage input to the amplifier. That is why it is called a volume control.

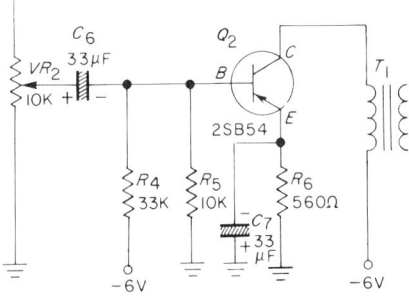

FIG. 17·5 Transistorized first audio-frequency amplifier stage. Part of Globe Patrol short wave radio receiver, Fig. 17·10. (*Allied Radio Shack, Electronic Division of Tandy Corporation, Fort Worth, Texas.*)

The true input circuit of the amplifier is composed of capacitor C_1 and resistor R_2. Voltages developed across resistor R_2 also appear across the input to the triode V_1. The vacuum-tube triode serves as a valve to control the electron flow through resistor R_3, and is *coupled* to the next stage through capacitor C_2. Terminals B+ and B— are the output terminals of the power supply. B+ is approximately 100 volts in most circuits of this nature.

The amplifier stage of Fig. 17·5 is essentially the same as that shown in Fig. 17·4. However, since Fig. 17·5 uses a transistor as the *active component*, some basic differences are essential. For example, a transistor requires that a *bias current* flow through the *base* element. Therefore, resistor R_4 in Fig. 17·5 serves to provide this bias. Another common variation is that a transformer (T_1) is used to couple the output signal to the following stage.

AMPLIFIER GAIN

The internal construction of a vacuum tube or transistor allows for a small signal on the control element to cause a large corresponding voltage change (signal) at the plate or collector of the same unit. This increase in signal voltage is considered a *voltage gain*, and the gain in effect is amplification. It should be noted that amplification is accomplished by the entire circuit, and not by the active component alone.

COUPLING CIRCUITS

An amplifier stage requires that the incoming signal be coupled into the control element circuit and that the amplified signal be coupled out to the next stage. Coupling must be accomplished with a minimum of disturbance to the circuits. Various types of coupling arrangements are employed in electronics. The two most widely used are *RC* coupling and transformer coupling (*RC* stands for resistor-capacitor). The circuit of Fig. 17·4 employs *RC* coupling. The input coupling circuit consists of C_1 and R_2. The output coupling circuit consists of R_3 and C_2. The amplifier of Fig. 17·5 makes use of *RC* coupling for the input and transformer coupling for the output. The transistor amplifier in Fig. 17·8 utilizes transformer coupling for both the input and the output.

TRANSISTOR AMPLIFIERS

Most circuit-analysis considerations that apply to electron tube circuitry also apply to transistorized circuits. The main difference is in the internal operation of a transistor as compared with the internal operation of an electron tube. This difference is covered in Chap. 14. Special attention should be given Figs. 14·10 and 14·11.

17·4 THE POWER OUTPUT STAGE

A loudspeaker or any other electrical device that does work requires large quantities of electrical power. Electrical power is obtained from the electrons that move along a conductor as a current. Therefore, when large amounts of power are required, a demand for heavy electric current is also necessary. However, energy for this power cannot be created. It must be obtained from an outside source.

The power output stage in an audio amplifier acts like a valve that allocates electrical power to the loudspeaker. The power handled by this stage is obtained from the power supply of the radio receiver or a similar device, utilizing an audio amplifier (see Fig. 17·6). The voltage signal from the first audio-frequency amplifier controls the valve.

A simple power output stage is illustrated schematically in Fig. 17·7. A voltage signal on the control grid controls the heavy current flow in the plate circuit. The plate current will fluctuate in step with the *drive* signal voltage on the grid. The

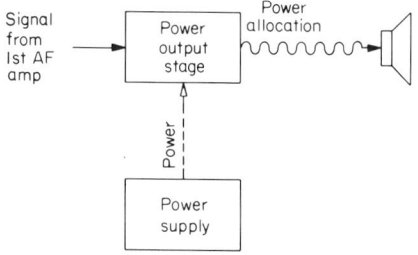

FIG. 17·6 Power transfer through the power output stage.

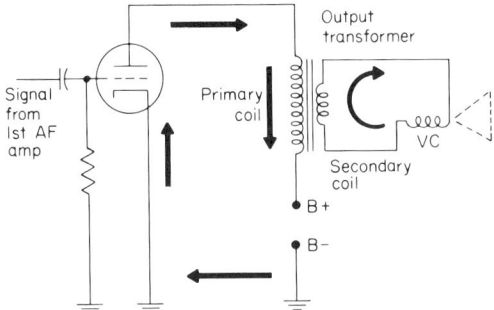

FIG. 17·7 The basic power output stage.

plate current flows through the primary winding of the *output transformer*. The loudspeaker requires a great deal of power as compared with the rest of the audio amplifier because it must move large volumes of air while it is reproducing sound. Figure 17·8 shows the circuit of a transistorized power output stage.

THE VOICE COIL

The current that is induced into the secondary winding of the output transformer also flows through a small coil cemented to the paper cone of the loudspeaker. Whenever an electric current flows through a coil of wire the coil is magnetized temporarily. The polarity of the magnetic field depends on which way the current flows through the coil. In the case under discussion, an alternating

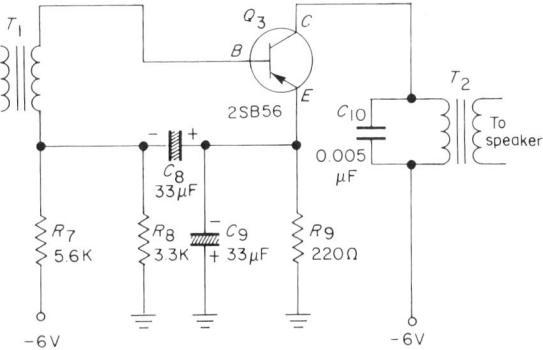

FIG. 17·8 Transistorized power output stage. Part of Globe Patrol short wave radio receiver, Fig. 17·10. (*Allied Radio Shack, Electronic Division of Tandy Corporation, Fort Worth, Texas.*)

current flows through the coil in the loudspeaker since the output transformer blocked the d-c component of the plate or collector circuit, allowing only the a-c signal to pass on to the secondary winding. Since an alternating current flows through the loudspeaker coil, this coil is alternately magnetized. Each end is alternately charged magnetically with a *north* and *south* pole. A small permanent magnet placed near the loudspeaker coil will cause the coil to be either attracted or repelled by the permanent magnet. Attraction will occur once every half cycle of the signal. Repulsion will occur during the other half cycle of the signal. As the coil moves, it carries with it the paper cone which is cemented to it. The paper cone thus sets the surrounding air in motion at the same rate as the a-c signal current. Because the loudspeaker coil is responsible for the reproduction of sound, it is called the *voice coil*.

17·5 THE COMPLETE AUDIO AMPLIFIER SUMMARIZED

The study of an audio amplifier is the study of low-frequency amplifiers. In addition, the study of amplifiers requires a working knowledge of the power supply since the voltages and currents necessary for this operation of the circuits involved originate at the power supply.

A complete schematic diagram of a phonograph audio amplifier, including the power supply, is shown in Fig. 17·9. In review, the pickup needle vibrates as it rides the grooves of a record. These vibrations are converted to varying voltages by the crystal Y_1. (Some pickup needles operate magnetic devices.) Potentiometer R_2 acts as a volume control by selecting the amount of signal voltage to be fed the first audio-frequency amplifier, acting as a voltage amplifier. The amplified signal voltage is coupled to the power output stage through an RC coupling network of which capacitor C_4 is part. The power output stage uses a pentode vacuum tube. The purpose of the power output stage is to deliver heavy currents to the voice coil of the loudspeaker.

A complete transistorized (solid-state) audio amplifier, including the power supply section is shown as part of the radio receivers in Figs. 4·2 and 17·10.

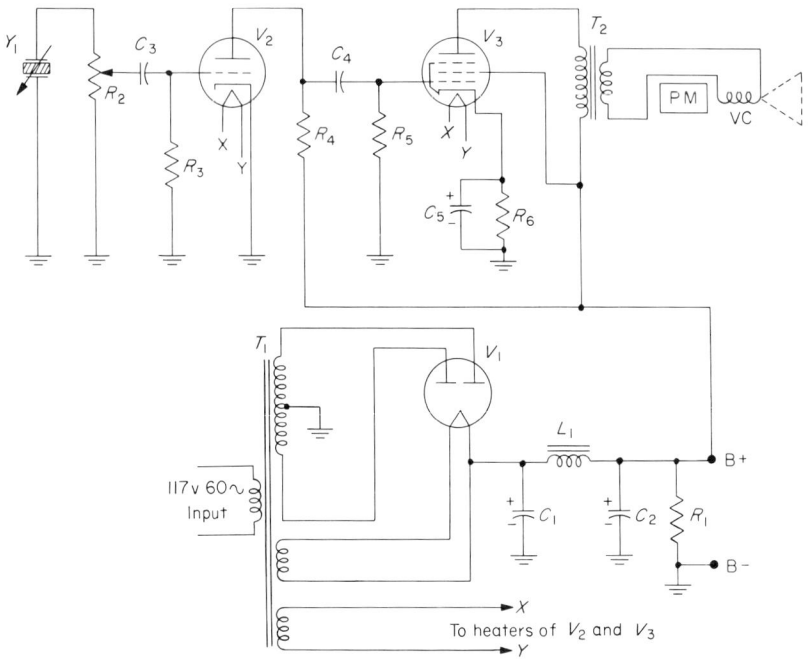

FIG. 17·9 A complete audio amplifier for use in a phonograph.

17·6 LOW-FREQUENCY OSCILLATORS

REGENERATIVE AMPLIFIERS

An amplifier that reamplifies its own output signal is called a *regenerative amplifier*. A regenerative amplifier is used to strengthen a weak signal by amplifying it over and over. However, excessive reamplification will cause the signal buildup to reach a point where distortion of the signal will take place. Therefore regenerative amplifiers are limited in the amount of reamplification they can exercise if they are to be useful.

An application of regenerative amplifiers is for the generation of alternating currents by electronic means. Advantage is taken of the fact that the signal within the regenerative amplifier *circulates* in a closed loop while being reamplified (see Fig. 15·5). If the signal is allowed to be reamplified beyond the distortion limits, it will periodically cut itself off, only to try again. This repeating action is called *oscillation*. An oscillator is therefore a regenerative amplifier used for the generation of alternating currents. The original signal can be any slight variation of current within the amplifier. From then on this signal is reamplified until oscillation is a continuing action.

OSCILLATOR-CIRCUIT ANALYSIS

A simple oscillator is shown schematically in Fig. 17·11. Notice that the plate current flows through the primary of the coupling transformer. This will allow plate-current changes to be coupled into the grid circuit of the same vacuum tube. Assume that this circuit has just had its power applied. As the plate current increases, a signal is produced. Any change in voltage or current is an electrical signal. Now this signal in the primary will induce a voltage across the secondary of the transformer, the voltage being applied across the input to the vacuum-tube triode. The polarity of the voltage on the control grid is dependent upon the secondary winding lead connections. If proper, the polarity will be positive, causing a further increase in plate current. As the plate current increases, the grid positive voltage also increases, further increasing the plate current.

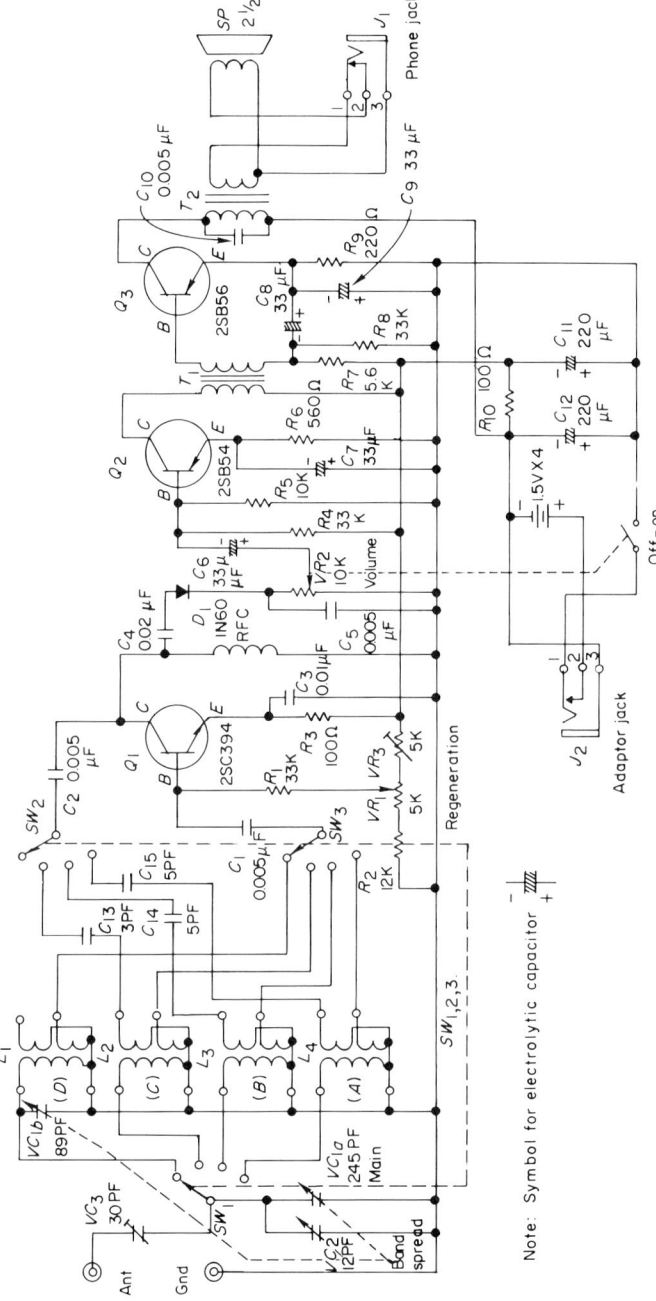

FIG. 17·10 Science Fair TM Globe Patrol short wave radio receiver. (Allied Radio Shack, Electronic Division of Tandy Corporation, Fort Worth, Texas.)

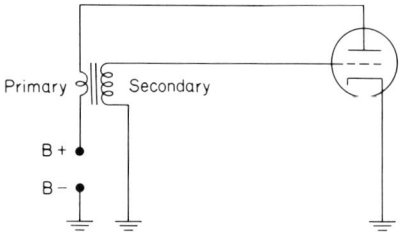

FIG. 17·11 A simple oscillator.

The point is finally reached where no further increase in plate current is possible. This point is known as plate current *saturation*. When saturation has been reached and there is no change in the plate current, the magnetic field in the transformer comes to a standstill. If the magnetic field does not move to cut the conductors in the secondary winding, no voltage is induced. With no positive voltage on the control grid, the plate current decreases. This in turn causes the magnetic field to retract into the core of the transformer. The magnetic field cutting the secondary winding in reverse will induce a voltage with opposite polarity. This now places a negative voltage on the control grid, which will further decrease the plate current. The plate current can thus be reduced until it is completely cut off. *Cutoff* refers to plate-current elimination.

By the elimination of the plate current, the transformer magnetic field is removed. Since no magnetic field exists, no voltage can be induced. The lack of voltage on the control grid allows the plate voltage to attract electrons and thus start the plate current again. The cycle of plate current flow is therefore repeated. The number of times this cycle is repeated in one second is known as the *frequency*. The frequency can be either at an audio rate or in the r-f range. The circuit of Fig. 17·11 is an audio-frequency oscillator since it makes use of an iron-core transformer.

SOLID-STATE AUDIO OSCILLATOR

The above circuit analysis relating to vacuum-tube oscillators is fundamental for all oscillators. Although the description dealt with vacuum tubes as the active component, and with transformer coupling of the feedback signal, in reality, any possible combination of circuit components and circuit arrangements is possible for successful oscillation. Just keep in mind that the feedback signal *must be positive*; that is, it must reinforce the original signal. To do this, it is necessary to use an active component and a transformer; two active components in series; or an active component and several resistors and capacitors arranged to gradually invert the signal polarity. Active components referred to in this section are limited to vacuum tubes and transistors. However, these tubes and transistors must be part of circuits that act to *invert* the signal polarity at the same time that they *amplify* the signal.

A solid-state (transistor) audio oscillator, with the possibility of tone (frequency) change and tone calibration is shown schematically in Fig. 17·12. Note that *RC* coupling is used at the input, and transformer coupling is used in the output circuit. Tone, or frequency change is made by the selection of either R_3 or R_4. After having selected the desired tone, adjustment of R_1 will alter the frequency slightly, therefore serving as a frequency calibration device.

The circuit of Fig. 17·12 is the basis for an electronic organ. To have a greater selection of tones, just increase the number of resistors to be selected by switch S_1. An audible output can be obtained by connecting a loudspeaker to points (1) and (2), as shown; or by connecting a pair of headphones

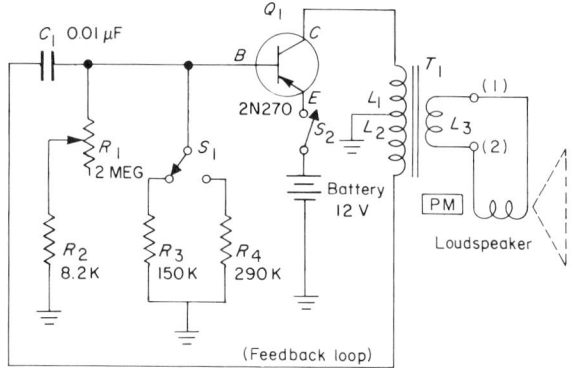

FIG. 17·12 A solid-state, variable tone, audio oscillator.

of proper impedance at the same terminals. Values of battery voltage and components are covered in summary Job 17·7 at the end of this chapter.

17·7 SUMMARY: AUDIO AMPLIFIERS AND OSCILLATORS

Audio amplifiers are low-frequency amplifiers. An amplifier is used to build up, or strengthen a signal-voltage or current. Amplifiers are classified according to the frequency they are designed to handle; whether they will primarily amplify voltage or current (power); and the type of coupling circuits or components used to transfer the signal from one stage to the next.

Oscillators are actually *modified* amplifiers; modified so that the amplifier feeds back part of the output signal. By so doing, it *regenerates* this signal. If the regeneration goes beyond certain limits, the signal recirculates over, and over. This condition is known as oscillation. Therefore, an oscillator is a regenerative amplifier. Oscillators find application in tone generators. These tone generators may be used as musical instruments, or as scientific test instruments.

The following jobs are designed to give a student a broad understanding of audio amplifiers and audio oscillators through practice. In addition, the use of basic test instruments and related testing and troubleshooting procedures are covered.

EXERCISES

JOB 17·1 How to check a loudspeaker with a dry cell

OBJECTIVES
1. To become acquainted with loudspeakers
2. To appreciate electromagnetism

MATERIALS REQUIRED

Power source
1.5-volt dry cell

Equipment
Ohmmeter

Supplies
5-inch loudspeaker, PM type

Miscellaneous
Two clip leads

PROCEDURE
1. Compare the loudspeaker supplied with the one shown in Fig. 17·2.
2. The schematic diagram illustrated in Fig. 17·3 applies to the loudspeaker supplied.
3. Apply 1½ volts direct current from the dry cell across the leads of the voice coil of the loudspeaker.
4. Place the loudspeaker near your ear and alternately open and close one lead connection from the dry cell to the voice coil. A definite sound should be heard upon opening and closing of the circuit.
5. The sound is produced when the voice coil is energized, thereby making it an electromagnet which will react against the permanent magnet of the loudspeaker. However, notice that a sound is generated when the circuit is broken as well as closed.
6. Repeat step 4, but instead of using the dry cell, use the ohmmeter. Set the ohmmeter at the $R \times 1$ range and proceed to make a series of continuity tests. Listen to the loudspeaker as in step 4. Repeat these continuity tests on all resistance ranges while listening to the loudspeaker.

QUESTIONS
1. Explain what the voice coil looks like.
2. What is the function of the paper cone?
3. What is the function of the permanent magnet?
4. Draw an exploded view of the loudspeaker tested. Identify all parts.
5. Explain the production of the sound generated when using the dry cell in step 4.
6. Why does an ohmmeter produce a sound in the loudspeaker in a continuity test?

7. Were you able to detect any difference in the intensity of the sound when the different resistance ranges were used in step 6? Explain why.

JOB 17·2 How to make an overall test of an audio amplifier

OBJECTIVES
1. To become acquainted with the audio-amplifier section of a radio receiver
2. To become acquainted with the function of a volume control
3. To appreciate the use of a potentiometer
4. To learn to use an isolation transformer

MATERIALS REQUIRED

Power source
117-volt, 60-cycle alternating current

Equipment
Radio receiver with a-c/d-c power supply
Isolation transformer
Filament transformer, 6.3-volt output

Miscellaneous
Two clip leads

PROCEDURE
1. Apply 117-volt a-c power to the isolation transformer. (See Fig. 16·6.)
2. Plug the line cord of the radio receiver into the isolated output receptacle of the isolation transformer. This will also apply 117 volts alternating current to the receiver since an isolation transformer has a 1:1 turns ratio. However, the isolation transformer is needed for safety purposes because the a-c/d-c power supply in the radio receiver does not utilize a power transformer.
3. Turn the power switch on the radio receiver to the ON position.
4. After a normal warmup period, tune in a station and listen to the output from the loudspeaker. Keep the volume down to a minimum listening level.
5. The fact that the radio plays is an indication that the power supply and audio amplifier are apparently functioning normally.
6. To make a more complete test, check the audio amplifier with a known test signal of constant amplitude. For such a test, remove the radio program being received by turning the tuning capacitor to an extreme position, which normally prevents a program from being heard at the loudspeaker.
7. Detune the radio receiver, using the information given in step 6.
8. Apply the 6.3 volts alternating current from the filament transformer *across* the volume-control potentiometer. Do not connect to the wiper terminal of the potentiometer.
9. Apply power to the filament transformer by inserting the a-c cord into a power outlet.
10. Rotate the volume control and listen to the 60-cycle note at the loudspeaker. Keep the volume down.
11. Slowly increase the volume and listen for any distortion of the note being heard.
12. The test made from steps 7 to 11 is only a check of the power supply and audio amplifier. In checking the audio amplifier a test is also being made of the condition of the volume control and loudspeaker.

QUESTIONS
1. Why is the test signal applied across the volume control?
2. Why does a normal output from the loudspeaker indicate a satisfactory power supply?
3. What kind of an output from the loudspeaker would you expect from an audio amplifier that had a dirty volume control?
4. Explain why an isolation transformer is needed when working with a radio receiver utilizing an a-c/d-c power supply.
5. Explain with the aid of diagrams how an ordinary radio receiver could be used as a record-player amplifier.

JOB 17·3 How to make a voltage analysis

OBJECTIVES
1. To learn to make voltage measurements on an operating radio receiver

2. To learn how to determine if a voltage measurement is acceptable

MATERIALS REQUIRED

Power source
117-volt, 60-cycle alternating current

Equipment
VTVM
Radio receiver with a power transformer and vacuum tubes
Schematic diagram for the radio receiver supplied
Vacuum-tube manual

PROCEDURE

1. Apply a-c power to the VTVM and allow it to warm up. Set the VTVM to function as a d-c voltmeter.
2. Remove the cabinet from the radio receiver.
3. Apply a-c power to the radio receiver and after a warmup period tune in a radio broadcast to ensure that the receiver is in good condition. No isolation transformer is needed since the radio receiver makes use of a power supply with power transformer.
4. Locate the power output tube on the radio receiver.
5. Locate the power output tube used in the vacuum-tube manual. Refer to Job 2•25 if necessary.
6. Look up the typical operating plate voltage of this tube in the tube manual. Keep in mind that the power supply being employed by the radio receiver makes use of a power transformer and that the output of this type of power supply is approximately 300 volts (see Fig. 16•15). With this as a guide select the appropriate column in the tube manual to obtain the plate voltage to be expected. (All other characteristics of this tube will be given in the same column.)
7. Compare the information obtained from the tube manual with voltages listed in the schematic diagram.
8. Knowing what amount of d-c voltage to expect at the plate of the power output tube, you can measure the actual voltage to see if it falls within the allowable tolerance. Tolerance for voltage measurements will be 20 percent unless otherwise noted on the schematic diagram.
9. To make the voltage measurement, set the voltmeter at the next higher range than the voltage expected.
10. Connect the negative lead of the voltmeter to B—. On a power-transformer type of power supply, the metal chassis is usually B—. Check to see if this is correct on your radio receiver.
11. Connect the positive lead of the voltmeter to the tube socket terminal attached to the plate of the power output tube. Remember to locate the correct terminal of the socket by counting *clockwise* from the key. Refer to Job 2•24.
12. Read the d-c voltage on the voltmeter. Record this information.
13. Compute to see if the measured plate voltage is within the allowable tolerance of the expected voltage.
14. Using the same procedure outlined, determine if the screen grid voltage of this same tube is within tolerance.
15. Continue to *analyze* the voltages at the plates of the remaining tubes. Do not attempt to measure d-c voltage at the plates of the rectifier tube since these plates have an a-c voltage applied.

QUESTIONS

1. Explain why an isolation transformer was not used in this job.
2. Explain the meaning of a *voltage analysis*.
3. Give an example of the computation required to determine if a voltage is within tolerance.
4. Explain the purpose of the screen grid.
5. Can a voltage analysis be made without a schematic diagram or tube manual? Explain.

JOB 17•4 How to use a signal tracer

OBJECTIVES

1. To become acquainted with signal tracers
2. To learn how to isolate trouble at a certain stage

MATERIALS REQUIRED

Power source
117-volt, 60-cycle alternating current

Equipment
Radio receiver (tube-type) with PM loudspeaker
Schematic diagram of the receiver
Signal tracer

Miscellaneous
Oscilloscope
Two clip leads

PROCEDURE
1. Apply power to the radio receiver and check to see if it is in good operating condition.
2. Disconnect the louspeaker leads to simulate a faulty loudspeaker.
3. Most signal tracers will provide terminals for their loudspeaker's voice coil (see Fig. 17·13). If the signal tracer you are using has these terminals, use your clip leads to temporarily replace the loudspeaker in the radio receiver. Under these conditions the loudspeaker in the signal tracer is being used as a *test loudspeaker*. When using the test loudspeaker the line cord to the signal tracer does not need to be connected to the power source. The presence of sound in the form of music or voice will indicate that the test loudspeaker is connected properly.
4. Disconnect the test loudspeaker.
5. Apply power to the signal tracer and allow a warmup period.
6. Connect the ground lead of the signal tracer to B— of the radio receiver. (Remember that an isolation transformer should be used if the receiver is employing an a-c/d-c power supply.)
7. Apply the tip of the probe to the control grid of the power output tube. If the radio receiver is tuned to a radio broadcast and the volume control of both the receiver and the signal tracer is properly adjusted, a radio program should be heard from the loudspeaker of the signal tracer. The volume control of the signal tracer is often identified as the *gain control*. If the signal tracer employs two probes, the audio probe should be used.
8. Analyze that, if no output were obtained from either the loudspeaker of the radio receiver or from the test loudspeaker, but an output was obtained from the control grid of the power output tube, then it can be determined that the power output stage is at fault. The rest of the radio receiver can be considered to be in good condition. Refer to your schematic diagram or to Fig. 17·9.
9. Obtain the audio signal at the three terminals of the volume control. Rotate the volume-control knob during each test.
10. Repeat this exercise using an oscilloscope as the signal tracer (refer to Job 13·1). Ask your instructor for assistance.

QUESTIONS
1. Assuming the loudspeaker of the radio receiver had an open voice coil, would it have been necessary to disconnect the faulty loudspeaker before using the test loudspeaker? Explain.
2. Explain how a signal tracer can be used to isolate trouble at a given stage in a radio receiver.

FIG. 17·13 Signal tracer. (*The Heath Company, Benton Harbor, Michigan.*)

3. Make a list of all the control knobs and terminals found on the signal tracer you used and give an explanation of their purpose.
4. State in summary form the results obtained when using the signal tracer at the volume control.
5. Write a summary of your experience using the oscilloscope as a signal tracer.

JOB 17·5 How to inject a test signal

OBJECTIVES

1. To become acquainted with audio signal generators
2. To learn how to use a signal generator to isolate trouble

MATERIALS REQUIRED

Power source
117-volt, 60-cycle alternating current

Equipment
Radio receiver (either tube or transistor)
Schematic diagram of the receiver
Sine wave audio signal generator (see Fig. 17·14)

Supplies
0.01-mfd 600-volt capacitor

Miscellaneous
Leads for the signal generator
Three clip leads

PROCEDURE

1. Review Job 17·2.
2. Apply power to the radio receiver and check to see if it is in good operating condition.
3. Detune the radio receiver with the large variable capacitor.
4. Apply power to the audio signal generator and allow a warmup period.
5. Set the signal generator to produce a signal of 1000 cps. On most generators this is done by setting the frequency dial at 100 and the frequency-range switch at the ×10 position.
6. Set the output attenuator at the midposition.
7. Apply the ground lead of the signal generator to the common (ground) terminal of the volume

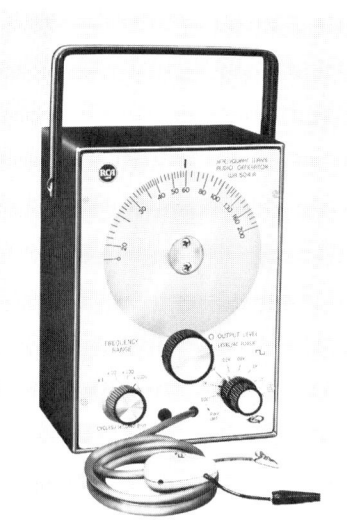

FIG. 17·14 Audio signal generator. (*Radio Corporation of America, Electronic Components Division, Harrison, New Jersey.*)

control in the radio receiver. (See Figs. 17·9 and 17·10 as applicable.)

8. Apply the 1000-cycle signal from the generator to the terminal of the volume control appearing at the opposite end from the common terminal. Do not apply the signal to the wiper terminal.
9. Adjust the volume control to produce a low-level output from the loudspeaker.
10. Remove the 1000-cycle signal from the "hot" terminal of the volume control. Maintain the ground lead connected to the common terminal.
11. Apply the 1000-cycle signal to the input control electrode of the first audio-frequency amplifier through a 0.01 mfd capacitor. Adjust the volume control of the receiver and the output level control (attenuator) of the signal generator to produce a low-level note from the loudspeaker. Record your observations.
12. To become familiar with the different frequency notes produced by various audio frequencies, produced by the signal generator, rotate the frequency dial and the frequency-range switch on the signal generator. Without increasing the volume, check which is the highest frequency you can hear.

13. Adjust the signal generator to produce a 400-cycle signal.
14. Remove the 400-cycle signal from the control electrode of the first audio-frequency amplifier. Maintain the ground lead at the common terminal of the volume control.
15. Maintain the capacitor (0.01-mfd 600-volt breakdown rating) in series with the "hot" signal lead of the signal generator. Use a clip lead to continue the overall signal lead.
16. Apply the clip lead at the end of the capacitor to the output electrode of the first audio-frequency amplifier. It may be necessary to adjust the signal generator attenuator to produce an audible signal.
17. Analyze that, if an output is heard when the signal is fed to the output electrode of the first audio-frequency amplifier, but is not heard when the signal is applied to the input control electrode of the same amplifier, then there must be a fault with the first audio-frequency amplifier stage. Refer to your schematic diagram or to Figs. 17·9 or 17·10.

QUESTIONS

1. How does this job compare with Job 17·2?
2. Did you use an isolation transformer? Why?
3. Describe the volume control, both physically and electrically.
4. What is an attenuator?
5. What was the highest frequency you could hear at step 12?
6. Why must the ground lead be maintained at the common terminal of the volume control?
7. Describe a signal generator.
8. How can a signal generator be used to isolate trouble?
9. Explain why a capacitor was used in steps 11, 15, and 16.

JOB 17·6 How to convert an amplifier to an oscillator

OBJECTIVES

1. To prove that an oscillator is a regenerative amplifier
2. To emphasize the importance of proper feedback

MATERIALS REQUIRED

Power source
117-volt, 60-cycle alternating current

Equipment
Radio receiver (tube type)
Schematic diagram of the receiver

Supplies
0.01-mfd 600-volt capacitor

Miscellaneous
Two clip leads

PROCEDURE

1. An oscillator will be made of the audio amplifier in Fig. 17·9. Compare this circuit with the audio amplifier in the schematic diagram of the receiver furnished.
2. Remove the cabinet of the radio receiver.
3. Apply power to the radio receiver and tune in a broadcast program. Adjust the volume control to a normal level.
4. With the aid of clip leads, connect one end of the capacitor supplied to the plate of the first audio-frequency amplifier and the other end of the capacitor to the control grid of the same tube. In effect, the output signal of the tube involved has been coupled back as an input to itself. (The tube will invert the phase of the grid signal by 180°.) Record your observations.
5. Move the clip lead that has been connected to the plate of the tube in step 4 to the plate of the power output tube. Maintain the control grid connection to the first audio-frequency amplifier. Since the power output tube will again invert the phase of the signal, the signal has now been inverted 360°. A note other than that transmitted as a broadcast signal should now be heard. Record any other observations.
6. Adjust the volume control up and down and record your observations.
7. Rotate the large variable capacitor in the r-f amplifier in order to tune in other radio broadcast programs and observe the note generated by feedback. Record the results.

QUESTIONS
1. During which step in the above procedure was the audio amplifier of the receiver converted to an oscillator? Why?
2. Explain your observations recorded in step 4. Is this what you expected? Why?
3. Describe in detail your observations of step 5.
4. What happened when the volume control was varied in step 6?
5. Was the oscillation note present at all times, or was it present only when a station was tuned in? Refer to step 7.
6. List any peculiar observations not covered in the preceding questions.
7. What is meant by positive feedback?
8. Explain how a transformer could have been used instead of the power output tube (step 5) to provide the proper feedback to the control grid of the first audio-frequency amplifier.

JOB 17·7 How to make a tone generator

OBJECTIVES
1. To better appreciate an oscillator circuit
2. To associate *tone, low-frequency,* and *audio* to each other
3. To better comprehend that useful activity can be derived when various electronic components are assembled into a circuit

MATERIALS REQUIRED

Equipment
Tools for electronics assembly
Tools for mechanical assembly

Supplies
Resistors, one 8.2K, one 150K, and one 290K; all ½-watt
Potentiometer, 2 Megohm, ½-watt
Capacitor, 0.01 mfd, 600 volt
Transformer, output type, 4K primary center-tapped, 3.2 ohm secondary
Transistor, 2N270
Loudspeaker, 4-inch, 3-watt, 3.2-ohm
Battery, 12-volt
Switches, one dpdt and one spst
Hook-up wire
Wire solder, 60/40 alloy
Miscellaneous hardware
Pegboard, 6 × 10 inches

PROCEDURE
1. Assemble the solid-state, variable tone, audio oscillator shown schematically in Fig. 17·12. Use the parts supplied. Mount all parts on the pegboard in "breadboard" fashion.
2. When completed, *double-check* that the polarity of the battery connection is as shown *before closing switch* S_2.
3. If the polarity is all right, close switch S_2. You should now hear an audio, low-frequency tone from the loudspeaker.
4. Open and close the switch S_2 repeatedly. If this switch was a telegraph key, you would have a code-practice oscillator.
5. Vary *rheostat* R_1. There should be a slight change in tone or frequency of the note being heard.
6. Change switch S_1 to the alternate position. The tone should change approximately one octave.
7. Again vary R_1. Note the slight change of tone once again.
8. Disassemble the circuit.

QUESTIONS
1. Write a report explaining the specific function of each component used to produce the tones you heard.
2. Why is the feedback loop necessary?
3. What does *low-frequency* mean?
4. Would it be possible to use this tone generator as a test signal generator? Explain.
5. Why was R_1 called a rheostat in step 5?

JOB 17·8 How to work with a low-frequency, integrated-circuit amplifier

OBJECTIVES
1. To minimize the mystery of integrated circuits
2. To give the student practical experience working with integrated circuits, including selection of associated parts

MATERIALS REQUIRED

Equipment
Tools for electronics assembly
Tools for mechanical assembly
Audio signal generator
Bench-type power supply

Supplies
Integrated circuit, CA3020
Various parts to make integrated circuit of Fig. 14•20 serve a useful function
Hook-up wire
Wire solder, 60/40 alloy
Pegboard, 6 × 10 inches

PROCEDURE

1. Assemble the audio amplifier shown schematically and symbolically in Fig. 14•20. Use the integrated circuit supplied. Ask your instructor for the additional parts you will need to complete the assembly. Mount all parts on the peg board in "breadboard" fashion.
2. Apply power to the circuit by using the bench-type power supply. Be sure to check the polarity and the amount of voltage applied.
3. Obtain a test audio signal from the signal generator. Apply this test signal where the diagram of Fig. 14•20 calls for *Ratio Detector Output*. This point is actually the input to the audio amplifier being assembled.
4. With the power supply connected and turned ON, and with the signal generator supplying the test signal, and, if everything is connected properly, you should hear an audible tone from the loudspeaker.
5. Vary the 5K potentiometer of Fig. 14•20 and record what happens.
6. Disassemble the circuit.

QUESTIONS

1. Describe an integrated circuit.
2. List the advantages and disadvantages of integrated circuits.
3. Summarize your experience in working with integrated circuits.
4. On the basis of your observations in step 5, what *functional name* would you assign the 5K potentiometer?

18

HIGH-FREQUENCY AMPLIFIERS AND OSCILLATORS

A high-frequency amplifier is used to strengthen a weak a-c signal. The circuits of a high-frequency amplifier must also select the signal to be amplified. This is necessary because most high-frequency amplifiers are subjected to various signals simultaneously. This is not the case with low-frequency amplifiers.

High-frequency amplifiers amplify a-c signals whose frequency of current variation may occur at 1 million cycles per second. A 1-million-cycle-per-second signal frequency is expressed as a 1-*megacycle signal*. The human ear cannot respond to this high rate, nor can a loudspeaker reproduce such a signal. However, a radio transmitter can radiate through an antenna a signal of this frequency. Because of this a 1-megacycle (Mc) signal is classified as a *radio-frequency* signal. A 1-megacycle signal is also written 1 MHz.

18·1 RADIO-FREQUENCY SIGNALS

Radio-frequency (r-f) signals are not confined to 1 Mc. They can be current variations occurring at a few hundred kilocycles per second, or the current changes can occur at several million megacycles per second. One receives an r-f signal upon tuning to a radio broadcast or to a television program. The center of the dial in a standard radio broadcast receiver lies at about 1 Mc (1 MHz).

When one tunes to a radio broadcast and listens to a person talking, an audio-frequency signal is heard. In order for the audio signal to arrive, it must be *carried* by an r-f signal from the radio transmitter through part of the radio receiver. From this it can be seen that an r-f signal is in reality a *carrier* of intelligence. *Carrier wave* is a name often given an r-f signal.

SELECTIVITY

The degree to which an r-f amplifier selects a chosen carrier wave for amplification is a measure of its *selectivity*. The more selective an r-f amplifier is, the less the interference from any additional carrier waves that may possess adjacent radio frequencies. The selection of carrier waves is often variable. In radio receivers this allows for the selection of radio programs. The variable capacitor of Fig. 18·1 is widely used for performing this function. However, the variable capacitor is used in conjunction with an inductor for the selection of a specific radio frequency.

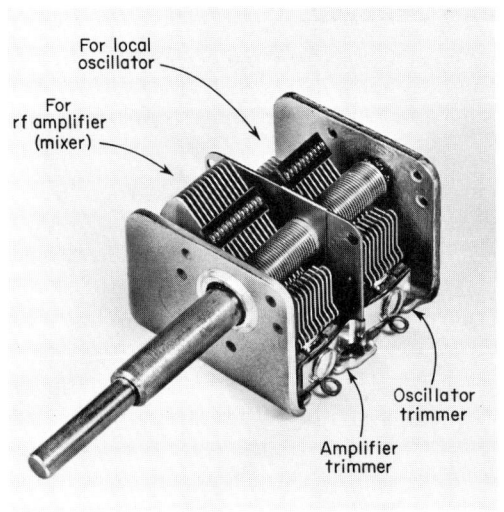

FIG. 18·1 A dual variable capacitor used in radio receivers.

FIG. 18·2 A resonant, or tuned, circuit.

18·2 TUNED CIRCUITS

Selection of a specific r-f signal is accomplished by a *tuned circuit*. The schematic diagram of a tuned circuit is shown in Fig. 18·2. This circuit is also known as a *resonant circuit*.

RESONANCE

A resonant circuit is one in which electrons circulate readily at one frequency within a circuit composed of an inductor and a capacitor in series. When the electrons circulate most easily, the circuit is said to be resonant. A circuit of this nature is resonant to one frequency only. Electrons trying to circulate at a different frequency encounter opposition other than that offered by the resistance of the wire. The arrows in Fig. 18·2 indicate that the electron current is alternating.

A winding of a transformer can be made part of a tuned circuit by the addition of a capacitor as shown in Fig. 18·3. Electrical energy is coupled into the tuned circuit by transformer action. The transformer may have a powdered-iron core or an air core. A regular iron core as used in low-frequency amplifiers and power supplies would make the transformer very inefficient when used at the high radio frequencies.

Both tuned circuits of Fig. 18·3 are variable. The resonant frequency of a tuned circuit can be changed by altering either the inductance or capacitance. In Fig. 18·3a the inductance of the coil is variable. In Fig. 18·3b the capacitance of the circuit is variable. While a variable capacitor may

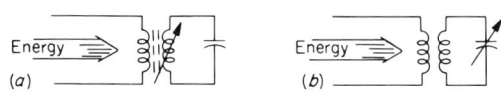

FIG. 18·3 Variable tuned circuits: (a) Inductor variable. (b) Capacitor variable.

be used in tuning a resonant circuit, the variation of inductance is accomplished by moving the powdered-iron core in and out of the coil.

Electrical energy can also be fed to a resonant circuit in a series manner, as illustrated in Fig. 18·4. Regardless of how energy is fed to the tuned circuit, the tuned circuit will resonate at one frequency. When resonance is obtained, the high circulating current will develop a high voltage across the inductor. The high voltage will also appear across the capacitor.

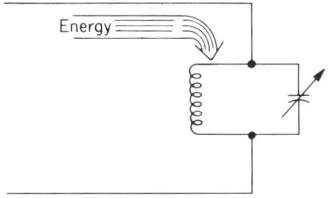

FIG. 18·4 Series-fed tuned circuit.

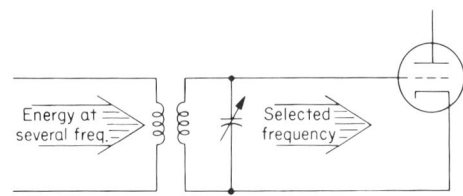

FIG. 18·5 Tuning the input to an amplifier.

18·3 APPLICATION OF TUNED CIRCUITS

A tuned circuit is made to resonate at one frequency in order to select that frequency for amplification. This is illustrated in Fig. 18·5. While several frequencies are applied to the transformer primary, only one is selected and presented to the vacuum tube or transistor for amplification. The input circuit of the active component is thus said to be tunable, or selective.

RADIO-FREQUENCY AMPLIFIER

A typical vacuum-tube, high-frequency (r-f) amplifier is shown in Fig. 18·6. Not all r-f amplifiers obtain input from an antenna. Sometimes they are given different designations and serve other purposes besides receiving and selecting radio programs. For example, the amplifier shown in Fig. 18·6 is used in radio and television receivers, but is termed the *intermediate-frequency amplifier*.

This amplifier must also select a carrier frequency and amplify it. An intermediate-frequency amplifier is not controlled by an operator, even though it is selective. The variable capacitors or variable inductors used are adjusted by the electronics technician by the process known as *alignment*. The variable capacitors of the type used in Fig. 18·6 are illustrated in Fig. 2·18b. Notice that, in the amplifier of Fig. 18·6, both the primary and secondary of the transformers are tuned.

For another illustration of high-frequency amplifiers and their related circuits, refer to Fig. 17·10 and the simplified diagrams or Figs. 18·7 and 18·8. Specifically, notice that a radio-frequency signal

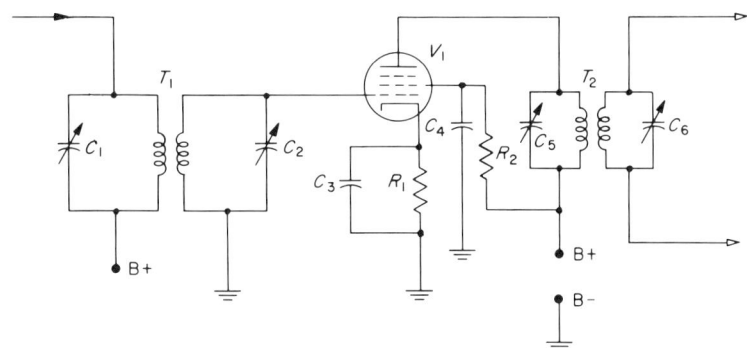

FIG. 18·6 High-frequency amplifier.

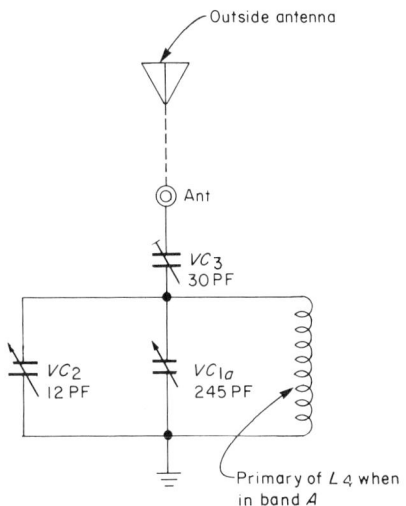

FIG. 18·7 Resonance in action. This variable resonant circuit is a simplified version of the tuning circuit for the short wave bands of the Globe Patrol radio receiver in Fig. 17·10. (Allied Radio Shack, Electronic Division of Tandy Corporation, Fort Worth, Texas.)

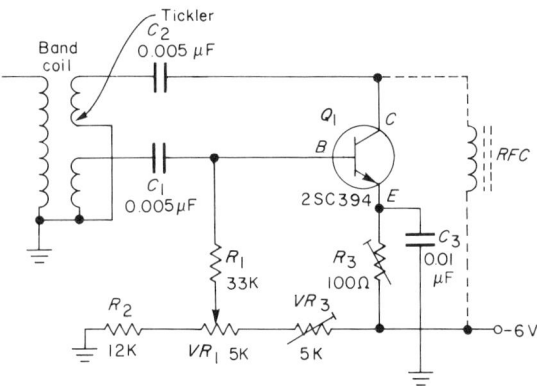

FIG. 18·8 A transistorized, high-frequency regenerative amplifier. Part of Globe Patrol short wave radio receiver, Fig. 17·10. (Allied Radio Shack, Electronic Division of Tandy Corporation, Fort Worth, Texas.)

that is received by an antenna is fed to the resonant circuit of Fig. 18·7 through variable capacitor VC_3. In this situation, *band switch* SW_1 has been set to select band A. Band switch SW_1 is necessary since the receiver of Fig. 17·10 can operate in any one of four bands. A *band* refers to a *band of frequencies*. Figure 18·9 locates the band switch on the front panel of the receiver in question. After having selected band A, variable capacitors VC_{1a} and VC_2 help the primary winding of the r-f transformer L_4 to *tune* in a station within that band.

Having *tuned in* a selected radio program, amplification of the signal is accomplished with the high-frequency (r-f) amplifier of Fig. 18·8. In this example, a *regenerative amplifier* is used to strengthen the signal. Capacitor C_2 feeds part of the output back into the input circuit for reamplification. However, this amplifier is *not supposed to oscillate*. Therefore, regeneration must be controlled. There must be enough regeneration to strengthen the signal; yet not so much that the amplifier will oscillate and give forth squeals. Potentiometer VR_1 serves as the regeneration control. The regeneration control is shown in Fig. 18·9. Comparing Fig. 18·7 and 18·9, capacitor VC_2 is the *bandspread* control, and capacitor VC_{1a} is the *main tuning* control. Loudness of the signal is controlled by the *volume control* of Fig. 18·9, which is potentiometer VR_2 of the first audio-frequency amplifier. See Figs. 17·5 and 18·9. Potentiometer VR_2 (volume control) gets the *audio* signal of the selected radio program through the diode rectifier, D_1 (Fig. 17·10) serving as a *detector* (demodulator). The *detector* is needed to separate the audio information from the *carrier* signal (r-f). (See Sec. 18·1.)

18·4 HIGH-FREQUENCY OSCILLATORS

High-frequency (r-f) oscillator circuits operate on the same principles as low-frequency oscillators, as discussed in Chap. 17. However, because of the extremely high frequencies involved, the inductors use air-cores, or powdered-iron cores. R-f oscillators also use tuned, resonant circuits to determine the frequency at which they will oscillate. These tuned circuits are the same as that shown in Fig. 18·2.

R-f oscillators find application in radio transmitters and in test-signal generating equipment. Fig. 18·10 shows a photograph of an r-f signal generator. This instrument is designed as an *r-f signal-substitution oscillator* for general television and radio

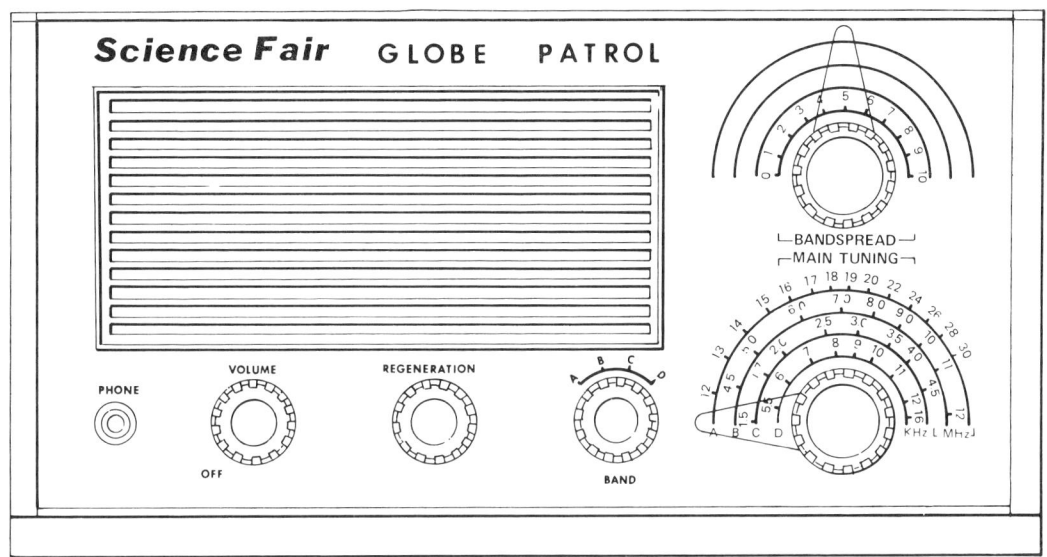

FIG. 18·9 Globe Patrol, short wave radio receiver. Front panel of receiver shown schematically in Fig. 17·10. (*Allied Radio Shack, Electronic Division of Tandy Corporation, Fort Worth, Texas.*)

servicing and other applications which require a continuous-wave or modulated r-f sine wave from 85 kHz to 40 MHz.

R-f oscillators also find application in radio receivers where it is advantageous to *convert* the incoming r-f carrier frequency to a lower *intermediate-frequency* carrier. When an oscillator is used in a radio (or television) receiver for this purpose, it is known as a *local oscillator*. Regenerative radio receivers, such as found in Fig. 17·10, do not use a local oscillator. A radio receiver that uses a local oscillator is known as a *superheterodyne* radio receiver. Although regenerative radio receivers have certain advantages, the superheterodyne radio receiver is the most popular. It is not within the scope of this text to enter into a discussion regarding the advantages and disadvantages of each.

FREQUENCY CONVERTERS

A radio-frequency stage that serves to convert an r-f carrier frequency to a lower intermediate-frequency carrier is known as a *frequency converter* stage. The term *converter* is often used by itself to designate this stage. Actually, a converter is composed of one or more *active* compo-

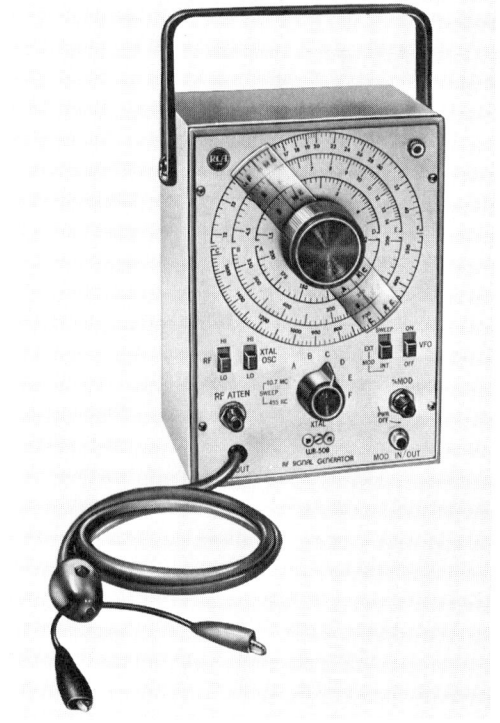

FIG. 18·10 The r-f signal generator. (*Radio Corporation of America, Electron Tube Division, Harrison, New Jersey.*)

nents, which together with surrounding *passive* components serve to amplify an r-f incoming signal; generate a local r-f signal; and mix the two together to produce a third, lower r-f signal (intermediate-frequency, *i-f*): *all at the same time*. The radio receiver circuits of Figs. 4·2 and 4·3 are superheterodyne circuits. In Fig. 4·2, transistor Q_1 is the converter. Notice that next to it is the oscillator coil (transformer T_2); and that the windings of T_2 are made part of a resonant circuit by capacitors C_1C and C_1D. Further, notice that the following stages are identified as i-f amplifiers. Again, as with the regenerative radio receiver discussed in Sec. 18·3, the diode detector *extracts* the audio signal from the i-f carrier and feeds it to the audio amplifier through the volume control, R_{12}.

In the case of Fig. 4·3, vacuum tube V_1 is the converter; T_2 is the oscillator transformer; tube V_2 is the i-f amplifier; the dual-diode plates of V_3 serve as the detector; R_6 is the volume control; and vacuum tubes V_3 (triode section) and V_4 are the audio amplifier section.

18·5 SUMMARY: R-F AMPLIFIERS AND OSCILLATORS

In summary, an r-f amplifier has the function of selecting and amplifying a high frequency. High-frequency amplifiers that make use of inductors require that those inductors use air or powdered-iron cores. Inductors used in r-f amplifiers are usually part of a resonant circuit. Resonant circuits are necessary in order to do the actual selection, or tuning of a desired frequency.

High-frequency oscillators operate on the same basic principles as low-frequency oscillators. R-f oscillators find application in radio transmitters and in test-signal generators. R-f oscillators also find application as local oscillators in superheterodyne radio receivers.

The following jobs will give the student practical experience with r-f amplifiers and oscillators. They will also serve to acquaint the student with the overall operation of a radio receiver, including the use of related basic test equipment.

EXERCISES

JOB 18·1 How to identify a radio-frequency stage

OBJECTIVES
1. To learn to recognize high-frequency components
2. To become familiar with the r-f section of a radio receiver

MATERIALS REQUIRED

Equipment
Radio receiver
Schematic diagram of the receiver

PROCEDURE
1. Study the schematic diagram of the high-frequency amplifier in Fig. 18·6. Notice that the coupling transformers do not employ an iron core; that several of the capacitors are variable; and that no electrolytic capacitors are used. High-frequency amplifiers are also known as radio-frequency (r-f) amplifiers.
2. Compare the observations made in step 1 with the audio-frequency amplifier shown schematically in Fig. 17·9. In this low-frequency amplifier will be found a transformer with an iron core and an electrolytic capacitor. Also note that the low-frequency amplifier of Fig. 17·9 does not make use of variable capacitors.
3. With the above comparisons in mind, locate the r-f stages in the schematic diagram of the radio receiver supplied.
4. Remove the cabinet from the radio receiver.
5. Locate the variable capacitors used in the receiver. Refer to Figs. 2·18 and 18·1.
6. Locate the high-frequency inductors used in the receiver. Refer to Fig. 2·24.
7. The small trimmer capacitor of Fig. 2·18b is often

found within the same metal case used to house a high-frequency transformer (see Fig. 18·11). The capacitor is adjusted through holes in the metal can.

8. Trace the wires leading to a high-frequency inductor and note that the same wires will also be attached to a variable capacitor. Do not remove the metal case *shielding* the inductor unless instructed to do so by your instructor.

9. At times a powdered-iron core is used with a high-frequency inductor. This powdered-iron core is often adjustable to provide changes in the inductance of the coil or transformer using it. When such a *variable inductor* is used, the capacitor employed (to make the circuit resonant) will be fixed. Check the radio receiver supplied to see if a variable inductor of this type is used.

QUESTIONS

1. Which of the amplifiers shown in Figs. 18·6 and 17·9 use tuned circuits? Why?
2. Why is an air or powdered-iron core used in a high-frequency inductor instead of laminated iron?
3. Explain the main function of the radio frequency received and amplified by the radio receiver.

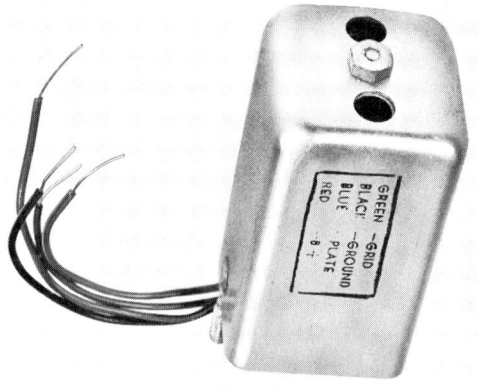

FIG. 18·11 A high-frequency transformer is located inside the case.

4. What is resonance?
5. Give an example of poor selectivity in a radio receiver.
6. Why are electrolytic capacitors associated with low frequency?
7. Define shielding.

JOB 18·2 How to make a resistance analysis

OBJECTIVES

1. To learn to analyze resistance in a circuit
2. To appreciate the resistance in high-frequency inductors

MATERIALS REQUIRED

Equipment
Ohmmeter
Radio receiver, superheterodyne, tube-type
Schematic diagram of the receiver

PROCEDURE

1. Remove the cabinet from the radio receiver. *Do not apply power to the receiver.*
2. Refer to Fig. 18·6 for the instructions that follow.
3. Locate on the schematic diagram of the radio receiver supplied the high-frequency amplifier that is similar to the one shown in Fig. 18·6. It is the intermediate-frequency amplifier.
4. In the control grid circuit of Fig. 18·6, one can expect to find a low resistance between the control grid and ground. This is because the wire in the secondary of transformer T_1 offers less resistance than most resistors. Capacitor C_2 can be disregarded because it is an open circuit to direct current.
5. Measure the resistance of the secondary winding of the comparable transformer in the radio receiver supplied. It is not necessary to disconnect any wires to make this measurement since the control grid acts as an open circuit to the direct current developed by the ohmmeter. Record the results.
6. The cathode circuit of Fig. 18·6 contains a resistor which will determine the resistance in this circuit. The capacitor is considered an open circuit once more since the problem at hand is one of resistance measurement.

7. Locate the *cathode resistor* in the i-f amplifier of the radio receiver supplied. Use the schematic diagram for this receiver as an aid.
8. From the schematic diagram and the color code on the resistor, determine how much resistance is to be expected in the cathode circuit of the i-f amplifier.
9. Measure the resistance of the cathode circuit by applying one lead of the ohmmeter to the cathode terminal of the vacuum-tube socket and the other ohmmeter lead to B− ground. Record the results.
10. To measure the plate circuit resistance in an amplifier, use the B+ terminal as the reference point. This is necessary because the electrolytic capacitors employed between B+ and B− will give a false and variable resistance reading due to their charging period (see Job 13·2). Therefore the resistance of the primary of transformer T_2 in Fig. 18·6 can be determined by measuring with an ohmmeter between the plate of the tube and B+.
11. Measure the resistance of the primary winding of the output i-f transformer in the radio receiver in your possession. Refer to step 10. Record the results.
12. The condition of resistor R_2 in Fig. 18·6 can be determined by comparing its measured value with the color-coded value or schematic-diagram value. (Both the color-coded value and the schematic-diagram value should be the same. However, at times it will be found that the colors have faded or have been burned off.) Locate the screen dropping resistor of the i-f amplifier in the radio receiver and schematic diagram given to you for practice.
13. Without disconnecting any wires, determine if the screen dropping resistor referred to in step 12 is within tolerance. Record the results.

QUESTIONS
1. When making a resistance analysis, why is it necessary to ensure that no power is applied to the chassis under test?
2. How much resistance did the secondary winding of the input i-f transformer possess? Is this resistance the *impedance* of this coil? Explain.
3. How much resistance did the primary winding of the output i-f transformer possess? Is this what you expected? Why?
4. Describe the cathode circuit of the i-f amplifier you tested. Give reasons for any deviations that existed between it and the amplifier shown in Fig. 18·6.
5. Explain why some circuits use B− ground as reference and others use B+ as reference when making a resistance analysis.
6. Explain how you determined the condition of the cathode resistor and the screen dropping resistor of the i-f amplifier you tested. Show how all values were obtained.

JOB 18·3 How to use an r-f signal generator

OBJECTIVES
1. To become acquainted with r-f signal generators
2. To appreciate that an r-f signal *carries* an audio signal

MATERIALS REQUIRED

Power source
117-volt, 60-cycle alternating current

Equipment
Radio receiver with power transformer
R-f signal generator

PROCEDURE
1. Tune in a radio broadcast program to ensure that the radio receiver is in good operating condition.
2. Remove the cabinet from the receiver.
3. Apply power to the r-f signal generator and allow several minutes for warmup.
4. Compare the r-f signal generator supplied with that shown in Fig. 18·10.
5. Tune the radio receiver to the extreme low end of the broadcast dial. This should be at the place where 550 kilocycles (KHz) of r-f energy can be received. The plates of the large variable capacitor should be fully closed.

6. Turn the volume control of the radio receiver clockwise to the maximum loudness position. A strong hissing sound may be audible from the loudspeaker. This is normal and represents background noise. However, if a radio program can be heard, detune the large variable capacitor until no station is heard.
7. Place the MODULATION selector on the r-f signal generator to the INTERNAL position. This provides for a 400-cycle note to be generated.
8. The signal generator output leads used are in the form of a shielded cable. The shield serves as the ground lead, and the center conductor is the "hot" signal lead.
9. Attach an output shielded cable to the A-F OUTPUT jack, or MODULATION OUTPUT jack as applicable.
10. Attach the loose ends of the output cable across the volume control of the radio receiver as directed in Job. 17•5. A 400-cycle note should be heard.
11. Maintain the volume control of the receiver at the maximum clockwise position.
12. Adjust the level of loudness with the A-F or MOD OUTPUT control at the signal generator.
13. No adjustment in frequency can be made of this audio signal.
14. Remove the output cable leads from across the volume control.
15. Apply the "hot" signal lead of the cable to the antenna terminal and the ground lead of the same cable to B− ground. Since a power transformer is used, B− is the metal chassis.
16. No audio signal should be heard from the loudspeaker since the signal generator is set up to produce a 400-cycle signal and the antenna requires a *radio frequency*, which is quite high.
17. Transfer the shielded output cable from the A-F OUTPUT jack to the R-F OUTPUT jack in the signal generator if applicable.
18. Maintain the antenna connection as in step 15, if the same cable is used. If not, make this antenna connection with the r-f cable.
19. Set the frequency dial of the signal generator at approximately 550 kc. Note on what *band* or scale it is located.
20. Select the band needed (see step 19) with the BAND SELECTOR on the front panel of the signal generator.
21. Increase the r-f output level by adjusting the R-F OUTPUT (attenuator) controls to a maximum clockwise position.
22. Slowly rotate the frequency dial back and forth at approximately 550 kc. Do not make any adjustments on the radio receiver. Maintain this *rocking* of the frequency dial, but gradually increase the frequency deviation from 550 kc. Eventually, a frequency will be reached where a 400-cycle note will be heard from the loudspeaker. If the radio receiver is in perfect condition, the frequency at which an output can be heard will be 550 kc.
23. Maintain the volume control of the radio receiver at the maximum clockwise position.
24. Adjust the level of audio loudness with the R-F OUTPUT controls.
25. Note that the 400-cycle signal was not able to enter the radio receiver at the antenna, but had to be *carried* in by a radio frequency. The radio frequency can thus be called an *r-f carrier*.

QUESTIONS
1. What are the functions that an r-f signal generator can perform?
2. Compare an r-f signal generator with the audio signal generator used in Job 17•5.
3. Why is a shielded cable necessary for the output leads of an r-f signal generator?
4. Why was it impossible to introduce a pure audio signal into the antenna and get a note from the loudspeaker?
5. Explain why it was necessary to rock the frequency dial of the signal generator to get an audio output from the loudspeaker.
6. Explain the role played by resonance in this assignment.
7. What is an r-f carrier?
8. Explain how an r-f signal generator can be used to check out the audio amplifier of a record player.

JOB 18·4 How to align an intermediate-frequency amplifier

OBJECTIVES

1. To appreciate tuned circuits
2. To learn the meaning of alignment

MATERIALS REQUIRED

Power source
117-volt, 60-cycle alternating current

Equipment
Radio receiver, superheterodyne type
Schematic diagram of the receiver
R-f signal generator
Nonmetallic alignment tool

PROCEDURE

1. Locate the i-f amplifier stage on the schematic diagram of the receiver supplied.
2. Locate the r-f stage immediately preceding the i-f amplifier in the schematic diagram of the receiver supplied.
3. Remove the cabinet from the radio receiver.
4. Locate all parts pertinent to the i-f amplifiers.
5. Apply power to the radio receiver and tune in a radio broadcast to ensure that it is in good operating condition.
6. Apply power to the r-f signal generator and allow several minutes for warmup.
7. Adjust the volume control of the radio receiver for maximum loudness. Maintain the volume control in this position throughout this assignment.
8. With the large variable capacitor detune the radio receiver so that no radio broadcast can be heard.
9. Set the r-f signal generator to deliver a carrier frequency of 455 KHz and an audible signal of 400 cycles as in Job 18·3. (Note that 455 KHz is a standard i-f frequency. However, if the manufacturer of the receiver you are using calls for a different frequency, *use it instead*.)
10. Apply the "hot" signal lead from the signal generator to the control electrode of the last i-f amplifier and the ground lead to the common reference point. A 400-cycle signal should be heard from the loudspeaker.
11. Adjust to a low level of loudness with the output attenuators of the signal generator.
12. The tuned circuit or circuits of the output i-f transformer will now be adjusted (see Fig. 18·4). With the alignment tool, adjust the capacitance of the parallel capacitors. These capacitors can be reached through holes in the i-f transformer case (see Fig. 18·11). If the transformer case has only one hole for adjustment, the variable capacitors have been replaced with fixed capacitors. When this is found, adjustment of the tuned circuits is made by varying the inductance of the i-f transformer. One of the two adjustments can be reached from the top of the chassis, while the second adjustment is reached from beneath the chassis. In any case, adjustment of the i-f tuned circuits is made to produce maximum audio signal from the loudspeaker.
13. Remove the "hot" signal lead from the control electrode of the last i-f amplifier. Maintain the ground lead at the common reference point.
14. Apply the "hot" signal lead from the signal generator to the control electrode of the stage immediately preceding the last i-f amplifier stage. The 400-cycle note should be heard.
15. Use the r-f attenuators at the signal generator to reduce to a low level the audio signal heard.
16. The tuned circuits of the interstage i-f transformer can now be adjusted. With the alignment tool adjust the variable parallel capacitors until a maximum audio note is heard from the loudspeaker. Everything stated in step 12 with respect to variable inductances applies here also.
17. Repeat steps 13 through 16 until all i-f stages are tuned.

QUESTIONS

1. Explain the fact that an i-f amplifier does not produce an i-f frequency.
2. Explain the fact that an i-f tuned circuit serves as a gate to pass an r-f signal.
3. Explain the role of the active component in an i-f amplifier stage.
4. Explain how a tuned circuit accomplishes the selection of one frequency over another.
5. Why is alignment of tuned circuits important?

JOB 18·5 How to check an r-f oscillator

OBJECTIVES
1. To become acquainted with r-f oscillators
2. To become aware of the bias produced by an operating vacuum-tube oscillator

MATERIALS REQUIRED

Power source
117-volt, 60-cycle alternating current

Equipment
Superheterodyne radio receiver, vacuum-tube type
Schematic diagram of the receiver
VTVM

Miscellaneous
A clip lead

PROCEDURE
1. Remove the cabinet of the radio receiver.
2. Apply power to the radio receiver and tune in a broadcast program to ensure that the receiver is in good operating condition.
3. Apply power to the VTVM and allow a warmup period.
4. Set the VTVM as a negative-reading d-c voltmeter.
5. Locate the local oscillator coil (transformer) in the radio receiver (refer to Fig. 2·24d).
6. Study the circuit of Fig. 4·3. The oscillator coil is transformer T_2.
7. Compare Fig. 4·3 with the schematic diagram of the radio receiver supplied. Specifically, locate on the diagram the local oscillator coil.
8. When an oscillator circuit oscillates it produces a negative voltage for its control grid. Measure this voltage across resistor R_1 in Fig. 4·3. Place the negative lead of the voltmeter at the control grid of the oscillator and the positive lead to B− ground. Record the voltage measured. (If you wish to continue listening to the program being received, it will be necessary to retune the large variable capacitor since part of it is C_3 in Fig. 4·3.)
9. Remove the leads of the voltmeter from the radio receiver.
10. Tune in a radio broadcast program.
11. Short out the primary winding of the oscillator transformer. Use a clip lead for this procedure. In Fig. 4·3 the primary winding of the oscillator transformer can be shorted out by simply connecting one end of the clip lead to the cathode of the converter tube and the other end to B− ground.
12. When the oscillator coil is shorted as indicated in step 11, the program being heard in step 10 will vanish.
13. With the oscillator coil shorted as in step 11, measure the voltage across R_1 again. Record the voltage measured.

QUESTIONS
1. What is bias?
2. Why is the voltage developed for the control grid of the oscillator negative?
3. Why is a local oscillator that does not oscillate responsible for the lack of an output from the loudspeaker?
4. Explain how a voltmeter can verify that an oscillator is oscillating.
5. Compare the voltages measured in steps 8 and 13.
6. Why did the application of the voltmeter leads in step 8 detune the radio receiver?

JOB 18·6 How to make tracking adjustments

OBJECTIVES
1. To learn about calibration
2. To appreciate how the i-f frequency is maintained constant
3. To appreciate ganged capacitors

MATERIALS REQUIRED

Power source
117-volt, 60-cycle alternating current

Equipment
Superheterodyne radio receiver, vacuum-tube type
Schematic diagram of the receiver
R-f signal generator
Alignment tool, nonmetallic

PROCEDURE

1. Remove the cabinet from the radio receiver.
2. Apply power to the radio receiver and tune in a broadcast program to ensure that the receiver is in good operating condition.
3. Apply power to the signal generator and allow several minutes for warmup.
4. Reduce the volume of the receiver to a barely audible condition.
5. Make sure that the pointer on the radio dial can travel to both extreme ends. To do this rotate the large variable capacitor to both extreme positions and check that the pointer can follow this motion without obstructions. If the pointer of the radio dial cannot reach the extreme ends, make the necessary mechanical adjustments to the dial string assembly. If no dial string is used, adjust the pointer directly.
6. Apply the r-f signal generator leads to the antenna terminal as directed in Job 18·3.
7. Set the signal generator to deliver 1500 KHz and an audible signal of 400 cycles as in Job. 18·3.
8. Rotate the large variable capacitor until the pointer on the radio dial indicates 150 (150 represents 1500 KHz).
9. Adjust the volume control on the radio receiver to maximum loudness.
10. Reduce the output from the loudspeaker with the r-f output attenuators on the signal generator.
11. Adjustments for proper tracking and calibration will be made to overcome physical variations in the large ganged variable capacitors. As shown in Fig. 18·1, one of the variable capacitors is smaller than the other. However, each of the variable capacitors has a trimmer capacitor to allow for fine adjustments (see also Fig. 2·18).
12. With the alignment tool adjust the trimmer capacitor on the side of the smaller of the large variable capacitors until the maximum signal is heard from the loudspeaker. By doing this you adjust the local oscillator frequency to the right value.
13. The smaller of the two large variable capacitors is C_3 in Fig. 4·3. Its trimmer capacitor is C_4 in the same diagram. Locate their respective counterparts in the schematic diagram of the receiver being worked on.
14. The correct adjustment for the trimmer in step 12 will be obtained at approximately one-half turn of the screw in a counterclockwise direction from a fully tight condition. A large deviation from this specification may require an adjustment of the oscillator coil inductance. However, not all oscillator coils provide such an adjustment. The oscillator coil shown in Fig. 2·24d will allow for an inductance change of the coil. When an oscillator coil of the type shown is used, it should be adjusted first in the same manner as indicated in step 12.
15. Having adjusted the oscillator to the correct frequency for the 1500-KHz position, check to see if it is also correct at the low end of the radio dial. To do this, rotate the large variable capacitors until the radio dial indicates 600 KHz (60 represents 600 KHz). Next set the carrier frequency of the signal generator at 600 KHz. A 400-cycle note should be heard from the loudspeaker. Rock the large variable capacitors in the receiver until the maximum signal is heard from the loudspeaker. The dial should be pointing to 600 KHz. If it is not pointing at exactly 600 KHz, consult your instructor.
16. Return the pointer of the radio dial to the 1500-KHz position. Readjust the r-f signal generator to this same frequency. The larger of the two large variable capacitors will now be adjusted in order to "bring in" as strong a signal as possible to the mixer-amplifier. This capacitor is C_1 in Fig. 4·3, and its trimmer is capacitor C_2. Locate their respective counterparts in the schematic diagram of the radio receiver being worked on.
17. Notice in Fig. 4·3 that capacitors C_1 and C_3 are ganged, as indicated by the broken lines. The two actual capacitors are rotated simultaneously by a common shaft. Therefore it is important that both capacitors be adjusted for a common maximum effort at each point of rotation.
18. With the radio dial pointer and the signal generator set for 1500 KHz, adjust the trimmer capacitor C_2 (Fig. 4·3) for 400-cycle signal output from the loudspeaker. No further check or adjustment is necessary at the low end of the radio dial.

19. Remove the signal generator leads from the antenna terminal of the radio receiver.
20. Tune in a radio station whose carrier frequency is known. When the radio program is adjusted by the large variable capacitors to produce the strongest signal, read the frequency on the radio dial to which the pointer indicates. This frequency should be exactly the same as the carrier frequency of the radio station being received.
21. The above tracking adjustments were necessary to ensure that a single i-f frequency is always produced by the frequency converter stage, regardless of the carrier frequency that the radio station is transmitting. Therefore alignment of the intermediate-frequency amplifier should always precede any tracking adjustments, unless the person performing the tracking adjustments is certain that the i-f amplifier is satisfactory.

QUESTIONS
1. What does tracking mean?
2. Explain how two parts are ganged together.
3. What is the purpose of a converter stage?
4. What is the relationship between local oscillator, r-f amplifier, mixer, and converter?
5. Does the frequency converter stage produce the i-f signal? Explain.
6. Why is a nonmetallic alignment tool necessary when making high-frequency adjustments?
7. Explain how calibration of the radio dial was accomplished.
8. Why was it necessary to listen to the 400-cycle note from the loudspeaker when making adjustments to the resonant circuits of the converter stage?
9. What is the purpose of a trimmer capacitor as used in this assignment?

JOB 18·7 How to produce a beat signal

OBJECTIVES
1. To observe the effects of two different frequencies mixing
2. To learn how to check the calibration of a signal generator

MATERIALS REQUIRED

Power source
117-volt, 60-cycle alternating current

Equipment
Radio receiver
R-f signal generator

PROCEDURE
1. Apply power to the radio receiver and tune in a broadcast program to ensure that the receiver is in good operating condition.
2. Apply power to the r-f signal generator and allow several minutes for warmup.
3. Tune the radio receiver to a radio station whose carrier frequency is known. The carrier frequency of the radio transmitter can be considered quite accurate since it is policed by the Federal Communications Commission, a governmental agency of the United States.
4. Set the r-f signal generator to deliver the same carrier frequency.
5. Set the MODULATION selector of the signal generator at the EXTERNAL position. This will remove the possibility of the 400-cycle note interfering with the observations.
6. Attach the shielded output cable of the signal generator to the R-F OUTPUT jack if this cable is not permanently attached.
7. Connect the loose ends of the output cable from the signal generator to the antenna terminal and the common reference point of the radio receiver.
8. Rock the frequency dial of the signal generator back and forth and notice that a variable note is produced by the loudspeaker as the rocking motion takes place.
9. Note also that, at a certain point, while the note produced is at a low-frequency level, the note will sharply disappear, only to reappear as the point is passed. This center point where the note reduces to a zero level is known as the *zero beat* point.
10. The audio-frequency note heard is produced by mixing two high frequencies together. The note heard is known as the *beat note*, and represents

the mathematical difference between the two high frequencies producing it. The closer the two high frequencies are with respect to each other, the lower will be the frequency of the beat note.

11. Zero beat indicates that the two high frequencies producing it are exactly alike; therefore there is no difference in frequency between them.

12. When zero beat is obtained, read the frequency dial of the signal generator. The frequency should be exactly the same as the carrier frequency transmitted by the radio station. In this manner the calibration of the signal generator can be checked.

QUESTIONS

1. What is a beat note?
2. How can a zero beat be used?
3. Was there any heterodyne action demonstrated during this assignment? Explain.
4. Would you say that the intermediate frequency produced by the frequency converter stage is a beat signal? Why?
5. Why is it important to check the calibration of a signal generator?
6. How can one accurate r-f signal generator be used to calibrate another?
7. Assuming zero beat was obtained in the manner outlined in this job, would it be possible to get out of the zero-beat condition by slightly detuning the radio receiver from the station being received? Explain thoroughly.

19

ELECTRONIC TESTING TECHNIQUES

Many methods and techniques are employed for testing electronic components and devices. Which method, and what technique is to be used depends upon what is to be tested, equipment available, and conditions under which the test is to be made. However, fundamentals of testing remain the same for most applications. This chapter will acquaint the student with these fundamentals.

19·1 COMPONENT TESTING

Testing of discrete components, such as resistors, capacitors, inductors, transistors, and vacuum tubes is done in order to determine if the component is faulty, as well as to determine the exact value of the part. For example, when testing a capacitor the technician may be only interested to learn if a *short* exists. In this case, a quick check with an *ohmmeter* may be sufficient. However, during troubleshooting procedures, the technician may have already determined that a shorted capacitor was not a possibility; but because of the way the circuit using the capacitor reacted, he may *suspect* that the value of capacitance may have changed and it was no longer saitsfactory for the particular circuit application. In this case, he would have to use a *capacity checker* in order to establish the *true value* of the capacitor. As can be seen, component testing may be a part of the troubleshooting activity. In addition, testing of components may precede their installation; either prior to the original installation at the factory, or prior to their installation as replacement components at the repair shop.

19·2 PRODUCTION TESTING

Testing of large quantities of discrete components prior to their being installed in new devices may be accomplished on a *sampling* basis. This is done by selecting a component at random from a large group of similar parts, and proceeding to test it thoroughly. If the part passes all tests satisfactory, the *odds* are pretty good that the rest of the components in that group are also good. This sampling test is repeated periodically, making the test on a different component every time.

If the sampling type of test is not permissible by the prospective user of the finished product, and if large quantities of components are to be tested, special *test panels* similar to that illustrated in Fig. 10·2 are often used. This type of test panel is also used to test completed subassemblies, or the finished product.

19·3 TROUBLESHOOTING

Testing of an electronic device in search for the cause of faulty operation of the device is known as *troubleshooting*. Troubleshooting requires a knowledge of how the circuit being tested should operate when no trouble exists, and it requires a knowledge of how each individual component is supposed to work within the circuit. In addition, it also requires the ability to use an assortment of test instruments, with emphasis relating to the proper time within the troubleshooting activity to use the proper instrument. Tests normally made during troubleshooting include: (1) signal tracing, (2) test-signal injection, (3) voltage measurements, (4) circuit-resistance testing, and (5) component testing. One of the summary jobs at the end of this chapter guides the student in coordinating many of the jobs already worked in this textbook into a troubleshooting series.

19·4 SUBSTITUTION TECHNIQUES

When troubleshooting, it is often better to substitute a *known* good component for a *suspected faulty* part. In the case of a capacitor, it is difficult to test a capacitor at the various *peak* voltages to which it may be subjected. The capacitor may test good on a checker, but *break down* in actual use. Component substitution presents a problem, however. It requires a large inventory of parts. If substitution of component parts is to be practiced, acquiring of a *component substitution box* may be better. See Fig. 19·1.

Substitution of complete subsections or stages in an electronic device during troubleshooting procedures is also good practice. For example, if trouble is suspected in the power supply of a radio receiver being repaired, the main lead feeding B voltage to the amplifiers may easily be unsoldered. Then, by temporarily connecting a spare power supply to this same junction, it can be quickly determined if the fault indeed is located in the receiver's power supply. This determination, of course, comes about if the receiver works properly with the spare power supply, but not with its own supply. Having verified that trouble does exist in the power supply, the technician can concentrate his further troubleshooting efforts to the power supply. This technique is quite useful since many times a short-circuit may exist in a stage other than the power supply,

FIG. 19·1 The component substitution box. (*Radio Corporation of America, Electronic Components Division, Harrison, New Jersey.*)

yet cause a drop in *B* voltage due to loading. Commercially produced bench-type power supplies are available that can serve the purpose of substitution. See Fig. 16·2.

19·5 SIGNAL TRACING

Most electronic devices are designed to generate or pass along some sort of signal from one stage to the next. As such, when the absence of a signal, or when the poor quality of the signal manifests itself to the technician, he needs to try to locate the *general area* of where the signal encountered difficulties. In order to locate this area, the technician must *trace* the signal from its point of origin to the point where it encountered difficulty. This process of troubleshooting is called *signal tracing*. A popular instrument used for signal tracing is the *oscilloscope*. See Fig. 19·2. Once the technician determines the area or stage where the trouble exists, he can then concentrate at this location with other troubleshooting procedures, such as making voltage measurements, making resistance analysis, and testing individual components.

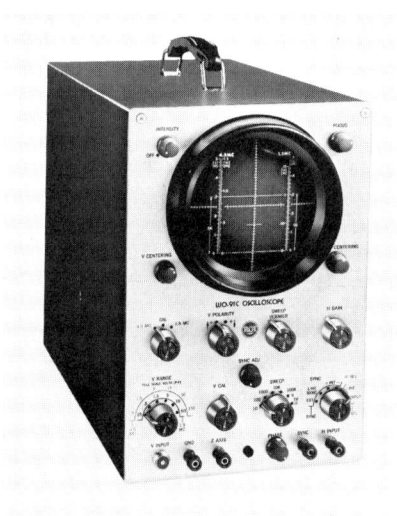

FIG. 19·2 The oscilloscope. (*Radio Corporation of America, Electronic Components Division, Harrison, New Jersey.*)

19·6 SIGNAL INJECTION

In the event that a signal is not present to be traced, it will require the *injection* of a test signal into the circuit. This requires the use of a *signal generator*, as shown in Fig. 18·10. A signal generator is also useful when making adjustments to high-frequency tuned circuits.

19·7 CIRCUIT LOADING

Use of test equipment while making measurements in an operating circuit may give misleading results. This is due to unavoidable disturbance of the circuit when the test leads from the equipment are applied to the circuit under test. This condition is known as *circuit loading*.

Circuit loading can be tolerated if the technician allows for errors that may be evident. However, sometimes circuit loading is so bad that it cannot be tolerated. In order to minimize circuit loading, special *test probes* may be used. Test probes are also available to extend the range, or to extend the usefulness of meters and equipment. One such probe can be seen in Fig. 19·3. Although test probes are fine, and very useful, whenever possible the technician should attempt to minimize circuit-loading problems by selecting test equipment which has been specifically designed to make the particular test in question.

For example, in order to minimize circuit loading in high-resistance circuits when making voltage measurements, voltmeters with a high *input impedance* rating (at least 11 megohms) should be used. The meter shown in Fig. 19·4 falls in this category. If however, voltage measurements are to be made across low-resistance circuits, a voltmeter having a rating of 20,000 *ohms-per-volt* may be sufficient. The meter shown in Fig. 14·22 has this type of rating.

19·8 SUMMARY: ELECTRONIC TESTING TECHNIQUES

In summary, the ultimate in electronics work is centered around methods and techniques used in testing. A person *who understands how to test* in

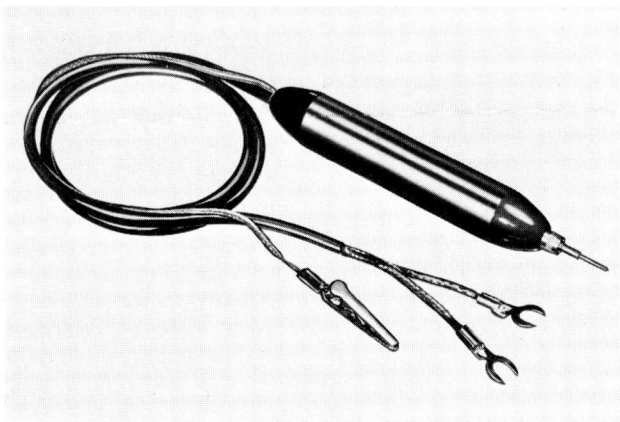

FIG. 19·3 A detector probe. (*The Heath Company, Benton Harbor, Michigan.*)

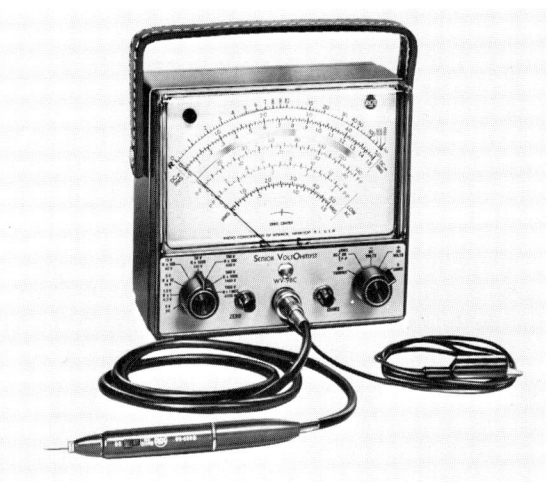

FIG. 19·4 The vacuum-tube voltmeter (VTVM). (*Radio Corporation of America, Electronic Components Division, Harrison, New Jersey.*)

the field of electronics is well versed on electronic component and circuit theory, and has a thorough knowledge of electronic test equipment. A person who has the above mentioned knowledge, also knows the limitations of such equipment. There is only one way to really acquire this broad spectrum of knowledge—*through practice*, following a well-rounded course in electronics theory. However, for the person who will work in production-type testing, basic concepts and fundamentals are all that are really necessary. The reason is that his supervisors will be responsible for the detailed operation, and as long as the production-test-technician is able to communicate and follow directions, he will do nicely at routine testing tasks.

The following jobs are designed to further acquaint the student with testing techniques and test equipment.

EXERCISES

JOB 19·1 How to make a troubleshooting procedures chart

OBJECTIVES
1. To provide a means for summarization of electronics knowledge acquired to date
2. To provide the student with a handy visual aid to be used when servicing radio receivers

MATERIAL REQUIRED

Supplies
Four sheets linear graph paper, 8½ × 11 inches each
One poster board, 17 × 22 inches
Rubber cement
Assorted colored pencils or pens

PROCEDURE

1. Rubber cement the four sheets of graph paper on the poster board so that the entire surface of the poster board is covered with graph paper. On this total ruled surface you will draw a large block diagram that will *program* you through a series of tests when troubleshooting a faulty radio receiver.
2. On the total ruled surface of the poster board neatly draw the block diagram of a typical radio receiver. Make the drawing as large as possible. However, put the identification of each block outside of the block in order to leave room inside of each block for guiding instructions. (Refer to Fig. 4·6.)
3. Using a sharp blue pencil (or pen), write within the appropriate blocks the *page numbers* of this textbook to be reviewed in case trouble is suspected to exist within that particular stage.
4. Using a sharp red pencil (or pen), write within the appropriate blocks the *job numbers* of exercises within this textbook to be reviewed in case trouble is suspected to exist within that particular stage.
5. Using a sharp green pencil (or pen), write within the appropriate blocks any *other information* that can quickly remind you how to make tests or adjustments applicable to that particular stage. This information should include what can be done next in a troubleshooting situation that requires further investigation of the receiver circuits.
6. Using a sharp purple pencil (or pen), write within the appropriate blocks *sources of information* available outside of this textbook which should also be consulted when troubleshooting a radio receiver.
7. Use additional colored pencils or pens to include in the blocks added information that you think would be helpful to you when troubleshooting.
8. Consult with your instructor about other things to include in this chart.

QUESTIONS

1. Are circuits normally found in most radio receivers representative of circuits found in more complex units of electronic equipment? Explain.
2. Are troubleshooting and testing procedures associated with radio receivers applicable to nonradio receiver devices? Explain.
3. How do you intend to use this chart?

JOB 19·2 How to establish priorities for test equipment

OBJECTIVES

1. To appreciate the relative importance of test equipment
2. To appreciate problems of deciding on the purchase of test equipment

MATERIALS REQUIRED

Supplies
Test-equipment catalogs

PROCEDURE

1. Assume that you are preparing to start a small radio and television repair shop and that your capital is limited.
2. The first test instrument to consider is a vacuum-tube voltmeter (VTVM), the reason being that the two most common measurements made are those of resistance and voltage.
3. An isolation transformer is an absolute necessity for safety purposes when working on a-c/d-c radio receivers.
4. The third piece of equipment to consider is an r-f signal generator with an audio test signal available. The audio signal is a fixed frequency output but is sufficient for testing the operation of the audio stages. The r-f signals are also necessary for alignment and tracking adjustments, besides their use in troubleshooting.
5. An oscilloscope should be included as the fourth most important instrument if repairs on television receivers is to be part of the business. An oscilloscope can be used to trace the signal in the audio section as well as in the r-f section. A detector probe is mandatory when the signal is to be traced in the r-f section. In a television receiver one will also find it necessary to observe the wave-forms of many of the cathode-ray-tube sweep and synchronizing signals.
6. A capacity substitution box is next in importance since many of the waveforms in television receivers are dependent upon the proper amount of capacitance in a capacitor. Analysis of the condition of a capacitor by voltage and resis-

tance measurements is often made by an experienced technician. However, as a beginner, and for the sake of expediency, obtain a capacity substitution box.

7. A vacuum-tube tester is not essential for locating a faulty tube. Normal block-diagram analysis, signal injection or signal tracing, and possibly a voltage analysis can easily reveal the condition of a vacuum tube in an operating circuit. However, for expediency and for customer service, a vacuum-tube tester should be part of an electronics repairman's test equipment.
8. For the repair of transistorized radio receivers one must definitely have a battery tester. A battery tester checks the voltage output of a battery *under load.* A simple voltage measurement with a VTVM is seldom enough to reveal the condition of a battery. The majority of customer complaints about faulty transistorized radio receivers can be traced to weak batteries. For example, the common complaint of a weak and distorted output signal is often found to be caused by a weak battery.
9. The statements made about vacuum-tube testers in step 7 also apply to semiconductor testers.
10. For the repair of automobile radio receivers, a battery eliminator is quite clean and handy as a power source. The minimum power source for this application is 6 volts and 12 volts.
11. For the repair of transistorized radio receivers and other similar equipment, a well-regulated, variable output power and bias supply is necessary. Output voltage of this power source may be from 0 to 30 volts.
12. For the rest of equipment selection one must evaluate one's knowledge of electronics and the extent and type of repair business to be set up. In addition, one must take into account available finances. There is no need to purchase a sweep-and-marker generator if a technician does not know about sweep alignment. Nor should money be spent for a color dot-bar generator if no repairs are to be made on color television.
13. Make a list of all the equipment mentioned in steps 2 through 11 above.
14. Use the catalogs supplied to obtain the current price of each instrument listed in step 13. Do not include in this list the price of any kit that needs assembling. Keep in mind the specification that will meet your needs and yet sells for a minimum price.
15. Make another list as called for in step 13.
16. Repeat step 14, but this time confine yourself exclusively to test instruments that are available in kit form. These kits will need to be assembled by you. Your knowledge of electronics assembly will be helpful here.
17. Expand the lists made in steps 13 and 15 to include pieces of test equipment which have been recommended by an experienced electronics technician.

QUESTIONS

1. Which jobs in this text taught you to use the test equipment listed in step 13?
2. Which jobs in this text will be helpful in assembling a piece of test equipment available in kit form?
3. Explain why a vacuum-tube tester is not as necessary as an r-f signal generator.
4. Explain why a VTVM is preferred over a portable voltohmmeter (VOM) as a basic instrument.
5. Describe how an oscilloscope can be used as a signal tracer in a radio receiver.
6. How much capital is needed to outfit a business with the fully-wired test equipment listed in step 14?
7. How much capital is needed to outfit a business with test equipment available in kit form? Refer to step 16.
8. List here the additional pieces of test equipment recommended by an experienced electronics technician in step 17.
9. To the list of question 8 add the reason for the necessity of including each unit.
10. Define a voltage under load.

20

ELECTRONIC CALCULATIONS

An attempt has been made to keep this text on a nonmathematical basis. This has been done because it has been the experience of the authors that students who are first drilled on concepts during introductory and survey-type courses are more apt to remain in the course during and after the critical period of career orientation. Detailed mathematical analysis seems to confuse the student in the beginning, especially if the student does not have a strong mathematical background. However, there is no doubt that quantitative analysis of circuit conditions is absolutely necessary for the electronics technician and engineer. Therefore, in the interest of continued motivation and preview, this chapter has been written. It is intended to give an insight to the *basic mathematical concepts* that the person who will continue with advanced studies in electronics may expect to encounter. The chapter will also help the person who will complete his electronics instruction at the level this book has been written, with some basic mathematical concepts relating to electrical and electronic calculations. This in turn will help him to appreciate, and possibly better comprehend some of the subjects covered in this book.

20·1 CALCULATING D-C CIRCUITS

SOLVING FOR TOTAL RESISTANCE IN SERIES CIRCUITS
Anytime that added resistance is introduced into a series-connected circuit, the total resistance to

electron movement will be *increased*. As a result, the amount of current flowing in that circuit will be decreased. These statements should be fairly well understood by the student thus far. However, this understanding should be further expanded by giving the student *tools* with which to better evaluate practical circuit conditions. These particular tools are *mathematical formulas*. A formula is sort of a mathematical *plan*. Its proper use will serve to direct the operator (student) as to what course of action needs to be taken in order to find a solution to a problem. Formulas, often referred to as *equations*, are often used repeatedly in order to compare results. *Comparison* itself is a tool. By comparing, one usually is able to detect a *trend*.

For example, when two resistors are connected in series, such as the circuit of Fig. 20·1a, the formula $R_T = R_1 + R_2$ tells us that in order to determine the total resistance of the circuit, it will be necessary to *add* the two separate values of the individual resistances. By applying this formula, or equation, to the specific circuit conditions of Fig. 20·1b, we find that the total resistance of the circuit is 2000 ohms. Then, by repeating the operation with different circuit conditions, *comparison* of the results reveals that the addition of resistance to a circuit which is less than one-tenth the original value will have very little effect on the total resistance of the circuit. However, if the degree of precision required for the circuit application is high, then absolute attention to detail should be enforced. Note the *rule* stated in Fig. 20·1. Such a rule can be safely assumed on the basis of repeated comparisons.

SOLVING FOR TOTAL RESISTANCE IN PARALLEL CIRCUITS

Anytime additional resistors are introduced into a parallel-connected circuit, providing the additional resistors are also connected in parallel to the existing network, the total resistance to electron movement will be *decreased*. As a result, the amount of current flowing out of the voltage source will be increased. To better comprehend these statements, it is necessary to get an insight to *trends* related to two-resistor parallel networks.

Complexity of mathematical solutions usually is related to the complexity of the circuit conditions. However, in order to *grasp* the trend, mathematical *approximations* are sufficient. For example, in Fig. 20·2 the total value of simple resistors connected in parallel can be determined by considering the value of resistor R_2 in terms of the value of resistor R_1. That is to say, R_2 is so-many-times greater than R_1. Circuit conditions and the basic formula are shown in Fig. 20·2a.

To continue with the example, notice that if R_2 has the same value as R_1, the total resistance will be *cut in half*. However, if the value of R_2 is twice the value of R_1, the total resistance will be *two-thirds* the value of R_1. And if the value of R_2 is five times the value of R_1, the total resistance will be *five-sixths* the value of R_1. Further, notice that as the value of R_2 increases, the total value of resistance becomes more nearly like R_1. From these examples two trends become evident: (1) The build-up, or evolution of the formula for simple-numbered resistors connected in parallel; and (2) repetition of this formula under different circuit conditions also reveals that the addition in parallel of resistance having values greater than ten times the value of the original resistance will have very little effect on the total resistance of the circuit.

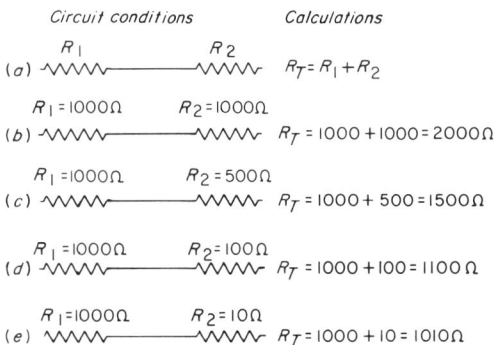

FIG. 20·1 Calculating resistors connected in series.

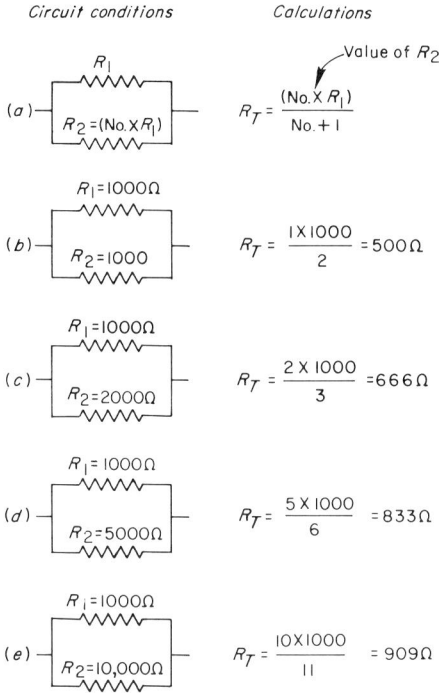

Rule: The total resistance of resistors connected in parallel will always be less than the value of the smallest resistor.

FIG. 20·2 Calculating resistors connected in parallel.

Again, the rule stated in Fig. 20·2 can be safely assumed on the basis of repeated comparisons.

By now it should be evident to the reader that through mathematics, a convenient method exists by which to *anticipate* electrical or electronic application results. Further, it should be evident that repeated solutions of related problems are necessary for purposes of comparison. Therefore, the student of advanced electronics will become involved with extensive application of mathematical tools.

A further example of the use of the calculation method illustrated in Fig. 20·2 is this: Assume that a 4700-ohm resistor is to be connected in parallel with a 1000-ohm resistor, and we wish to get *an idea* of how much total resistance to expect. By quickly *comparing* our two resistors with the circuit conditions of Fig. 20·2, we notice that we can use Fig. 20·2d for an easy *approximation* to our problem. After all, 4700 is approximately 5000. From this, we should expect a total resistance of *about* 800 ohms. Since resistors, test equipment, and voltage sources all have tolerance factors, it becomes evident that an error of a few ohms during approximation procedures is acceptable—especially since the time saved can better be used during actual measurement.

Another method that can be safely used to get an approximate answer regarding the total resistance of two more complex-valued resistors connected in parallel is illustrated in Fig. 20·3. Notice, however, that this method is really an extension of the method used in Fig. 20·2. Practice in solving two-resistor, parallel-connected circuits is essential since multiple-resistor parallel circuits can easily be solved using the two-resistor method by reducing such complex networks into two-resistor combinations. This will become more apparent with solution experience.

For situations where exacting calculations are required, the methods shown in Fig. 20·4 should be used. Notice how close these answers are to the approximate answer obtained in Fig. 20·3 where the same problem was solved. The answers in Fig. 20·4 were obtained by using long-hand multiplication and division. Under routine conditions, a *slide rule* would be used as a mathematical aid, and this slide rule would then introduce an error

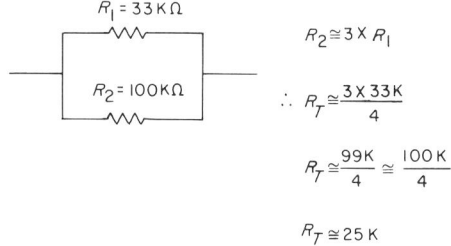

Notes: (1) The symbol $\cong$ indicates that the answer will be approximate.
(2) The symbol $\therefore$ is a short way of stating *therefore*.

FIG. 20·3 Method for finding the *approximate* total resistance of two resistors connected in parallel.

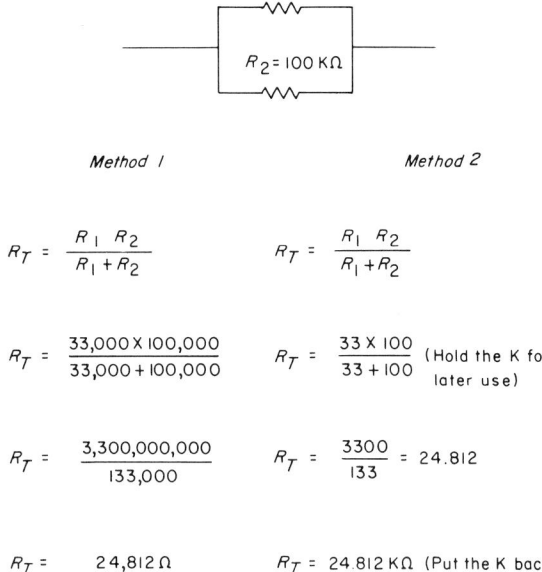

FIG. 20·4 Methods for finding the *exact* total resistance of two resistors connected in parallel.

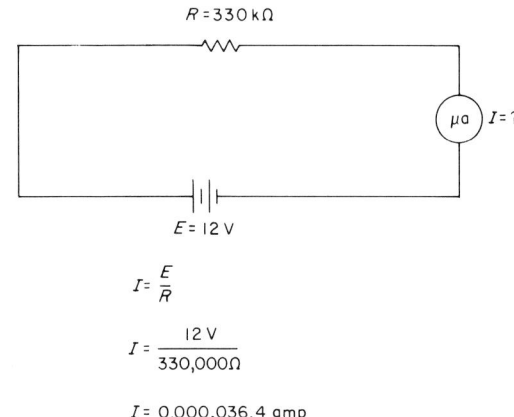

$$I = \frac{E}{R}$$

$$I = \frac{12\text{ V}}{330{,}000\,\Omega}$$

$$I = 0.000{,}036{,}4 \text{ amp}$$

The current value can also be expressed as follows:

$$I = 36.4 \text{ microamps, or}$$

$$I = 36.4 \text{ μa}$$

FIG. 20·5 Basic Ohm's law relationship between current, voltage, and resistance.

which would be tolerable. All of this proves that approximations have merit.

OHM'S LAW FOR D-C CIRCUITS

An electric current is dependent upon how much electrical pressure is applied to electrons, which causes them to move along a conductor, and upon how much resistance these electrons encounter in their *journey*. The more pressure (*voltage*) there is, the *more* current will flow, and conversely, the more resistance there is, the *less* current will flow. These concepts should be well-imparted in the student by now. However, the student should also be acquainted with the mathematical tool that aids the technician and engineer in anticipating circuit operation. Such a tool is known as *Ohm's law*.

Ohm's law is actually a mathematical expression that relates current in a circuit to the voltage causing that current and to the resistance encountered by the current. Figure 20·5 gives an example of the application of Ohm's law. Other variations of this law exist.

As with all electronic calculations, Ohm's law serves only to alert the technician or engineer of what to expect. The actual test is in the measurement. For example, the solution of the problem in Fig. 20·5 tells us that 36.4 microamperes of current are supposed to flow in that particular circuit. Therefore, we select and set up our test equipment accordingly, expecting the meter-pointer to deflect to the point indicated in Fig. 20·6. If everything is perfect, the meter will give the expected reading. However, if a circuit-fault exists, or if we are using defective equipment, the meter reading will be different. At that time, the technician will have to start troubleshooting, that is, making an attempt to locate the problem.

20·2 CALCULATING A-C CIRCUITS

INDUCTIVE REACTANCE

All of the circuit conditions and techniques used to solve the exampled resistive problems which were discussed in Sec. 20·1, also apply when the resistors are applied to an a-c voltage source. However, in a-c type circuits, one encounters inductors

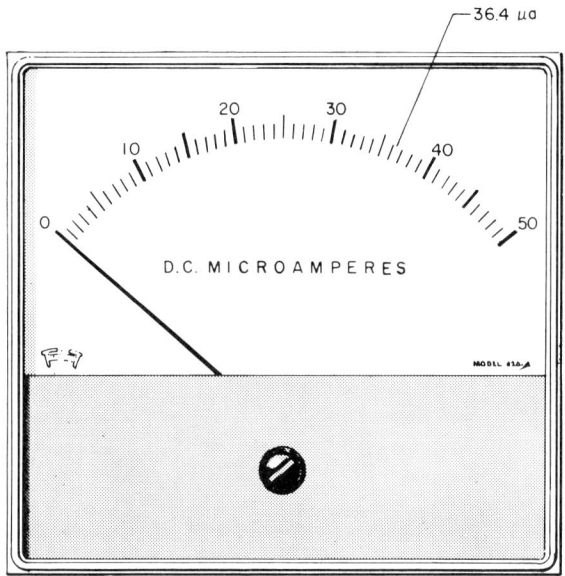

FIG. 20·6 The linear scale of a current measuring meter.

Inductive reactance is similar to resistance in that it too opposes current flow, but its opposition is *only to alternating currents*. Further, whereas resistance is *frictional* in nature, reactance is a build-up of opposing forces; in the case of an inductor, *magnetic* in nature. The amount of inductive reactance depends upon the physical properties of the coil (such as the number of wire-turns in the coil, spacing between the turns, etc.), and at the frequency of current alteration. Although the physical properties of the coil (inductance, rated in henries) does not normally change, the frequency of current-change does alter regularly; therefore, one can expect inductive reactance to change accordingly.

Figure 20·7 gives three examples of what the inductive reactance will be at different audio-frequencies. Notice in the diagram that the value of inductance (L) is fixed, and so is the value of resistance. It is normal to illustrate the value of resistance offered by the wire that makes up the coil (primary winding) as a separate resistor. Inductive reactance is expressed as X_L. Notice should

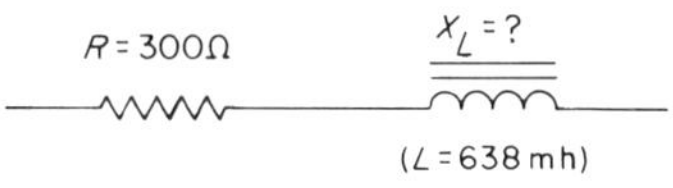

and capacitors which *react*, at times violently, to the current changes associated with alternating currents. Once again, this chapter will only cover the basic concepts related to a few examples of a-c circuit problem solution. The study of a-c problem-solving is extensive, and certainly beyond the scope of this book. For our discussion, an inductor will be used to give examples of a-c circuit problem-solving.

Consider the primary winding of an audio-amplifier coupling-transformer to be used in a transistor device. Let us say that the winding has an inductance of 638 millihenries, and that the wire that makes up the coil has a total resistance of 300 ohms. We wish to get an idea regarding how much current will flow through the primary winding of this transformer if we connect it to a 12-volt power source at various audio frequencies. To find our answer, it will first be necessary to determine what will be the *combined opposition* to current flow. This combination must take into account the wire resistance (300 ohms) and the *inductive reactance* of the loops of wire. The combined opposition is known as *impedance*.

$R = 300\,\Omega \qquad X_L = ?$

$(L = 638\,\text{mh})$

at 100 Hz

$X_L = 2\pi f l$

$X_L = 2 \times 3.14 \times 100 \times 0.638 \qquad (638\,\text{mh} = 0.638\,\text{h})$

$X_L = 400\,\Omega$

at 1000 Hz

$X_L = 2\pi f l$

$X_L = 2 \times 3.14 \times 1000 \times 0.638$

$X_L = 4000\,\Omega \qquad (4000\,\Omega = 4\,\text{K})$

at 10,000 Hz

$X_L = 2\pi f l$

$X_L = 2 \times 3.14 \times 10{,}000 \times 0.638$

$X_L = 40{,}000\,\Omega \qquad (40{,}000\,\Omega = 40\,\text{K})$

FIG. 20·7 Relationship of inductive reactance to inductance, frequency, and resistance. (*Note: π is the mathematical constant 3.14.*)

be taken that as the frequency of current-alteration increases, so will the inductive reactance increase.

IMPEDANCE

Having calculated the inductive reactance, it is next possible to combine it with the resistance of the wire to determine the amount of *impedance.* However, it is not possible to simply add them together, even though they are considered in series with each other. The combining of resistance and reactance is done by adding them *vectorially.* To do this, one of several methods can be employed. This text will use the *graphical method* since it requires very little mathematical background. However, even though it may appear to be rather awkward, it is surprisingly easy to use, and sufficiently accurate. The graphical method for solving for the impedance of an inductive circuit is illustrated in Fig. 20·8. The sample solution of Fig. 20·8 deals with the same circuit of Fig. 20·7 when it is connected to a 100 Hz signal source. (Notice that the frequency of the signal source is the important factor here, not the voltage.)

When a changing current is forced to go through an inductor or capacitor, a slight change in *timing* relation between the current and the voltage takes place. In the case of an inductor, the current is *delayed,* or made to *lag the voltage.* The amount of lag is dependent upon the amount of resistance in the circuit, and upon the inductive reactance also present. The amount of lag (in capacitive circuits it would be *lead*) is expressed in degrees. By using the graphical method to solve for the impedance, it is only a simple matter of using a protractor to measure the *angle of current lag,* as shown in Fig. 20·8. It should be stated that current is never made to lag by more than 90 degrees by an inductor. In the example of Fig. 20·8, when the primary of the transformer is connected to a 100 Hz power source, the resultant current will lag the voltage by 53.1°. This angle is also known as the *phase angle.*

OHM'S LAW FOR A-C CIRCUITS

An Ohm's law relationship between current, voltage, and *impedance* also exists. However, notice

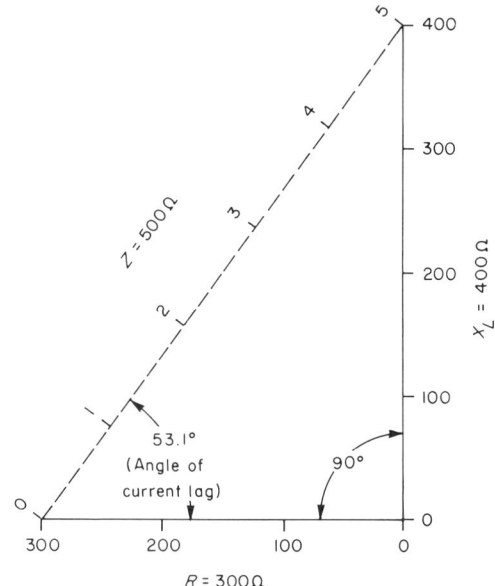

Step 1: Plot X_L and R on graph paper, 90° from each other. (Use the same scale for both. R should be the horizontal line.)

Step 2: Use ruler to measure length of dotted line. (Use the same scale as in step 1.)

Step 3: Use protractor to measure phase angle (between R and Z).

FIG. 20·8 Graphical method for solving for impedance and phase angle.

that for a-c circuits, it is the total combined opposing forces (impedance) that restrict electron movement. Figure 20·9 illustrates the use of Ohm's law for a-c circuits. The example given in Fig. 20·9 is a continuation of our problem with the audio coupling-transformer mentioned above. As can be seen, after having solved for the impedance of the primary winding, we can apply Ohm's law to help give us the idea of how much current may flow through the coil when it is connected to a 12-volt power source operating specifically at 100 Hz. Having this idea, or insight, as to what to expect in terms of current flow, we can proceed to verify it with the aid of test equipment.

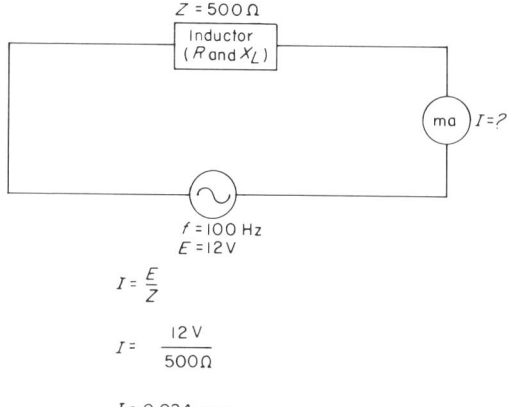

FIG. 20·9 Ohm's law for a-c circuits.

20·3 SUMMARY: ELECTRONIC CALCULATIONS

In summary, it is seldom necessary for an electronics assembler to attempt the solution of electrical or electronic mathematical problems. The electronics test technician may have need to make a few simple calculations, depending on the nature of his assignment, but the electronics repair technician and the engineer will definitely have need to solve many mathematical problems. This need will manifest itself extensively during their advanced training phases, as well as on the job.

However, everyone connected with electronics in one way or another should have a basic knowledge of electrical-electronic calculation *concepts* in order to better understand and appreciate what they do on the job. The following jobs are designed to impart a better understanding of these concepts through practice. Recall that through repetition of circuit-problem solution, and comparison of these solutions, trends in circuit analysis can be established.

EXERCISES

JOB 20·1 How to calculate for total resistance in series

OBJECTIVES
1. To give the student an insight as to what happens to the total resistance of a circuit when additional resistors are added in series to those already in the circuit
2. To give the student practice in computing the total resistance of resistors connected in series

MATERIALS REQUIRED

Supplies
Notebook paper and pencil

PROCEDURE
1. Study Fig. 20·1 thoroughly. Pay special attention to the rule.
2. Substitute the following resistor combinations for R_1 and R_2 in Fig. 20·1 and solve for the total resistance of the circuit in each case:

   a. R_1 = 1,000 ohms
      R_2 = 3,300 ohms
   b. R_1 = 5,600 ohms
      R_2 = 68,000 ohms
   c. R_1 = 39,000 ohms
      R_2 = 470 ohms
   d. R_1 = 1K
      R_2 = 2.2K
   e. R_1 = 4,700 ohms
      R_2 = 33K
   f. R_1 = 27K
      R_2 = 2.2 MEG

3. Solve other resistor combinations as directed by your instructor.

QUESTIONS
1. Explain what is meant by the *rule* as stated in Fig. 20·1.
2. Explain the procedures necessary to solve resistor combination e in step 2.
3. Write a summary statement about resistors connected in series.
4. For extra credit, obtain the materials and equipment necessary to measure these resistor combinations. Compare results obtained through com-

putation and through measurement. Explain any differences.

JOB 20·2 How to calculate for total resistance in parallel

OBJECTIVES

1. To give the student an insight as to what happens to the total resistance of a circuit when additional resistors are added in parallel to those already in the circuit
2. To give the student practice in computing the total resistance of resistors connected in parallel

MATERIALS REQUIRED

Supplies
Notebook paper and pencil

PROCEDURE

1. Review the text material on parallel circuits in Sec. 20·1.
2. Study Figs. 20·2, 20·3, and 20·4 thoroughly. Pay special attention to the *rule* stated in Fig. 20·2.
3. Substitute the following resistor combinations for R_1 and R_2 in Fig. 20·4, and solve for the total resistance of the circuit in each case: (*Use any method that is easy for you.*)
   a. R_1 = 1,000 ohms   d. R_1 = 1K
      R_2 = 3,300 ohms R_2 = 2.2K
   b. R_1 = 5,600 ohms   e. R_1 = 4,700 ohms
      R_2 = 68,000 ohms R_2 = 33K
   c. R_1 = 39,000 ohms f. R_1 = 27K
      R_2 = 470 ohms R_2 = 2.2 MEG

4. Solve other resistor combinations as directed by your instructor.

QUESTIONS

1. Explain what is meant by the *rule* as stated in Fig. 20·2.
2. Explain the procedures necessary to solve resistor combination f in step 3.
3. Notice that the resistor combinations used in step 3 of this job are identical with the resistor combinations used in step 2 of Job 20·1. Compare the results obtained when you used each combination, first series-connected and then parallel-connected.
4. Write a summary statement about resistors connected in parallel.
5. For extra credit, obtain the materials and equipment necessary to measure the resistor combinations of step 3 connected in parallel. Compare results obtained through computation and through measurement. Explain any differences.

JOB 20·3 How to calculate d-c Ohm's law problems

OBJECTIVES

1. To stress that voltage changes or resistance changes will alter the amount of current flowing in a circuit
2. To teach the student how to anticipate current changes in a d-c circuit through computation

MATERIALS REQUIRED

Supplies
Notebook paper and pencil

PROCEDURE

1. Review the text material on *Ohm's law for d-c circuits* in Sec. 20·1.
2. Study Fig. 20·5 thoroughly. Notice the different ways to express current quantities.
3. Notice that since the current flowing in a circuit is dependent upon the voltage applied to the circuit, and upon the total resistance offered to this current flow, that any change in said voltage or resistance will alter the amount of current flowing. The formula for computing this *relationship* between voltage, resistance, and current is shown in Fig. 20·5. The sample problem in Fig. 20·5 is worthy of imitation in order to learn to anticipate current quantities when the values of voltage and resistance are known.
4. In order to become better-acquainted with the above-mentioned relationship (*known as Ohm's law*), substitute the following values of voltage (E) and resistance (R) in the circuit of Fig. 20·5 and determine how much current (I) will flow in the circuit by computing each relationship with the aid

of the formula $I = E/R$: (*State your answers in any current-quantity you wish to use.*)

a. $E = 12$ volts
   $R = 270K$
b. $E = 6$ volts
   $R = 330K$
c. $E = 120$ volts
   $R = 270K$
d. $E = 9$ volts
   $R = 4,700$ ohms
e. $E = 300$ volts
   $R = 33K$
f. $E = 12$ volts
   $R = 6$ ohms

5. Solve other d-c Ohm's law problems as directed by your instructor.

QUESTIONS

1. In your own words, explain the use of Ohm's law.
2. Explain the relationship between *amperes*, *milliamperes*, and *microamperes*.
3. If you knew the amount of resistance in a circuit and the amount of current flowing in that same circuit, would it be possible to calculate the amount of voltage applied to the circuit? Explain.

JOB 20·4 How to calculate inductive reactance

OBJECTIVES

1. To become acquainted with opposition factors to electric currents other than resistance
2. To become acquainted with inductive reactance
3. To acquaint the student that a change in inductive reactance is the result of a change in the frequency of alternations of the current that flows through an inductor

MATERIALS REQUIRED

Supplies
Notebook paper and pencil

PROCEDURE

1. Review the text material on inductive reactance in Sec. 20·2.
2. Study Fig. 20·7 thoroughly. Notice that as the frequency increases from 100 Hz to 10,000 Hz, the inductive reactance also increases from 400 ohms to 40,000 ohms. Also notice the organization and use of the formula to arrive at the respective solutions.

3. In order to become better acquainted with inductive reactance and the methods to be used when solving reactance type problems, substitute the following values of inductance (L) and frequency (f) for those used in Fig. 20·7, and proceed to solve for the inductive reactance. (*Notice that the value of resistance has nothing to do with inductive reactance.*)

a. $L = 165$ mh
   $f = 60$ Hz
b. $L = 427$ mh
   $f = 60$ Hz
c. $L = 17$ μh
   $f = 2.5$ MHz
d. $L = 17$ μh
   $f = 25$ MHz
e. $L = 300$ mh
   $f = 400$ Hz
f. $L = 300$ μh
   $f = 1500$ KHz

4. Solve other reactance-type problems as directed by your instructor.

QUESTIONS

1. In your own words, explain what inductive reactance is.
2. Explain what happens to inductive reactance when the frequency is increased.
3. Explain what happens to inductive reactance when the inductance is increased.
4. How can you increase inductance?

JOB 20·5 How to determine the impedance of an inductor

OBJECTIVES

1. To become acquainted with the total opposition offered by an inductor to an electric a-c current
2. To become acquainted with impedance and phase angle
3. To acquaint the student with the graphical method for solving impedance problems

MATERIALS REQUIRED

Equipment
One 1-foot rule
One 6-inch protractor

Supplies
Notebook paper and pencil
Graph paper, 8½ x 11 inches (linear, any size grids)

PROCEDURE

1. Review the text material on impedance in Sec. 20·2.
2. Study Fig. 20·8 thoroughly. Pay close attention to the procedure steps called out in Fig. 20·8.
3. In order to better comprehend the graphical method for solving impedance-type problems, solve the following inductor-resistor combinations for total opposition (*impedance*) and for the angle by which current will lag the voltage (*phase angle*): (*Follow the example and procedures as shown in Fig. 20·8.*)

   a. L = 165 mh (Note: X_L must be
       f = 60 Hz   obtained from L
       R = 60 ohms and f. See Job
   b. L = 427 mh   20·4.)
       f = 60 Hz
       R = 400 ohms
   c. L = 17 µh
       f = 2.5 MHz
       R = 80 ohms

4. Solve other impedance-type problems as directed by your instructor.

QUESTIONS

1. Define *impedance*.
2. What does the *phase angle* represent?
3. What would happen to the impedance of an inductor if the frequency of current changes increased? Explain.
4. Suggest other methods that can be used to solve for impedance and the phase angle. Give examples.

JOB 20·6 How to calculate a-c Ohm's law problems

OBJECTIVES

1. To illustrate that Ohm's law relationships apply to a-c circuits as well as to d-c circuits
2. To stress that in an a-c circuit, the current is dependent upon the frequency of current changes as well as on the applied voltage and the resistance in the circuit

MATERIALS REQUIRED

Supplies
Notebook paper and pencil

PROCEDURE

1. Review all of Sec. 20·2.
2. Study Fig. 20·9 and compare it with Fig. 20·5. Make a note of similarities and differences.
3. Using the problem of Fig. 20·9 as an example, solve for the resultant current when an a-c voltage source of 12.6 volts operating at 60 Hz is applied to a 165 mh inductor connected in series to a resistor having 60 ohms of d-c resistance. Also determine by what phase angle this current will lag the 12.6 volts applied. (*Hint: Review all jobs in this chapter.*)
4. Solve other a-c Ohm's law problems as directed by your instructor.

QUESTIONS

1. List the similarities and differences which you noticed by comparing Figs. 20·5 and 20·9 to each other.
2. Write a summary statement which compares *resistance*, *reactance*, and *impedance* in an a-c circuit.
3. Compare a-c Ohm's law to d-c Ohm's law.

APPENDIXES

I ELECTRON TUBES
II STANDARDS OF INDUSTRY
III BIBLIOGRAPHY
IV AUDIO-VISUAL AIDS AND TECHNIQUES

ELECTRON TUBES

An electron tube is a device used to regulate or control electron movement. In this respect it has a function similar to that of the variable resistor. However, the electron tube and the variable resistor are radically different in construction, and they perform their task by different methods.

An electron tube is constructed as a hollow, airtight chamber containing various electrodes within. Most electron tubes maintain a vacuum inside the chamber, and are therefore known as *vacuum tubes*. Vacuum tubes find extensive application in radio, radar, television, etc. Some electron tubes contain a gas in the chamber, instead of a vacuum. Such an electron tube is known as a *gaseous tube*. Gaseous tubes find application in industrial electronics. An assortment of electron tubes is seen in Fig. 2·7.

ELECTRON TUBES VERSUS RESISTORS

Recall that resistors are used primarily to control the amount of current that will flow in a simple circuit. The higher the resistance, the less the current, and likewise, the lower the resistance, or opposition, the more current will flow. Electron tubes, in effect, have a similar task, especially the vacuum tube. An electron tube can control the amount of current flowing through it, and therefore the current flowing in its associated circuit. One difference between resistors and vacuum tubes, however, is that resistors have one fixed value of resistance, while vacuum tubes can exert any amount of opposition, similar to a rheostat.

The vacuum tube introduces opposition to current flow into a circuit in a manner different from that of the resistor. Also, the vacuum tube can change the amount of opposition it offers to current flow much faster than can a rheostat.

AI·1 PRINCIPLES OF THE VACUUM TUBE

A vacuum tube has an electrode that *boils* electrons off its surface. It also has an electrode that attracts, or collects, these electrons and absorbs them into its structure. This collecting electrode is a fraction of an inch away from the electrode that boils off the electrons. There is therefore a gap between these two electrodes. Air cannot exist in this gap since it would obstruct the electrons moving from the hot electrode to the collecting electrode. To remove the air, a glass or metal covering is placed around the electrodes and the air pumped out, leaving a vacuum.

VACUUM-TUBE NOMENCLATURE

Standard identification of the electrodes improves communication. Therefore the electrode that boils off electrons is called the *cathode*. The electrode that collects the electrons is called the *plate*. The glass or metal covering that surrounds these electrodes is referred to as the *envelope*. Further, instead of saying that the cathode boils off electrons, it is said that the cathode *emits* electrons.

What has been described thus far is a *vacuum-tube diode*. The term diode is derived from the words "two electrodes." In a vacuum-tube diode there are two important electrodes, the cathode and the plate. This simple diode is the basis for all other electron tubes. Figure AI·1 shows the schematic symbol for a vacuum-tube diode.

Since the vacuum-tube diode is the foundation for other electron tubes, it is necessary to investigate how the cathode emits electrons and how the plate collects these electrons.

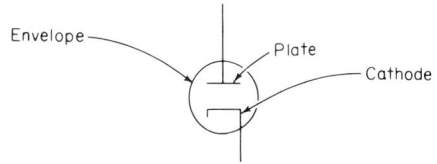

FIG. AI·1 The schematic symbol of a vacuum-tube diode.

AI·2 THE THEORY OF THE VACUUM-TUBE DIODE

THE CATHODE

The cathode is a metal sleeve coated with a claylike substance that can easily emit electrons when heated. The life of any electron tube depends on how active this coating is in terms of electron emission.

THE HEATER

Heating of the cathode is usually accomplished with a built-in electric heater. The heater element is placed within the cathode metal sleeve, and the connecting leads brought out of the envelope. By connecting the leads of the heater element to a source of electrical energy, such as a battery or a transformer, the heating element will become hot and transfer its heat to the cathode. The heater element is also known as the *filament*. Figure AI·2 shows the schematic symbol for a vacuum-tube diode with the heater included.

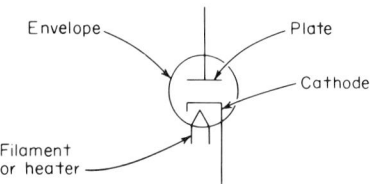

FIG. AI·2 The schematic symbol of a vacuum-tube diode utilizing an indirectly heated cathode.

Notice that this vacuum tube is still called a vacuum-tube diode even though the heater has been added. The heater is not considered important to the basic operation of the vacuum tube. In fact, in some tubes heating elements such as have been described are not used at all.

Sometimes the claylike substance is coated on the heating element, and this unified element is usually also referred to as the *filament*. However, for purposes of instruction or circuit analysis, the unified element can be referred to as the cathode, but never as the heater. Figure AI·3 shows the schematic symbol for a vacuum-tube diode employing the filament type of cathode.

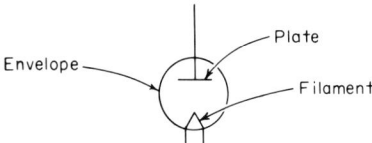

FIG. AI·3 The schematic symbol of a vacuum-tube diode utilizing a directly heated cathode.

THE PLATE

The plate collects the electrons emitted by the cathode. It therefore must have a large area and be able to withstand the bombardment of electrons. The plate is made of metal and is usually dark in color. Although the plate is shown as a straight line in the schematic symbol, it physically surrounds the cathode (see Fig. AI·8).

For the emitted electrons to be collected by the plate, it must have a positive polarity with respect to the cathode. The plate is therefore connected to an external circuit that takes electrons from its structure. This leaves a deficiency of electrons on the plate, resulting in negative electrons being attracted to the plate. However, the emitted electrons originated in, and left, the cathode. They must therefore be replaced.

For this replacement, the cathode is also connected to an external circuit that can supply additional electrons. A device that can meet both requirements, that of the plate and that of the cathode, is a battery. Therefore a battery is connected across the plate and cathode. Figure AI·4 shows such an arrangement in schematic diagram form. Current flow is indicated by the arrows. The B-battery designation is standard when a battery is connected in the plate circuit of a vacuum tube.

Opening statements in this appendix compared the vacuum tube to a resistor. To verify this, com-

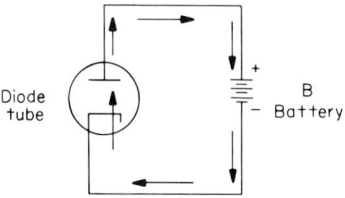

FIG. AI·4 Plate current flow.

pare the basic resistive circuit of Fig. AI·5 with that of Fig. AI·4. Again, the arrows indicate current flow.

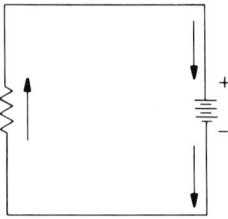

FIG. AI·5 The equivalent circuit when an electron tube conducts.

When a battery is used to supply the energy for the heater, it is designated the A battery. Figure AI·6 shows a vacuum-tube diode with a separate

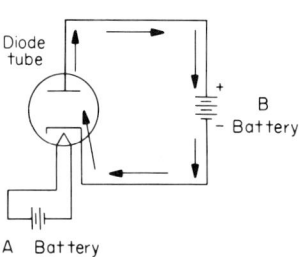

FIG. AI·6 The plate circuit is independent of the heater circuit when a separate cathode is used.

cathode and heater. Figure AI·7 shows a vacuum-tube diode using the filament-type cathode, with the A and B batteries connected. The unbroken arrows indicate plate current flow, while the broken arrows indicate the heater current flow.

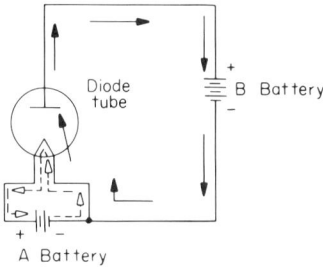

FIG. AI·7 Plate and filament current in an electron tube utilizing a directly heated cathode.

AI·3 THEORY OF THE TRIODE

The next major objective in the study of electron tubes is to investigate how the vacuum tube can regulate or control the flow of current in the plate circuit. As seen in Fig. AI·4, the current that flows in the plate circuit also flows through the vacuum tube. Therefore, by introducing an opposition to electron movement within the vacuum tube, the plate current in the external circuit is also reduced.

No physical object can be employed to interfere with the flow of electrons within the vacuum tube since no physical movement within the envelope is practical. Therefore an alternative method is employed. This method introduces an electrostatic interference to the flow of electrons as they travel through the gap between the cathode and the plate.

THE CONTROL GRID

The electrostatic force is introduced between the cathode and the plate by a coil of fine wire. The turns of this coil are widely separated thereby providing very little physical interference. The cathode is placed in the center of the coil, and the plate wraps around the coil. This coil is called the *control grid*.

The control grid is considered a third major element in a vacuum tube. Such a three-element vacuum tube is called a *triode*, derived from the words "three-electrodes." Figure AI·8 is a simplified cutaway view of a triode, showing the physical arrangement of the three basic electrodes.

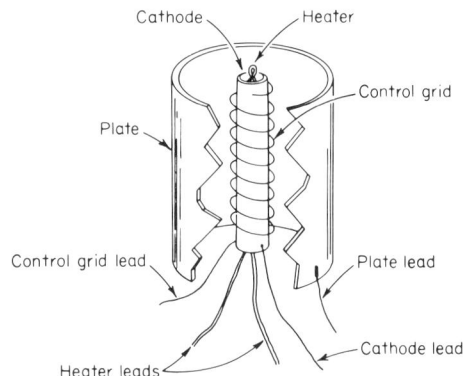

FIG. AI·8 Simplified cutaway view of a triode electron tube.

The control grid is connected on one end only, and no current flows through it. An electrostatic field can exist around a wire when a voltage is applied and does not require current flow. The electrostatic field is negative and acts to repel the negative electrons. Figure AI·9 shows the schematic symbol of a vacuum-tube triode.

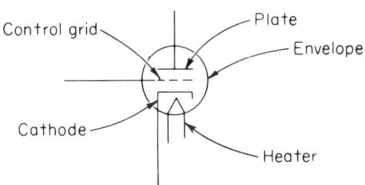

FIG. AI·9 Schematic symbol of a triode vacuum tube.

THE C BATTERY

The creation of a negative electrostatic force is accomplished by connecting the control grid to the negative terminal of a separate battery. In connecting the *grid battery*, the cathode is once more used as the reference point. In order to have the grid negative with respect to the cathode, the positive terminal of the grid battery must be connected to the cathode of the triode. The grid battery is also identified as the C battery. Figure AI·10 shows the vacuum-tube triode in a schematic diagram. The various batteries appear in their place, with the proper polarities applied.

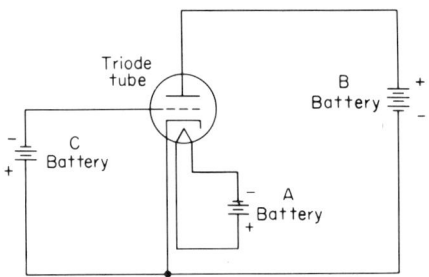

FIG. AI·10 The triode vacuum tube with the necessary batteries.

AI·4 BATTERY REQUIREMENTS

The requirements of the A, B, and C batteries govern their physical and electrical size. The requirements in turn are governed by their individual task.

THE A BATTERY

The A battery supplies heavy currents to the heater. This is accomplished with low voltages since the resistance of the heaters is quite low. Typical A-battery voltages are 6.3 and 12.6 volts. The A battery is physically the largest of the batteries used in electronics.

THE B BATTERY

The B battery has the highest voltage requirement. Typical B-battery voltages are 90 and 135 volts. Sometimes this voltage requirement is greater, and sometimes less. The current drain on the B battery is less than on the A battery. For this reason, the B battery is physically smaller than the A battery.

THE C BATTERY

The C battery applies a low negative potential to the control grid. Typical C-battery voltages are 4.5 and 7.5 volts. No current is required from a C battery. Therefore its physical size is quite small.

AI·5 THE TRIODE TUBE IN ACTION

The control of plate current in a vacuum-tube triode can best be explained by using a simple triode tube circuit. The following discussion refers to the circuit of Fig. AI·11.

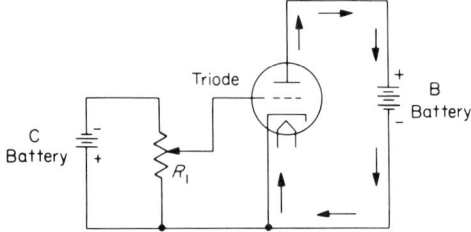

FIG. AI·11 The triode vacuum tube in use as a valve.

Potentiometer R_1 (which is a type of variable resistor used to control voltage application) is placed across the C battery. The center arm of the potentiometer is connected to the control grid. The positive terminal of the C battery is also connected to the cathode. This enables the potentiometer to control the amount of negative voltage that will be applied to the control grid, with respect to the cathode.

The arrows indicate the plate current flow. When a higher negative voltage is applied to the control grid, a higher negative electrostatic field will surround the control grid. Because of this, the electrons traveling from the cathode to the plate will encounter a repelling force in the vicinity of the control grid. This repelling force adds to any other opposition the plate current encounters. The negative voltage applied to the control grid can be made high enough to cut off completely all plate current.

As can be seen, the triode tube in effect acts as a valve that governs the amount of plate current. For this reason, in many countries the vacuum tube is referred to as a *valve*.

AI·6 SUMMARY OF ELECTRON TUBE FUNDAMENTALS

Electron tubes provide a means for an electric current to travel through a vacuum or gas for the primary purpose of exerting control over this current.

The basic electron tube is a two-electrode tube known as a *diode*. A diode tube consists of a *cathode* that emits electrons and a *plate* that collects these electrons.

The control of electron flow within a vacuum tube is accomplished by an electrostatic field introduced between the cathode and plate by an additional electrode called a *control grid*. Such a three-element tube is known as a *triode*.

STANDARDS OF INDUSTRY

Industry has for many years attempted to establish and abide by a series of self-imposed standards designed to protect the consumer pertinent to products he buys. These standardization efforts have been strengthened by the further establishment of consumer-imposed demands that, as condition to acceptance of products purchased, certain manufacturing criteria be met to ensure quality, reliability, and maintainability. Therefore, it is not unusual to see manufacturers' associations and large consumer organizations meeting to explore cooperatively ways and means by which their common objectives can be attained.

AII·1 MILITARY AND NASA STANDARDS

Two large-scale consumers of electronic products are the Department of Defense, with its various military services, and the National Aeronautics and Space Administration (NASA). Therefore, when one is employed by a firm producing electronic products for use by the military or by the aerospace branches of government, one will be subject to military standards (MIL-STD) or to NASA standards, respectively. Information on these standards can be obtained from the United States Government Printing Office in Washington, D.C., for a nominal fee.

AII·2 PROTECTING THE PUBLIC INTEREST

In general, the civilian, public population must rely upon ethics and self-policing policies of industry regarding manufacturing processes that contribute to produce quality, reliability, and maintainability. This becomes apparent when one considers that an individual normally buys one electronic product at a time, and as such does not have the same bargaining power as a large-scale consumer. Although it is true that some protection is afforded the public through various consumer bureaus, nevertheless, it is through the efforts of industrial-type associations that a large portion of the public interest is overseen.

ELECTRONIC INDUSTRIES ASSOCIATION

The Electronic Industries Association (EIA), headquartered at Washington, D.C., is the national industrial organization of electronic manufacturers in the United States. As such, it looks after the public welfare while at the same time supporting the interests of the electronics industry. It further cooperatively supports the standardization needs of the military as well as of NASA.

AMERICAN STANDARDS ASSOCIATION, INCORPORATED

Another public service organization is the American Standards Association, Incorporated (ASA), with headquarters in New York City. This organization strives to promote the development of standards nationally. Although the ASA adopted standards are not mandatory, they are looked upon favorably and respectfully by industry as a whole. The efforts of this association scan the mechanical as well as the electrical fields.

OTHER CONTRIBUTING FORCES TOWARDS STANDARDIZATION

Although it is beyond the scope of this text to completely cover all efforts being made towards standardization, the reader should be aware that the basis for such standards often originate within numerous technical societies and professional organizations found throughout industry. Therefore, students of electronics and fabrication techniques should make an effort to join such societies and organizations in an attempt to keep up with the state-of-the-art.

BIBLIOGRAPHY

AIII·1 ELECTRONICS ASSEMBLY AND FABRICATION

Autonetics Training Department: "Electronic Fabrication Handbook," Autonetics, Space Division of North American Rockwell Corporation, Anaheim, California, 1961.

Baer, Charles J.: "Electrical and Electronic Drawing," McGraw-Hill Book Company, Inc., New York, 1960.

Beason, Robert G. (editor): "How to Build Electronic Kits," 3rd edition, Allied Radio Corporation, Chicago, Illinois, 1966.

Heller, J. D.: "Soldering Fundamentals, Course Number 751 AX," Autonetics, Space Division of North American Rockwell Corporation, Downey, California, 1968.

Lockheed Training Department: "Handbook for Electrical Fabrication and Installation," Lockheed, Burbank, California, 1964.

National Aeronautics and Space Administration: "Quality Requirements for Hand Soldering of Electrical Connections," (NASA Quality Publication NPC 200-4) Superintendent of Documents, U.S. Government Printing Office, Washington, D.C., 1964.

Patrick, Owen G.: "Creative Electronics Fabrication," Holt, Rinehart and Winston, Inc., New York, 1966.

Shiers, George: "Electronic Drafting," Prentice-Hall, Inc., Englewood Cliffs, New Jersey, 1962.

AIII·2 BASIC ELECTRONICS

Frost, John S.: "Applied Electronics," Buck Engineering Company, Inc., Farmingdale, New Jersey, 1967.

Grob, Bernard: "Applications of Electronics," 2nd edition, McGraw-Hill Book Company, New York, 1966.

———: "Basic Electronics," 3rd edition, McGraw-Hill Book Company, New York, 1971.

Heller, J. D., J. M. Simoney, and R. E. Battershill: "ABC's of Electronic Components: A Programmed Course," Autonetics, Division of North American Rockwell, Anaheim, California, 1969.

Romanowitz, Alex H., and Russell E. Puckett: "Introduction to Electronics," Buck Engineering Company, Inc., Farmingdale, New Jersey, 1969.

Zbar, Paul B.: "Electricity-Electronics Fundamentals," McGraw-Hill Book Company, New York, 1969.

AIII·3 SOLID-STATE ELECTRONICS

Electronic Components and Devices: "RCA Linear Integrated Circuits," Technical Series IC-41, Radio Corporation of America, Harrison, New Jersey, 1967.

Gerrish, Howard H.: "Learning Experiences in Transistors and Semiconductors," Buck Engineering Company, Inc., Farmingdale, New Jersey, 1965.

Turner, Rufus P.: "Integrated Circuits," Allied Radio Shack, Fort Worth, Texas, 1971.

Parks, M. S.: "The Story of Microelectronics," Autonetics, Division of North American Rockwell Corporation, Anaheim, California, 1966.

AIII·4 COMPUTATIONS FOR ELECTRONICS

Cooke, Nelson M.: "Basic Mathematics for Electronics," 2nd edition, McGraw-Hill Book Company, New York, 1960.

Porter, Harold W., Charles H. Lawshe, and Orville D. Lascoe: "Machine Shop: Operations and Setups," American Technical Society, Chicago, Illinois, 1960.

Riggenbach, Frank: "The Slide Rule," The Technical Education Press, Seal Beach, California, 1970.

Smartt, James L.: "Length Measurement," Go-Power Corporation, Palo Alto, California, 1969.

AIII·5 ELECTRONICS DICTIONARIES

Beam, Robert E. (editor): "A Dictionary of Electronic Terms," 9th edition, Allied Radio, Chicago, Illinois, 1968.

Cooke, Nelson M. (editor): "Allied Electronics Data Handbook," 6th edition, Allied Radio Shack, Fort Worth, Texas, 1966.

Graf, Rudolf F.: "Modern Dictionary of Electronics," 3rd edition, Howard W. Sams and Company, Inc., Indianapolis, Indiana, 1968.

Todd, Alva C. (editor): "Encyclopedia of Electronics Components," 2nd edition, Allied Radio Shack, Fort Worth, Texas, 1967.

IV

AUDIO-VISUAL AIDS AND TECHNIQUES

AIV·1 EDUCATIONAL TECHNOLOGY

Modern teaching strategies encompass a detailed analysis of the learning characteristics of students in both group and individualized teaching environments. In order to best meet the learning needs of the student, the teacher should be acquainted with all phases of audio-visual technology. By having knowledge of audio-visual equipment and materials that can be used to supplement material contained in a textbook, the teacher will be able to organize his teaching so as to improve both quality and quantity of instruction. This, of course, is done with the student foremost in mind.

While the teacher may want to use audio-visual equipment and materials in a group situation as an *aid to instruction*, the student may want to use the same equipment and materials as an *aid to learning*. Therefore, a good teacher will provide for both instruction and learning needs. Traditionally, large-class instruction is conducted in regular classrooms. However, individualized instruction which caters to the learning needs of a student is conducted in *study carrels*, either in a special classroom or dispersed throughout a shop or laboratory area.

It is the responsibility of a teacher to search for,

obtain, and utilize audio-visual equipment and materials (making them when necessary) in order to provide all of the possible tools for learning to his students. Since the task is of giant proportions, this section has been prepared by the authors in order to give the instructor some assistance with his research for suitable materials in the area of audio-visual aids. Although the list is not complete, it should serve adequately as a basis upon which to expand.

AIV·2 AUDIO-VISUAL AIDS

The audio-visual aids listed below may be used to supplement material found in this book. They have been grouped by major areas as outlined in this book. Some of the aids recommended may apply to more than one area. It is therefore suggested that they be examined before use in order to determine their suitability for a particular group of students.

Titles of motion pictures (marked "MP"), filmstrips (marked "FS"), filmloops (marked "FL"), slide sets (marked "SS") and overhead transparencies (marked "OT") are included in the list. Each title is followed by the name of the producer and the distributor, unless one firm performs both functions. Abbreviations are used for names of producers and distributors, and these abbreviations are identified in the list of sources at the end of this bibliography. In many instances, the films can be borrowed or rented from local or state 16mm-film libraries. (A nationwide list of these sources is given in "A Directory of 2002 16mm Film Libraries," available for 35 cents from the Supt. of Documents, Washington, D.C., 20402.) Unless otherwise indicated, the motion pictures are 16mm sound black-and-white films and the film-strips are 35mm, black-and-white, and silent. The length of the motion picture is given in minutes (min), that of filmstrips in frames (fr).

This bibliography is a selective one, and film users should examine the latest annual edition and quarterly supplements of "Educational Film Guide," a catalog of 11,000 films published by The H. W. Wilson Company, New York. The Guide, a standard reference book, is available in most college and public libraries.

It is recommended that instructors contact the public relations office of large electronics manufacturing and service concerns regarding visual aids, which are often available to educational institutions.

I. ELECTRONICS ASSEMBLY

On Solder (MP, Scope/Class, 30 min color). Describes in detail the theory of the soldering process, the basis for developing an understanding of the requirements of good soldering techniques.

The Art of Soldering (MP, BOE, 20 min color). Explains the principles of proper soldering techniques as applied to precision electronics work.

10-9-8-Hold (MP, BOE, 20 min color). Illustrates the importance of reliability in the manufacturing process of a guided missile.

Wire Sizes and Voltage Drop (MP, UWF, 13 min). Covers factors influencing the ability of conductors to carry electron flow; measurement of wire sizes; wire area in circular mils; voltage drop; and Ohm's law.

Electronic Schematic Symbols and Resistor-Capacitor Color Code (Chart, EICO, 8½ x 11 inches). Notebook-size charts, prepared as a service to electronics students, providing a comprehensive reference for electronics symbols and hardware. Also includes a complete resistor-capacitor color-code chart.

A Tip On Irons (MP, CLASS, 20 min color). Covers the principles of operation and the use of electrical soldering irons. It explains heat transfer for purposes of soldering.

Soft Soldering (MP, EBF, 11 min color). Illustrates the three basic steps involved in soft soldering: cleaning, heating, soldering. Shows the uses of electric soldering tools and explains various types of fluxes used.

Basic Soldering Skills (SS, HIC, set of 90). A set of 90, 35 mm slides for use in teaching basic soldering skills. Lecture notes accompany this set. The slides were prepared by NASA, lecture notes prepared by S. R. Duarte.

II. FUNDAMENTALS OF ELECTRONICS

Principles of Electricity (MP, GE, 20 min color). Introduces the principles of electricity, using animation extensively. Breaks down the structure of matter into atoms to show the relationship of the electron. Treats separately the volt, ampere, and ohm.

Basic Electricity (MP, USAF/UWF, 20 min color). An animated cartoon explaining the fundamentals of electricity, including voltage, current, resistance, magnetic fields, induction, primary and secondary coils, and series and parallel circuits.

Basic Electronics (MP, USAF/UWF, 17 min color). An animated cartoon explaining the meaning of atoms and electrons, vacuum tube, cathode, rectifier tube, amplifier tube, grid, and bridge circuits.

Alternating Current (FS, USAF/UWF, 50 fr). An elementary introduction to the principles of alternating current. Demonstrates and explains Lenz's law, simple wave alternator, frequency, effective value, voltage-current-time relationship, and power.

Basic Electricity (FL, AEF, Set of 9). A set of Super-8mm motion picture filmloops covering: Resistors in Series; Resistors in Parallel; Capacitor Construction and Actions; Inductance Principles; Inductor Construction and Actions; The RL Time Constant; The Undamped Oscillation; The Damped Oscillation; The RC Time Constant.

Electronic Transparencies (OT, LAB, Set of 34). A set of transparencies suitable for every electricity-electronics classroom. Covers most of the fundamentals of electronics. Designed for use with standard 10″ x 10″ overhead projectors.

Capacitance (MP, USN/UWF, 31 min). Demonstrates electron flow through a circuit, the charging and discharging of capacitors, variations of a charge on a capacitor in relation to time, and the behavior of capacitance with alternating current.

Inductance (MP, USN/UWF, 35 min). Shows how a magnetic force reacts around a coil, the nature of self-inductance, and how to increase the inductance of a coil.

RCL: Resistance, Capacitance (MP, USN/UWF, 34 min). Explains current and voltage in relation to time; voltage and current curves; the relationship of current and voltage; the measurement of voltage at source; the addition of phase components; and the effect of impedance on resonance.

Sylvania Educational Aids: Vacuum Tube Construction (Wall charts, SYLV, set of six). Wall charts showing the internal construction of electron tubes and identifying the internal elements of a variety of tubes, including cathode-ray tubes.

Vacuum Tubes (MP, EBF, 11 min). Explains the three functions of the vacuum tube—amplifier, rectifier, and oscillator.

The Triode: Amplification (MP, USOE/UWF, 14 min). Principles of the diode and triode; electric fields; a triode amplifier circuit; amplification of d-c voltage changes; alternating voltages; distortion; amplification of audio-frequency signals.

Electric-Electronic Concepts (Study Guide, MTC, Single Sheet). A study guide in the area of electricity-electronics for persons in need of a brief layman's language survey of concepts. Eighteen pages of information photographically reduced to fit one page.

III. SEMICONDUCTORS

Principles of the Transistor (MP, McGraw, 21 min). Explains how the germanium diode and the transistor function, with details concerning the crystal lattice, P- and N-type germanium, and hole conduction. Describes some of the present applications of the transistor and its advantages over the thermionic tube.

The Transistor (MP, McGraw, 16 min color). Explains the working of a PNP transistor. Shows how the transistor may be used to obtain amplification. Analyzes the amplification process.

Transistors: Switching (MP, USN/UWF, 14 min). Shows examples of switching circuits in transistorized computers, explaining briefly the concept of digital computation and how transistors are used and, in more detail, how a simple transistor switch works.

Transistors and Transistor Circuits (FL, AEF, Set of 5). A set of Super-8 mm motion picture filmloops covering: Transistor Characteristic Curves; Audio

Voltage Amplifier; Direct-coupled Amplifier; Multivibrator; Tuned Collector Oscillator.

IV. MECHANICAL ASSEMBLY

School Shop Safety (MP, BFA, 15 min color). Explains that safety is the first lesson taught in school shops. This film emphasizes basic safety practices to be observed when handling various materials, hand tools, power tools, heated materials, and electricity.

Drill Press (MP, BFA, 17 min color). This film shows the procedures for positioning, clamping, and drilling. Also discussed are use of the pilot hole and large diameter drilling.

Precision Measuring, Parts I and II (MP, BFA, each part 8 min color). Part I explains in detail how to read the fractional scale. Part II explains in detail how to read the decimal scale.

Portable Power Tools (MP, BFA, 17 min color). The electric drill, belt sander, orbital sander, and the saber saw are demonstrated, and safety practices emphasized.

Hand Tools For Metalworking (MP, BFA, 25 min color). Proper use of specific tools for specific jobs are explained. Tools covered include: hammers, screwdrivers, pliers, wrenches, files, and hacksaw.

VISUAL AID SOURCES

AEF	Animated Electronic Films, P.O. Box 2036 Eads Station, Arlington, Virginia 22202
BFA	Baily-Film Associates, 11559 Santa Monica Blvd., Los Angeles, California 90025
BOE	The Boeing Company, Public Relations Department, Seattle, Wash.
CLASS	Classroom Film Distributors, Inc., 5620 Hollywood Blvd., Hollywood 28, Calif.
EBF	Encyclopaedia Britannica Films, Wilmette, Ill.
EICO	EICO, Long Island City 1, New York.
GE	General Electric Company, 1 River Road, Schenectady 5, N.Y.
GEN TEL	General Telephone Company of California, 2020 Santa Monica Blvd., Santa Monica, Calif.
HIC	Hickok Teaching Systems, Woburn, Massachusetts 01801
HUGHES	Hughes Aircraft Company, Culver City, Calif.
LAB	Lab-Volt Educational Systems, Buck Engineering Company, Inc., Farmingdale, New Jersey
MCGRAW	McGraw-Hill Book Company, Text-Film Dept., 330 W. 42d St., New York, N.Y. 10036.
MTC	Manpower Training Consultants, P.O. Box 2190, Seal Beach, California 90740
PHOTO	Photo and Sound Company, 5525 Sunset Blvd., Hollywood 28, Calif.
SCOPE	Scope Film Associates, Los Angeles, Calif.
SYLV	Sylvania Electric Products, Inc., 1740 Broadway, New York, N.Y. 10036.
USA°	U.S. Department of the Army, Washington, D.C. 20310
USAF°	U.S. Department of the Air Force, Washington, D.C. 20330
USC	University of Southern California, Department of Cinema, University Park, Los Angeles 7, Calif.
USN°	U.S. Department of the Navy, Washington, D.C. 20350
USOE°	U.S. Office of Education, Washington, D.C. 20202
UWF	United World Films, Inc., 1445 Park Ave., New York, N.Y. 10029

°*Films distributed by United World Films, Inc.*

INDEX

Active circuits, 216–223
Active components, 24
Alignment, 253
   procedures, 260
Alligator clip, 152
Alternating current (a-c), 186–195
   inductor, 22
   production by motor-generator units, 187
   (*See also* Circuits; Inductor)
Alternation, 186
American Standards Association (ASA), 289
Ammeter, 175
Ampere, 14

Amplification, 217
Amplifier, 235, 251
   audio-frequency, 236
   circuits, 237, 239, 253
   gain, 238
   high-frequency, 251–264
   low-frequency, 235–250
   power, 217, 238
   radio-frequency, 253
   regenerative, 218, 240, 254
   triode, 201, 202, 286
   voltage, 217
   (*See also* Audio amplifier)
AND gate, 219, 220
Anode, 197

Anode (continued)
   grid, 28
Antenna, loop, 23
   rod, 23
Anti-wicking tool, 96
Arsenic, 198
Assembly, mechanical:
   practices, 153–161
   workstation, 7
Assembly-line processes, 6
Audio amplifier, 236
   function, 236
   integrated-circuit, 207
   overall test, 244
   phonograph, 240

Audio frequencies, 236
Audio signal generator, 208, 247
   radio-frequency, 255, 258
Audio signals, 236
Audio-visual aids, 292–295

B supply output, 226
Base, 200
Battery, 14
   A, 286
   automobile, 173
   B, 286
   C, 286
Beat, note, 263
   zero, 263
Beat signal, production of, 263
Bias voltage, 197
Bistable multivibrator, 220
Binary information, 220
Bleeder resistor, 230
Block diagram, 61, 62
Blueprints, 64
   care, 66
   content, 65
   detail, 65
   final assembly, 66
   subassembly, 65
   unit assembly, 65
Bolt, 150
Box wrench, 81
Bracket, 151
Braid, flat, 111
Breakdown voltage, 19
Bridge, impedance, 195

Cable, clamp, 151
   coaxial, 111
   connectors, 95, 120, 123, 124, 129
   interconnecting, 112
   lacing, 119, 121
   multiple wire, 111
   shielded, 112
   symbols, 124
   ties, 119, 121
   wrapping tape, 120
Calculating a-c circuits, 274
Calculating d-c circuits, 271
Calibration, frequency, 263
Capacitance, 15, 19, 191
Capacitive reactance, 192
Capacitor, 19–21, 191
   checker, 266
   color code, 37
   construction, 38
   disk, 19

Capacitor (continued)
   electrolytic, 20
   ganged, 21
   hermetically sealed, 20
   identification, 19
   marking, 19
   microminiature, 19
   molded tubular, 19
   ratings, 19, 37
   symbols, 21
   test, 194
   trimmer, 21
   types, 19
   variable, 20, 21, 252
Carrier wave, 252
Catalogs, use of, 157
Cathode, 283
Center punch, 83, 153
Chassis, 6, 148, 153
   ground, 148
   wiring diagram, 58, 61
Choke, 22, 190
   construction, 39
   tubular, insulated, 22
Circuit breaker, 152
Circuit leading, 267
Circuits, 174
   a-c, calculation of, 274
   active, 216–223
   amplifier, 237, 239, 253
   cable, flat, 124
   coupling, 218, 238
   d-c, calculation of, 271
   electrical, 174
      closed, 174
      open, 174
   etched (see Circuits, printed)
   flip-flop, 219
   integrated, 25, 27, 51, 53, 205
      hybrid, 27, 206
      monolithic, 27, 206
      symbols, 29
   MOS-FET, 204
   parallel, 181, 182, 272
   printed, 51
      advantages, 51
      applications, 52
      connectors, 123
      mounting components, 138
      replacing components, 109
      soldering, 96
   resonant, 252
   series, 181, 271
      test, 183
   series-parallel, 182

Circuits (continued)
   switching, 219
   transistor, 200
   tuned, 252
      applications, 253
      series-fed, 253
Clamps, 151
Clips, alligator, 152
   Fahnestock, 152
Coil, primary, 191
   secondary, 191
   voice, 239
Collector, 200
Color code, 32
   capacitor, 37
   diode, 42
   resistor, 32
   transformer, 231
   wire, 118, 127
Component, active, 24
   discrete, 50, 139
   identification, 13–49
   passive, 24
   substitution, 266
   support, 139
   testing, 265
Conductor, 14
   cleaning with ink eraser
Connection, electrical, 91
   mechanical, 91
   terminal, 133–147
Connector, cable, 95, 120, 123, 124, 129
   pins, 95
   polarity, 122
   printed circuit, 122, 123, 124
   rack-and-panel, 123
   soldering of, 95
   solderless, 113, 115
   symbols, 124
Contact pins, 95
Continuity, checker, 163
   tester, 166
Contour soldering, 92–93
Control grid, 285
Control shaft, cutting, 155
Converter, 255
   frequency, 255
Copper wire, 110
Core, 190
   inductor, 23
   iron, 23
Cotter pin, 150
Crimping tool, 115, 116
Current, 14, 172
   control, 14, 172, 173

Current (continued)
  distribution, 181
  electric, 172
  electron, 172
  forward, 197
  hole, 199
  induced, 191
  leakage, 194
  measurement, 178
  plate, 185
    control, 186
    cutoff, 186
  reverse, 197
  units, 14, 175
  vacuum tube, 285, 286

Degrees, electrical, 189
Desoldering, 97
Detector probe, 268
Detectors, 236
Diagonals, 77
Diagrams, 57–74
  block, 61, 62
  interconnection, 58, 62
  layout, 63, 64
  pictorial, 63, 67
  printed-circuit, 63, 66
  schematic, 58, 59
  symbolic, 62, 63
  wiring, 58, 61
Diodes, 24, 28
  color code, 42
  identification of semiconductor, 42
  rectifiers, 25
  silicon, 198
  solid-state, 196
  special application, 204
  symbols, 197
  tester, 211
  theory, 197, 283
  tunnel, 205
  vacuum tube, 283
  zener, 204–205
Direct current (d-c), 15
  inductor, 22
  (See also Circuits; Inductor)
Discrete components, 50, 139
Doped germanium, 198
Drawing, pictorial, 63, 67
Drawpunch, 155
Drill, hand, 85
Drill press, 155, 156
Drill-size gauge, 154
Dry cell, 58
Dykes, 77

Dynamotor, 188

Effective value, 189
Electric current (see Current)
Electric motors, 22
Electrical chassis (see Chassis)
Electrical circuit (see Circuits)
Electrical concepts, 13
Electrical conductor, 14
Electrical connection, 91
Electrical degrees, 188, 189
Electrical-support hardware, 151
Electrical symbols, 172
Electrical wire, 14
Electricity, origin, 172
  principles, 171–179
Electrolytic capacitor, 20
Electromotive force, 172
Electron current, 172
Electron flow, 172
  in amplifiers, 239
  in rectifiers, 217, 228
Electron movement, 14
Electron tubes, 15, 28–30, 282–287
  classifications, 28
  construction, 48
  designations, 30
  as electronic valves, 286
  gaseous, 28, 282
  sockets, 28
  testing, 48
  versus resistors, 282
Electronic assembly workstation, 7
Electronic calculations, 271–280
Electronic fabrication, 3–12
Electronic Industries Association, 289
Electronic packaging, 50–56
  integrated-circuits, 52
  printed-circuits, 51
  3-D packaged circuits, 51
  wired circuits, 50
Electronic test station, 7
Electronic testing techniques, 265–270
Electronics, solid-state, 196–215
Electronics technician, 7
Electronics tester, 7
Electrostatic force, 191
Emitter, 200
Employment conditions, 3
  evaluation, 9–12
  safety, 4
  work practices, 5
Environmental testing, 164
Etched circuits (see Circuits, printed)
Exploded illustration, 63

Fabrication, electronics, 3–12
Fahnestock clip, 152
Farad (unit), 19
Filament, 283
Files, 85
Filter, 228
Flip-flop circuit, 219
Force, electromotive, 172
  electrostatic, 191
Forward bias, 197
Forward current, 197
Frequency, 186
Frequency calibration, 263
Frequency converters, 255
Front-to-back ratio, 198, 209
Full-wave rectifier, 229
Fuses, 152

Gallium, 198
Ganged capacitors, 21
  motor units, 187, 188
  tone, 242, 249
Gaseous tubes, 282
Generator, electrical, 188
Germanium, 198
  doped, 198
  N-type, 198
  P-type, 198
Graphical representation, 187
Grid, anode, 28
  battery, 286
  control, 285
  suppressor, 28
Grinder, 86
Grommet, 152
Ground wire, 117

Hacksaw, 85
Half-wave rectifier, 228
Hammer, 153
Hand drill, 85
Hand tools, 75, 79
Hardware, 148–153
  electrical support, 151
  physical support, 148
Harness, assembly procedures, 118–132
  protection, 120, 122
  termination, 120
Heat, damage, 94, 96
  problems, 101
  requirements, 90
Heat shunt, 161
Heat sink, 94
Heater, 283
Heating effect, 189

Henry (unit), 21
High-frequency amplifiers, 251–264
High-frequency oscillators, 254
Holding devices, 83, 84
Hole current, 199
Hole cutting, 154, 155
Hole flow, 199
Hole movement, 199
Hook splice, 92

Impedance, 21, 192, 276
Impedance bridge, 195
Impurities, 198
Induced current, 191
Induced voltage, 191
Inductance, 15, 21, 190
Inductive reactance, 190, 274
Inductor, 21–24, 190
   a-c, 22
   cores, 23
   coverings, 23
   d-c, 22
   high-frequency, 23
   identification, 21
   leads, 21
   low-frequency, 23
   marking, 21
   rating, 21
   symbols, 23, 24
   types, 22
Inspection, 3, 162–168
Instruments, testing, 266
Insulating aids, 152
Insulator, 172
Integrated circuits, 25, 27, 51, 53, 205
   classifications, 27
   hybrid, 27, 206
   monolithic, 27, 206
   packages, 27
   symbols, 29
Intermediate-frequency signal, 256
Iron core, 23
Isolation transformer, 227

Junctions, PN, 198

Key, 47
Kilohm, 175
Knot, square, 120

Lacing, 119
   procedures, 121
Language of symbols, 57
Lead, bending, 140
   identification, 118

Lead (continued)
   wrapping, 91
Leakage current, 194
Line cord, 76
Linear measurement, 68
   devices, 68
   English, 68
   metric, 68
Local oscillator, 255
Long-nose pliers, 77
Loop antenna, 23
Loudspeaker, 236
   symbol, 237
   test, 243
Low-frequency amplifier, 235–250
Low-frequency oscillator, 240
Low-frequency transformer, 23

Machine screw, 148
   fillister head, 149
   flathead, 149
   Phillips head, 149
   roundhead, 149
Magnetic field, 190
Magnetism, 190
Mathematics, for electronics, 271
Measurement, linear, 68
   English, 68
   metric, 68
Measuring devices, 68–69
Mechanical assembly practices, 153–161
Mechanical assembly workstation, 7
Mechanical connection, 91
Megohm, 175
Metal-oxide semiconductor field-effect transistors (*see* MOS-FET)
Military standards, 288
Milliammeter, 178
Milliampere, 175
Miniaturization, 50
Module, 52
MOS-FET, 203
MOS-FET circuits, 204
Motor-generator units, 187, 188
Mounting controls, 154
Multivibrator, bistable, 220

N-type germanium, 198
NASA standards, 288
Nondestructive testing, 163
NPN transistor, 201
Nut, 149
   driver, 80
   hexagon, self-locking, 150
   speed, 150

Ohm (unit), 175
Ohmic values, 17
Ohmmeter, 175
   use, 33
Ohm's law, 274, 276
Oscillation, 218
Oscillator, 240–243
   audio, 242
   circuit analysis, 240
   coil, 23
   high-frequency, 254
   local, 255
   low-frequency, 240
   radio-frequency, 254
   testing, 261
Oscilloscope, 193, 267
   signal tracer, 267

P-type germanium, 198
Parallel circuit, 181, 182, 272
Passive components, 24
Peak value, 189
Peel-pull test, 166
Pentode, 28
Period, 188
Pins, connector, 95
   contact, 95
   cotter, 150
Planning and layout, 153
Plate, 284
Plate current, 185
   control, 186
   cutoff, 186
Plated-through holes, 140
Pliers, diagonal-cutting, 77
   long-nose, 77
PN junctions, 198
PN materials, 198
PNP transistor, 201
Polarity, 173
   cable connectors, 129
Porcelain standoff, 152
Potentiometer, 18
Power amplifier, 217, 238
Power cord, 16
Power output stage, 236, 238
Power output transformer, 239
Power rectifier, 25, 26, 216
   markings, 43
Power source, 225
Power supplies, 224–234
   A, B, and C, 225
   bench-type, 226
   component identification, 231
   functions, 226

Power supplies (continued)
    need, 224
    outputs, 225
    transistorized, 230, 231
    types, 225
Power transformer, 226, 227
    lead identification, 231
Power transistors, 26
Printed-circuit cable, flat, 124
Printed circuits, 51
    advantages, 51
    applications, 52
    connectors, 122, 123, 124
    mounting components, 138
    replacing components, 109
    soldering, 96
Printed wiring, 124
Probe, detector, 268
Production testing, 266

Quality control, 162–168

Radio frequencies, 251
Radio-frequency (r-f) amplifier, 253
Radio-frequency oscillator, 254
    testing, 261
Radio-frequency signal generator, 255, 258
Radio-frequency signals, 251
Radio-frequency stage identification, 256
Radio receiver (see Receiver, radio)
RC coupling, 238
Reamplification, 240
Receiver, radio, 59, 60, 241
    kit, 9
    panel, front, 255
Rectification, 216, 227
    defined, 216
    filtering output, 228
Rectifier, diodes, 25
    dry-metal, 27
    full-wave, 229
    half-wave, 228
    polarity marking, 217
    power, 25, 26, 216
        markings, 43
    signal, 217
    silicon controlled, 205
Regeneration, 218
Regenerative amplifier, 218, 240, 254
Relay, cases, 23, 142
    identification, 23
    operation, 40
    symbol, 22
Reliability, 162

Reproducer, 236
Resistance, 173
    analysis, 257
    measurement, 176
    rating, 15
    soldering, 99
    units, 172, 175
Resistor, 14–19
    bleeder, 230
    color code, 32
    identification, 15–19
    markings, 17
    networks, 142
    parallel, 272
    power, 17
    precision, 17
    rating, 15
        power, 17
        wattage, 17
    series, 271
    series-dropping, 182
    symbols, 18
    tapped, 18
    variable, 17, 142
    testing, 35
Resonance, 252
Resonant circuits, 252
Reverse bias, 197
Reverse current, 197
R-f (see Radio-frequency)
Rheostat, 18
Rivet, 150
Rms voltages, 230
Rod antenna, 23
Rotary switch, 151

Schematic diagram, 58, 59
Scissors, 77
Screw:
    machine, 148–149
        fillister head, 149
        flathead, 149
        Phillips head, 149
        roundhead, 149
    sheet-metal, 150, 151
Screwdriver, blade, 80
    Phillips, 79
Selectivity, 252
Semiconductor, 24–28
    classifications, 24
    designations, 30
    diode identification, 42
    life expectancy, 207
    limitations, 207
    purpose, 24

Semiconductor (continued)
    symbols, 29
Series circuit, 181, 271
    test, 183
Series-dropping resistor, 182
Series-parallel circuit, 182
Signal, production of beat, 263
Signal generator, audio, 208, 247
    radio-frequency, 255, 258
Signal injection, 247, 267
Signal rectifier, 217
Signal tracer, 245, 267
    oscilloscope, 267
Silicon controlled rectifier, 205
Silicon diode, 198
Sine waves, 188
Sleeving, plastic, 122, 123
Sockets, 28, 142
    transistor, 44
    vacuum tube, 28
    wrench, 81
Solder, 90
    alloy, 90
    application, 92
    removing, 73
    requirements, 89
    results, 93
Solder cups, 95, 96
Solder lug, 152
Solder pot, 95, 96, 114
Soldering aid, 76
Soldering, 89–109
    bit, 98
    on connectors, 95
    contour, 92–93
    gun, 141
    iron, 75, 90
        tip, tinning, 91
    printed-circuit boards, 96
    process, 92–95
    resistance, 99
    terminals, 134
    wave, 98–99
Solderless connectors, 113, 115
Solderless, terminal lug, 115
    terminals, 138
Solid-state diode, 196
Solid-state electronics, 196–215
Speed nut, 150
Splice, 92
    hook, 92
    tap, 92
    Western Union, 92
Spot ties, 118
Square knot, 120

Standards of industry, 288–289
  American Standards Association, 289
  Electronic Industries Association, 289
  military, 288
  NASA, 288
Substitution techniques, 266
Superheterodyne radio receiver, 255
  block diagram, 62
  schematic diagram, 59, 60
Supply (see Power supplies)
Suppressor grid, 28
Switches, 151
  identification, 151
  rotary, 151
  toggle, 151
Switching circuits, 219
Symbolic diagram, 62, 63
Symbols, cable, 124
  capacitor, 21
  connector, 124
  diode, 197
  electrical, 172
  for inductors, 23, 24
  for integrated circuits, 29
  language of, 57
  loudspeaker, 237
  for relays, 22
  for resistors, 18
  for semiconductors, 29
  transistor, 29, 45, 200
  for vacuum tubes, 29

Tap splice, 92
Terminal connections, 133–147
Terminals, bifurcated, 135
  boards, 133
  flat, 135
  hook, 136
  screw-type, 134
  soldering, 134
  solderless, 138
  turret, 133
  use, 133
Test equipment, 175
  selection, 269
Test panels, 163, 164, 167
Test technicians, 3
Testing:
  components, 265
  electron tubes, 48
  environmental, 164
  instruments, 266
  nondestructive, 163
  production, 266
  techniques, 265–270

Testing (continued)
  variable resistors, 35
Tetrodes, 28
Tinning, 90
  for heat transfer, 91
  for penetration, 91
  with solder pot, 114, 117
  with soldering iron, 90, 91
Toggle switch, 151
Tone generator, 242, 249
Tools, 75–88
  crimping, 115, 116
  for electrical assembly, 75, 98
  hand, 75, 79
  for measuring and layout, 81
  for mechanical assembly, 79
  for microelectronics, 79
Toroidal choke, 22
Tracer (see Signal tracer)
Tracking adjustments, 261
Transformer, 190
  color code, 231
  construction, 227
  coupling, 218
  encapsulated, 227
  high-frequency, 257
  isolation, 227
  low-frequency, 23
  power, 226, 227
    lead identification, 231
  power output, 239
  test, 194
Transistor, 15, 25, 200
  bias, 200
  circuits, 200
  currents, 201
  field-effect, 202
  identification, 26
  leads, 44
  NPN, 201
  PNP, 201
  power, 26
  principles, 200
  radio receivers, 59, 241
  sockets, 44
  symbols, 29, 45, 200
  tester, 211
Transistorized power supply, 230, 231
Triode, 28, 29, 201, 285
  amplifier, 201, 202, 286
  theory, 200, 285
  tube as a valve, 286
Troubleshooting, 266
Tubes (see Electron tubes)
Tuned circuits, 252

Tuned circuits (continued)
  applications, 253
  series-fed, 253
Tunnel diode, 205
Turret terminals, 133

Units, current, 14, 175
  resistance, 173
  voltage, 14, 175
Units of quantity, 175

Vacuum tube, 282–287
  current, 285, 286
  diode, 283
  manual, 47
  nomenclature, 283
  principles, 283
  socket, 28
  symbols, 29
  tests, 48
Vacuum-tube voltmeter (VTVM), 268
Vise, 83, 84, 153
Voice coil, 239
Voltage, analysis, 244
  bias, 197
  breakdown, 19
  current relations, 189
  distribution, 180
  divider, 180
  drop, 181
  gain, 238
  induced, 191
  rns, 230
  source, 14, 173, 174
  units, 14, 175
Voltage amplifier, 217
Voltmeter, 175
  use, 178
  vacuum-tube (VTVM), 268
Voltohmmeter (VOM), 210
Volume control, 237

Washers, 150
  external-teeth binding, 150
  flat, 150
  internal-teeth binding, 150
  shoulder, 152
  spring-lock, 150
Wattage rating, resistor, 14
Wave, carrier, 252
  rectifier, 228, 229
  sine, 188
  soldering, 98–99
Wavesoldering, 98
Welding, 100

301

Welding (continued)
  electronic, 100
  percussion, 100
Western Union splice, 92
Wicking, 97
Wire, braids, 111, 114
  cleaning, 112
  color code, 118, 127
  copper, 110
  diameters, 116
  electrical, 14
  enamel-covered, 110
  flexible, 111
  ground, 117
  harness, 119
    protection, 122

Wire (continued)
  termination, 120
  jig board, 131
  insulation, 111
  preparation, 110–118
  shields, 111, 114
  size, 115, 116
  solid, 116
  strands, 116
  stripper, 77
    mechanical, 78, 113
    thermal, 78, 114
  stripping, 112
  tining, solder pot, 117
    soldering iron, 91
  types, 110

Wire (continued)
  wrapping specifications, 133–147
Wire-Wrap®, 113, 115
Wiring, diagram, 58, 61
  printed, 124
Wrapping, lead, 91
  tape, 122
  of wire specifications, 133–147
Wrench, adjustable, 81
  box, 81
  open-end, 81
  socket, 81

Zener diode, 204–205
Zero beat, 263